THE GLACIERS OF EQUATORIAL EAST AFRICA

SOLID EARTH SCIENCES LIBRARY

STEFAN HASTENRATH

The University of Wisconsin, Madison

THE GLACIERS OF EQUATORIAL EAST AFRICA

D. REIDEL PUBLISHING COMPANY

A MEMBER OF THE KLUWER ACADEMIC PUBLISHERS GROUP

DORDRECHT / BOSTON / LANCASTER

Library of Congress Cataloging in Publication Data

Hastenrath, S.
The glaciers of equatorial East Africa.

(Solid earth sciences library)
Bibliography: p.
Includes indexes.
1. Glaciers—Africa, Eastern. I. Title. II. Series.
GB2573.E27H37 1983 551.3'12'09676 83–21144
ISBN 90–277–1572–6

Published by D. Reidel Publishing Company,
P.O. Box 17, 3300 AA Dordrecht, Holland.

Sold and distributed in the U.S.A. and Canada
by Kluwer Academic Publishers,
190 Old Derby Street, Hingham, MA 02043, U.S.A.

In all other countries, sold and distributed
by Kluwer Academic Publishers Group,
P.O. Box 322, 3300 AH Dordrecht, Holland.

Printed in The Netherlands

Dedicated to Nayan Bhatt, Bob Caukwell, and Frank Charnley, and to the memory of Kamau Mwangi

Frontispiece: The Kibo summit of Kilimanjaro from the Northeast. Photograph taken by Alan Root during his hot air balloon flight on 10 March 1974.

PREFACE

In the course of a decade's work on the mountains of East Africa, I met some of the most wonderful people on Earth. It is impossible to record all those who have helped me in this study in one way or other.

Glacier research in East Africa has some history. Nearly half a century ago, Carl Troll completed the first detailed mapping of Lewis Glacier. I had the good fortune of exchanging ideas with him at his home in Bonn in 1974, shortly before his death. Paul C. Spink, Ulceby, North Lincolnshire, England, shared with me his photographs and observations on Kilimanjaro and Mount Kenya in the 1940s. When I joined the University of Nairobi in 1973, several members of the 1957–58 IGY Mount Kenya Expedition were still there. I received generous advice and help from Igor Loupekine, John Loxton, but especially from Robert A. Caukwell and Frank Charnley. Their continuous cooperation was a great encouragement over the years. Heinz Loeffler, University of Vienna, informed me about the depth of Lewis Tarn. Helmut Heuberger, University of Munich, provided me data on his measurements of glacier terminus positions. Peter Gollmer of Geosurveys and Alan Root, Nairobi, gave me aerial photographs of Kibo from the early 1970s.

I acknowledge support from various other colleagues at the University of Nairobi: Raouf Rostom of the Department of Surveying and Photogrammetry; Neville Skinner of the Department of Physics; G. C. Asnani, John Ng'anga, J. K. Patnaik, Hassan Virji, of the Department of Meteorology; Edmund Rodrigues and Nayan V. Bhatt of the Department of Geology. The latter five joined me on one or more working sorties to Lewis Glacier. My association with Nayan Bhatt extended over many years.

I recall with gratitude my other comrades of the Mountain Club of Kenya. John Youngs accompanied me on two hard working weekends at Lewis Glacier. To John Temple I owe the memory of a safari to West Kibo and discussions of ice extent on Kilimanjaro. Phil Snyder helped me in many ways, including with the organization of regular observation programs and the often difficult logistics on Mount Kenya. Communication half around the globe would have been more difficult without Susan Morris and John Hull. Iain Allan and Ian Howell always offered me freely their advice and cordial friendship.

In 1975, Uwe Radok brought me in contact with a small group of researchers in Australia. Their work, especially the exploration of the glaciers of New Guinea, was a source of inspiration. Subsequently, Phillip Kruss of the Melbourne group joined our Department of Meteorology at the University of Wisconsin. He accompanied me on various of the field expeditions to Lewis Glacier in the course of 1977–83. His work in the field and his theoretical studies of ice dynamics and glacier retreat are important contributions towards the understanding of glacier-climate relationships in the tropics.

Lonnie Thompson of the Institute of Polar Studies of Ohio State University, with whom I spent many field seasons in the Peruvian Andes, joined us for the retrieval of ice cores in February–March 1978. Michael Kuhn of the University of Innsbruck provided me with a steam drill for the 1977–78 and 1981–82 expeditions. Kamau Mwangi of Naro Moru participated in all our field seasons on Lewis Glacier during 1977–80, and performed glaciological observations in the intervening intervals. John Omirah Miluwi of the Mount Kenya National Park took an active interest in my work on Lewis Glacier over many years. His brother Martin Otieno assumed the responsibility for the monthly observations since 1980. The staff and students of Hillcrest Secondary School, Nairobi, especially Lindsay Berners-Price, Fay Ellyat, Patrick O'Dwyer, Phil Lewis, Kenneth Chapman, David Bowan, also carried out valuable glacier observations on their episodic trips to Mount Kenya.

At the University of Wisconsin, Richard Kleinheinz, Nivana Cheng, Paul Lobue, David Harding, and David Porter drafted most of the maps and figures. Mrs. Eva Singer again typed the various generations of the manuscript.

I am indebted to Paul Temple, University of Birmingham, John Whittow, University of Reading, and Karl Butzer, University of Chicago, for reviews of the manuscript.

For permission to reproduce illustrations, I thank Brian Baker, Karl Butzer, Frank Charnley, J. Coetzee, Mrs. Myrtle Dutton, John Flenley, Olov Hedberg, Paul Mohr, Alan Root, George Salt, Paul C. Spink, Paul Temple, L. A. J. Williams; the Alpine Club, American Association for the Advancement of Science, Deutscher Alpenverein Sektion Stuttgart, Geological Society of America, Mountain Club of Kenya, Royal Geographical Society; Balkema Publishers – Rotterdam, Bell and Hyman Ltd. – London, Butterworths Publishers – London, Jonathan Cape Publishers – London, Orell Füssli Verlag – Zürich; Geografiska Annaler, Journal of Glaciology, Svensk Botanisk Tidskrift.

It is above all people, not institutions, that matter in high mountain research. However, this study would not have become a reality without the material support I received over the years. The 1973–74 pilot project on Mount Kenya was aided by Vote 670–226 from the Deans' Committee of the University of Nairobi. The field programs on Lewis Glacier during 1977–83 and the work at the University of Wisconsin were supported by U.S. National Science Foundation Grants EAR76–18881, EAR77–13130, EAR79–23897, and INT79–19841, and National Geographic Society Grant 2239–80. For the 1977–78 and 1981–82 field seasons, I furthermore acknowledge Grants 180113 and 120108 from the Research Committee of the Graduate School, University of Wisconsin, Madison. Support by the Global Environment Monitoring System of the United Nations Environment Programme aided in the continuation of the monitoring project on Mount Kenya beyond 1981. Equipment was loaned by the Departments of Civil and Environmental Engineering, and Geology and Geophysics of the University of Wisconsin, Madison. In East Africa, equipment was placed at our disposal by the Kenya Water, the Mines, and the Meteorology Departments.

The 1974 field research in the Ruwenzori was authorized by the Office of the President of the Republic of Uganda, and the 1977–83 field program on Mount Kenya by the Office of the President of the Republic of Kenya. The Kenya Air Force flew airborne photogrammetric surveys of Lewis Glacier in 1974, 1978 and 1982, and the Air Survey and Development GmbH in 1982. The Kenya Meteorological Department, the Kenya Water Department, and the Administration of the Mount Kenya National Park joined in support of regular precipitation measurements on Mount Kenya. The continued monitoring of climate and glacier variations in equatorial East Africa requires some material commitment and will depend on the realization of priorities at the national and international levels. At the time of this writing, funding for the continuation of the glacier observation program on Mount Kenya is uncertain.

STEFAN HASTENRATH
Nairobi, January 1983

TABLE OF CONTENTS

LIST OF FIGURES

LIST OF PHOTOGRAPHS

LIST OF MAPS

Asterisks indicate maps larger than page size (back pocket)

LIST OF TABLES

CHAPTER 1

INTRODUCTION

"I endeavoured to explain to my people the nature of that 'white thing', for which no name exists even in the language of Jagga itself. . . . It made a singular impression on my mind in the view of the beautiful snow mountain so near to the Equator . . ."

(Johannes Rebmann, Narrative, 1849)

Glaciers near the Equator are found in three regions of the World: the Ecuadorian Andes, the cordilleras of New Guinea, and the high mountains of East Africa. The ice extent in any of these regions has never amounted to more than a small fraction of the total land surface. However, in terms of their response to atmospheric forcing, the glaciers are highly sensitive, though complex, indicators of long-term climatic variations. It is therefore not surprising that the inventory of present ice extent and the monitoring of glacier fluctuations have been identified as tasks of high priority in international endeavors directed at the understanding of large-scale environmental change (UNESCO, 1970; World Meteorological Organization – ICSU, 1975, pp. 7, 11, 16; International Association of Hydrological Sciences – UNESCO, 1977; Temporary Technical Secretariat for World Glacier Inventory, UNESCO–UNEP–IASH–ICSI, 1977). The glaciers of Irian Jaya have been investigated by the Australian Universities Expeditions in the early 1970s (Hope *et al.*, 1976; Allison and Kruss, 1977). The glaciation of the Ecuadorian Andes from Pleistocene to present has been the topic of a recent monograph (Hastenrath, 1981). The present publication attempts to update the knowledge on the glaciers of equatorial East Africa.

The existence of snow and ice in the equatorial Andes was known since the 1500s and historical accounts extend over more than four centuries (Hastenrath, 1981). The ice-capped Mount Carstensz (Jaya) in Western New Guinea was first sighted in 1623, but this report was disbelieved until around 1900. The peak region was first reached in 1912, and visits since then have remained rare (Hope *et al.*, 1976).

By comparison, the glaciated mountains in East Africa (Map 1:1) were discovered rather late. Rebmann's claim in 1848 of snow on Kilimanjaro and Krapf's observation in 1849 of ice-clad Mount Kenya were met with scepticism (Rebmann, 1849a, b, 1855; Krapf, 1849, 1858, 1860; Gumprecht, 1849, 1853; Petermann, 1853; Barth, 1862). Kersten and von der Decken visited Kilimanjaro in 1861–62 (Kersten, 1863; von der

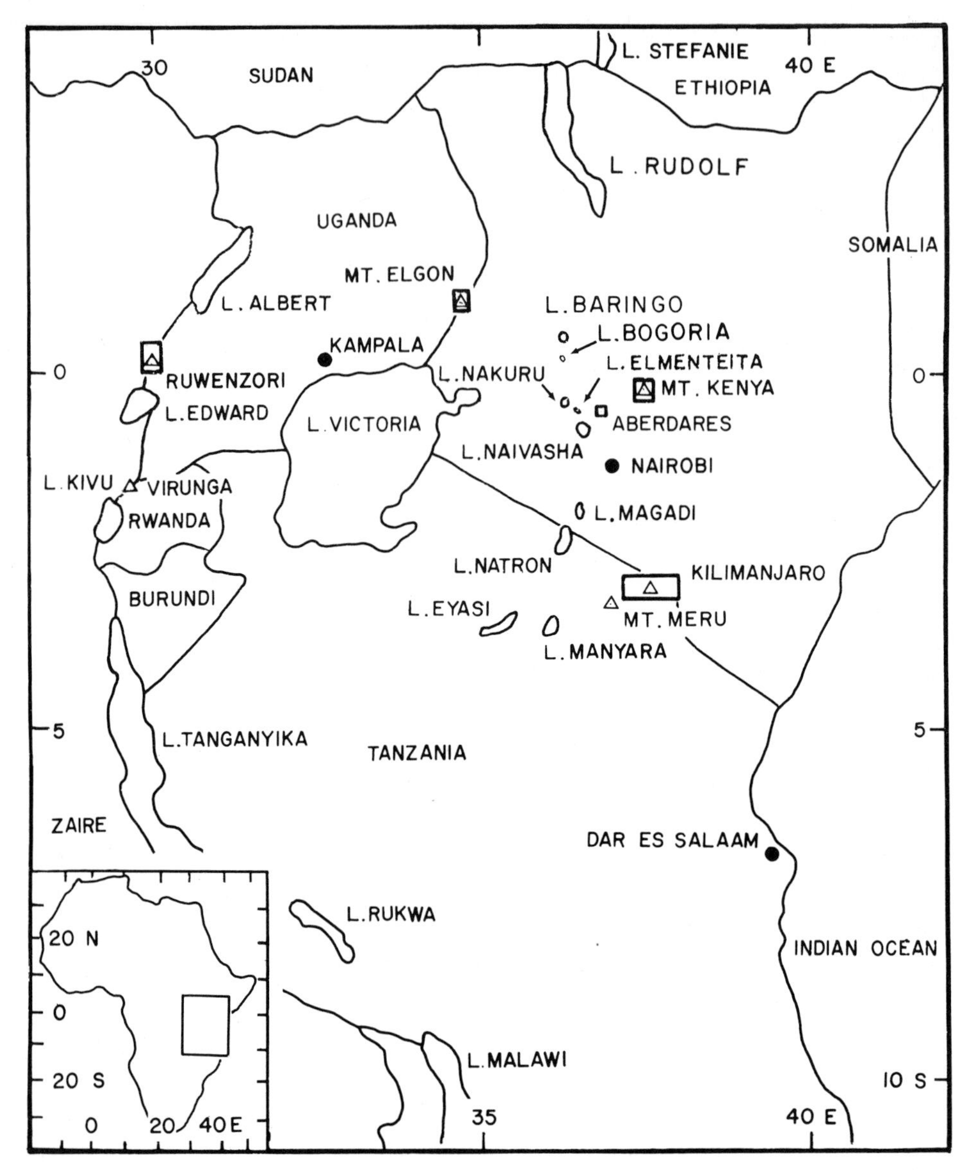

Map 1:1. Orientation map. Boxes indicate the location of Maps 3.1:1, 3.2:1, 3.3:1, 3.3:2, 3.4:1, 3.5:1, 4.1:1, 4.2:1, 4.3:1, 4.3:8. Insert illustrates area depicted in Maps 1.1 and 2.1:1.

Decken, 1869; Methner, 1932). Hildebrandt (1878a, b), travelled in Tanganyika in the 1870s. Reports of snow in the Ruwenzori date from the 1870s and 1880s, but the first visits to the peak regions did not materialize until much later (Stuhlmann, 1894; Kassner, 1911; Johnson, 1912; Weiss, 1917; see also Osmaston and Pasteur, 1972, p. 148–149).

Hans Meyer led expeditions to Kilimanjaro in 1887, 1889, and 1898, and in 1889 succeeded in first climbing the summit of Kibo (Meyer 1887a, b, 1888a, b, 1890a, b, c, d, e, 1891a, b, 1898, 1899, 1900a, b, 1901, 1928; Wichmann 1889, 1912; Reischel 1899;

Anonymous, 1900). He gathered the first systematic observations on the Pleistocene and recent glaciations. His early investigations were expanded by the expeditions of Jaeger (1909, 1931, 1968) in 1906–7, Klute (1914, 1920, 1921, 1929; Klute and Oehler, 1920; Reck, 1921–22; Gillman, 1923a) in 1912, and Nilsson (1931, 1935, 1940, 1949) in 1927–28. The University of Sheffield expeditions performed extensive field work in 1953, 1957, and 1967 (Downie and Wilkinson, 1972). Results regarding the Pleistocene and recent glaciation have been presented by Downie *et al.* (1956), Downie (1964), and Humphries (1959, 1972). Visits to the peak region of Kilimanjaro have been numerous since the end of the last century (von Eberstein, 1887; Ehlers, 1889; Volkens, 1894, 1897; Lent 1894a, b; Johannes, 1899a, b; Chanel, 1899; Uhlig, 1904, 1905, 1908; Lange, 1912; West 1915; Gillman, 1923b, 1944; Latham, 1926; von Salis, 1926; Mosterz, 1929/30; Geilinger, 1930, 1936, 1937; Anonymous, 1937; Methner, 1932; Rice, 1932a, b; Flückinger, 1934; Wyss-Dunant, 1935, 1938; Spink, 1943, 1944, 1945, 1947, 1949, 1952a, b; Hunt, 1947; Jeannel, 1950; Salt, 1951). Photographs of Kibo from the air have been taken on a few occasions (Anonymous, 1931, 1932; Mittelholzer, 1930, 1932; Light, 1941; Kurz, 1948; Tuckett, 1931; Nicol, 1964; Root, 1974).

The ice regions of the Ruwenzori were reached by several parties towards the end of the past and in the early years of this century (Osmaston and Pasteur, 1972, p. 149–150; Stanley, 1890; Stuhlmann, 1894; Scott-Elliot, 1895; Moore, 1901; Anonymous, 1904, 1912; Thomas, 1947; David, 1904, 1905; Revelli, 1906; Forster, 1901, 1902, 1907; Freshfield, 1906; Wollaston, 1908; Wauters, 1905, 1909; Behrens, 1906; Bright, 1908; Dawe, 1906; Maury, 1912; Alluaud, 1910; Henrici, 1911–12; Kmunke, 1913; see also Freiberg, 1929). The Duke of the Abruzzi led a mountaineering and scientific expedition to the Ruwenzori from the Uganda side in 1906. In addition to the biological and geological exploits, the expedition accounts (Cora, 1906; Roccati, 1907, 1909; Almagià, 1908; Filippi, 1908a, b, 1909a, b; Abruzzi, 1907a, b, 1909a, b) provide a rich photographic documentation of ice extent of several glaciers around the beginning of the century.

In 1927–8, Nilsson (1931) studied the Pleistocene moraine morphology on the East side of the mountain range. A Belgian scientific expedition explored the Ruwenzori from the Zaïre side in 1932 (Grunne, 1933; Grunne *et al.*, 1937; La Vallée Poussin, 1933; Meersch, 1933; Michot, 1933, 1937). Synge (1937) visited the mountain in 1934. A stereo-photogrammetric mapping of the peak region at scale 1:25 000 was performed by an expedition of the Deutscher Alpenverein in 1938 (Eisenmann, 1939a, b; Stumpp, 1952). Belgian and British expeditions in 1952 were concerned with the geology, biology, and glaciology of the Ruwenzori (Menzies, 1951a, b; Bergström, 1953, 1955; Kennedy, 1953; de Heinzelin, 1951, 1952, 1953, 1962; Ross, 1955a, b; Osmaston, 1965). Five scientific expeditions were undertaken by Makerere University in the course of 1957–60, with emphasis on the glaciers of the Ruwenzori (Whittow, 1959, 1960, 1966; Whittow and Shepherd, 1959; Whittow *et al.*, 1963; Whittow and Osmaston, 1966); In 1960 Livingstone (1962) studied lake sediments and moraine chronology on the East side of the Ruwenzori. Useful photographic documentation on varying ice extent has been created by the frequent visits to the peak regions of the Ruwenzori from the 1920s to the 1960s (Chapin, 1927; Humphreys, 1927, 1933; Shipton, 1932; Hodgkin, 1941, 1944; Firmin, 1945; Hicks, 1946, 1947; Eisenmann, 1939a, b; Stumpp, 1952; British Museum, 1939; Hicks, 1946, 1947; Bere, 1946; Busk, 1954a, b; Temple, 1968; Fantin, 1968), while access has become more difficult in the 1970s.

Gregory (1893, 1894, 1900) undertook the first scientific expedition to the peak region of Mount Kenya. Mackinder (1900) achieved the first ascent of the highest peak in 1899. During 1927–8 and 1932 Nilsson (1931, 1935, 1940, 1949) searched for evidence of Pleistocene glaciation in the mountains of East Africa, and carried out field research on Mount Kenya, as well as at the Ruwenzori, Kilimanjaro, Mt. Elgon, and the Aberdares. A comprehensive survey of the Pleistocene geology of Mount Kenya was completed by Baker (1967) in the 1960s. Visits to the peak regions with occasional accounts have become abundant since the 1920s (Anonymous, 1906, 1929; McGregor Ross, 1911; Arthur, 1921; Dutton, 1929; Harris, 1929; Mountain Club of East Africa, 1932; Shipton, 1931, 1956; Tilman, 1937; Douglas-Hamilton, 1941–2; Firmin, 1945–6; Hicks 1945–6; Cameron and Reade, 1950; Nicol, 1954; Howard, 1957). Troll and Wien (1949) produced the first detailed map of Lewis Glacier. The next map was produced in the course of the IGY Mount Kenya Expedition in 1958 (Charnley, 1959). In 1963 E. Schneider mapped the entire peak region at a scale of 1:5000 (Forschungsunternehmen Nepal-Himalaya, 1967).

Investigations on the vegetation and vegetation history in the highlands of East Africa and their climatic significance have been carried out by Hedberg (1951, 1964), Salt (1954), Van Zinderen Bakker (1962, 1964), Coetzee (1964, 1967, 1978), Van Zinderen Bakker and Coetzee (1972), Livingstone (1967, 1975, 1976a, b, 1980), Kendall (1969), Livingstone and Kendall (1969) Kendall and Livingstone (1967), Morrison (1961, 1966), Hamilton (1972, 1973, 1974), Morrison and Hamilton (1974), Hecky (1978), and others. Observations on patterned ground and soil frost phenomena have been reported by Flückinger (1934), Troll (1944), Zeuner (1949), Hastenrath (1973, 1974a, b, 1978), Furrer and Freund (1973), Furrer and Graf (1978), Winiger (1979, 1981). Long-term lake level variations have been studied by Richardson (1966), Washbourn-Kamau (1967); Kendall and Livingstone (1967), Livingstone and Kendall (1969), Butzer (1971, 1980), Butzer *et al.* (1972), Richardson and Richardson (1972), Grove *et al.* (1975), Livingstone (1975, 1976a, b, 1980), Street and Grove (1976, 1979). Loeffler (1968) studied the modern limnology of East African mountain lakes.

The present monograph is based primarily on field research during the course of the 1970s and early 1980s. The first visits to the peak regions of Kilimanjaro and Mount Kenya materialized in June–July 1971. During my association with the University of Nairobi in 1973–74, I undertook field trips to various high mountains of East Africa, and started the field program on Lewis Glacier. Field expeditions to Lewis Glacier were further undertaken in December 1977 to March 1978, in December 1978 to January 1979, December 1979 to January 1980, July–August 1980, December 1980 to January 1981, December 1981 to January 1982, and December 1982 to January 1983. A basic observation program related to the mass and heat budget characteristics was maintained throughout the intervening intervals.

Evidence of continued and drastic glacier recession in various high mountains of East Africa (Hastenrath, 1975), as well as in other parts of the tropics, invited a survey of current ice extent and an assessment of long-term glacier behavior. These endeavors broadly coincide with the objectives of the international programs of a World Glacier Inventory and of the monitoring of the Fluctuations of Glaciers (Temporary Technical Secretariat for World Glacier Inventory, UNESCO–UNEP–IUGG–IASH–ICSI, 1977; International Association of Hydrological Sciences – UNESCO, 1977). The survey of current ice extent and the reconstruction of glacier variations since the end of the last century combines own field observations with the evaluation of earlier photographs and of other historical sources.

The ultimate aim of such efforts is the quantitative inference of the climatic variations from the apparent glacier response. To this end, a detailed study of a few selected glaciers is called for. In all of the tropics, Lewis Glacier on Mount Kenya seems most suited for elucidating the relationship between climatic forcing and glacier response. The Lewis is the largest glacier of Mount Kenya; its catchment area is well defined; it is of comparatively easy access; but most importantly, the detailed mappings of the ice surface topography accomplished in 1963, 1958, and 1934, along with the historical photographs from earlier decades and dating back to Mackinder's expedition in 1899, provide a documentation on the secular changes in ice extent and volume that is unparalleled in the entire tropics.

Therefore a multi-annual field program was undertaken on Lewis Glacier, including the monitoring of surficial ice movement, measurements related to the mass and heat budget, total β radioactivity, microparticle and oxygen isotope analysis of ice cores, determinations, of the bedrock topography, and repeated mappings of the ice surface topography. This field program along with the earlier, unique historical documentation forms the observational basis for the theoretical study and the numerical modelling of climatic forcing and glacier response. This novel approach leads to a quantitative reconstruction of climate variations from the observed glacier behavior. In view of the similarity of fluctuations of other glaciers on Mount Kenya, in East Africa at large, and in other high mountain regions of the tropics, the results from the detailed study of Lewis Glacier are expected to be of general importance.

Furthermore, an attempt was made to obtain a picture of glaciations in Pleistocene and early Holocene times by combining own observations in the field with evaluation of air photographs and the study of the pertinent literature. Comprehensive studies have been published on the Pleistocene morphology of Kilimanjaro (Downie and Wilkinson, 1972) and Mount Kenya (Baker, 1967), although absolute dating of glaciations is lacking. Extensive work on the Ruwenzori is not readily accessible (Osmaston, 1965). Following the pioneering investigations of Nilsson (1931, 1935), the recent work of Perrott (1982a) for the Aberdares, and of Hamilton and Perrott (1978, 1979) for Mount Elgon is of interest. Evaluation of air photography in conjunction with own field observations further contribute to the survey of the Pleistocene glaciation in these mountain regions. The synopsis of geomorphic evidence of Pleistocene to early Holocene glaciations is expected to complement the study of the recent glaciation and of climate–glacier relationships.

As a background to the study of the glaciers, the environmental setting of East Africa is reviewed in Chapter 2. The Pleistocene to early Holocene glaciation is discussed in Chapter 3, and the recent ice extent in Chapter 4 and Appendix 3. The study of Lewis Glacier is the topic of Chapter 5. Topographic charts, air photographs, and satellite imagery, are listed in Appendix 1, historical documentation in Appendix 2. Chapter 6 presents an appraisal of East African glaciers in perspective with environmental change throughout the global tropics.

CHAPTER 2

THE ENVIRONMENTAL SETTING

" . . . the thing about the geology of Africa that strikes one as especially significant is, that throughout this vast area just opening up to science there is nothing new."

(Henry Drummond, *Tropical Africa*, 1890)

"Et comme toutes ces montagnes . . . sont caractérisées par des flores très semblables entre elles, mais d'un type unique sur la planète . . . les problèmes les plus intéressants se présentent . . . sur . . . ces formes végétales si particulières."

(Lucien Hauman: in *Grunne, Ruwenzori*, 1937)

"The epochs of probably both colder and moister climate . . . have . . . extended their effects to the areas between those mountains, and we find everywhere evidence of such epochs in the numerous . . . ancient lake basins now wholly or partly dried up."

(Erik Nilsson, *Pluvial Lakes*, 1931)

The purpose of this chapter is to provide the framework of the physical environment that is necessary for the study of East African glaciers. To this end, the gross physiographic structure, large-scale circulation and climate, altitudinal zonation of vegetation and subnival soil forms, and vegetation and lake history, are discussed in particular.

2.1. Physiographic Structure

Flanked as they are by the Indian Ocean and the Congo Basin, the highlands of East Africa carry the highest mountains of the continent. The gross physiography (Map 2.1:1, and Figs. 2.1:1 and 2.1:2) is dominated by two broadly meridionally oriented rift systems which are part of tectonic structures of continental dimensions (Krenkel, 1925, pp. 41, 227–240; Furon, 1963, pp. 283–293, 317–327; Baker *et al.*, 1972, pp. 7, 25–31, 40–41; Morgan, 1973, pp. 4–27). From the bifurcation in the Lake Malawi (Nyasa) area of Southern Tanzania, the Eastern Rift extends through Central Kenya into Ethiopia. Embedded in this trough are the Lakes Eyasi, Natron, Magadi, Naivasha, Elmenteita, Nakuru, Baringo, Rudolf (Turkana), and several smaller lakes in Southern Ethiopia. The Western Rift harbors a chain of lakes extending from Lake Malawi northward into the Sudan, namely Lakes Tanganyika, Kivu, Edward, and Albert (Mobutu).

The gross tectonic structure of the Eastern Rift system is illustrated in Map 2.1:2. The Eastern Rift (Baker *et al.*, 1972, pp. 25–41; Skinner, 1977) consists of a complex of grabens and depressions which separate the East Kenya–Somalia crustal block from the remainder of Africa. Grabens are well-defined on the central parts and South slopes of the Kenyan and Ethiopian uplifts (Gregory and Ethiopian rifts). Splay faults across

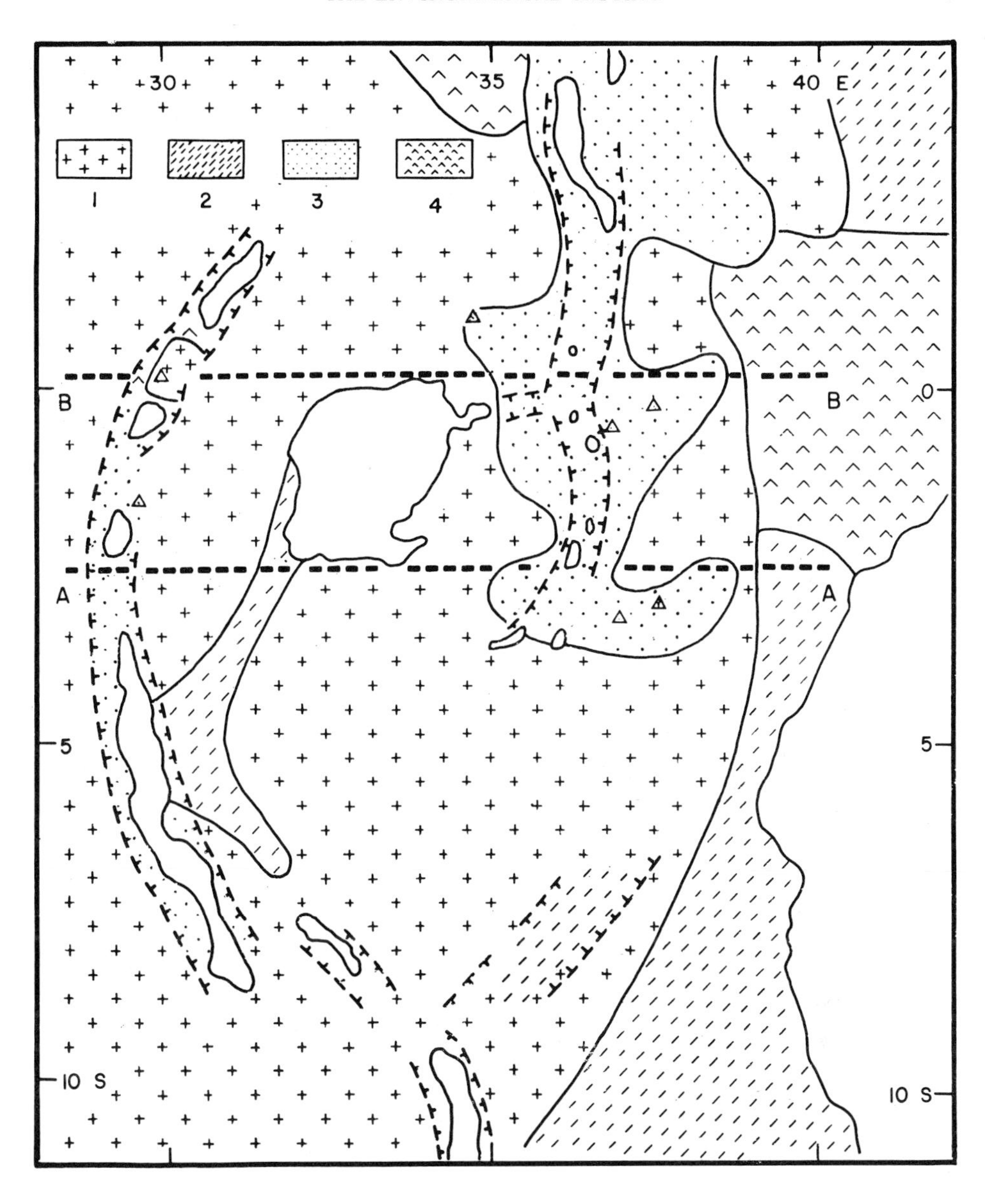

Map 2.1:1. Physiographic sketch. 1. Precambrian gneiss, schists, and quartzite; 2. Palaeozoic and Mesozoic sediments; 3. Tertiary and Quaternary volcanics; 4. Other Quaternary sediments. Dented broken lines indicate fault lines with dents pointing downslope. Broken lines labelled A–A and B–B denote position of transects in Figs. 2.1:2 and 2.1:1.

broad depressions characterize the less uplifted areas in North Tanzania and North Kenya. At the North end of the Eastern Rift the main faults diverge. Proceeding from South to North five main sections of the Eastern Rift are distinguished: the North Tanzania Divergence, the Gregory and Kavirondo Rifts, the Turkana Depression, the main Ethiopian Rift, and the Afar Depression. The North Tanzania Divergence is located in the transition

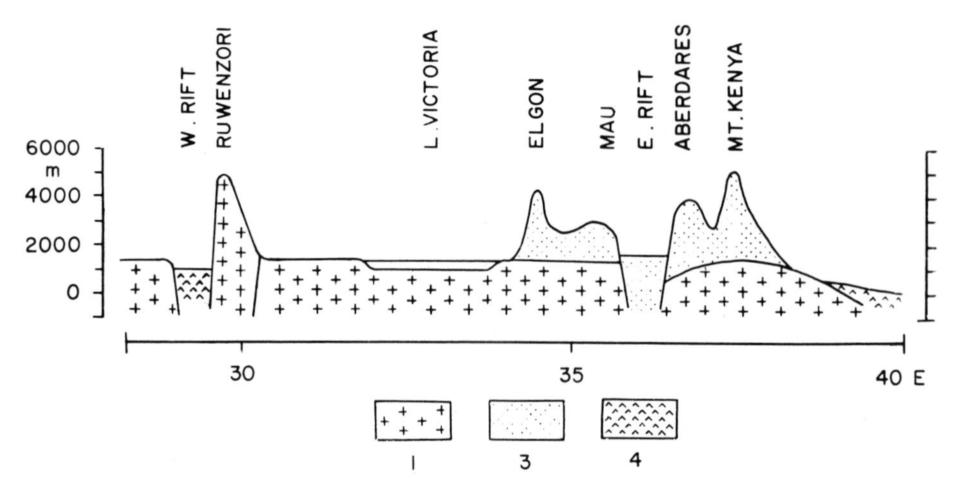

Fig. 2.1:1. Schematic zonal transect at 0–1°N. Symbols as Map 2.1:1. Vertical exaggeration of topography is fivefold. See Map 2.1:1 for location of transect B–B.

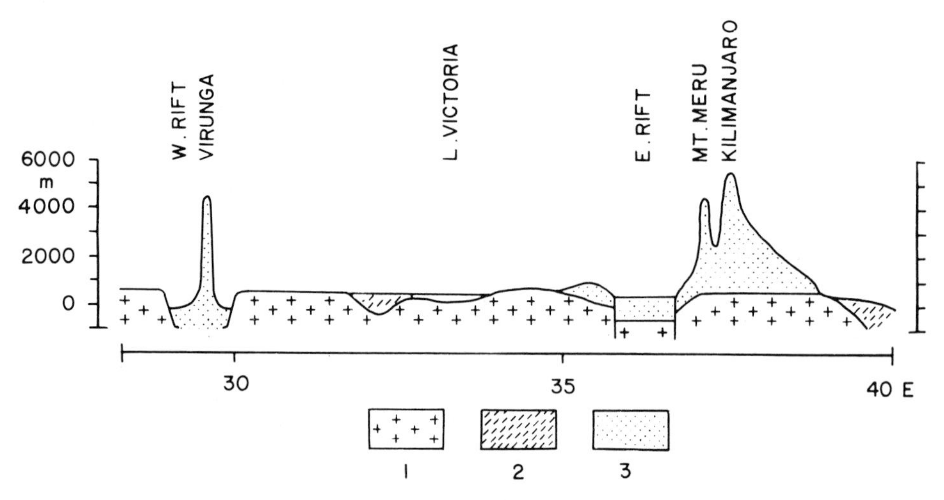

Fig. 2.1:2. Schematic zonal transect at 2–3°S. Symbols as Map 2.1:1 and Fig. 2.1:1. Vertical exaggeration of topography is fivefold. See Map 2.1:1 for location of transect A–A.

(around 2–3° S) from the interior plateau of North Tanzania to the South flank of the Kenya domal uplift. The Gregory Rift extends from the Natron–Magadi basin northward, and bisects the Kenya domal uplift. At the center of the Kenya uplift, the Kavirondo Rift branches from the Gregory Rift, trending South of West into Lake Victoria. North of 1° N, the Kenya Rift widens and passes into the Turkana Depression. The main Ethiopian Rift extends between the Kenyan and Ethiopian domes. According to Baker *et al.* (1973) the Eastern Rift was formed by repeated warpings of the crust, preceded and accompanied by faulting and volcanism. The margins of the domal uplifts, such as Turkana sandwiched between the Kenyan and Ethiopian uplifts and North Tanzania, are characterized by divergent fault systems crossing broad depressions.

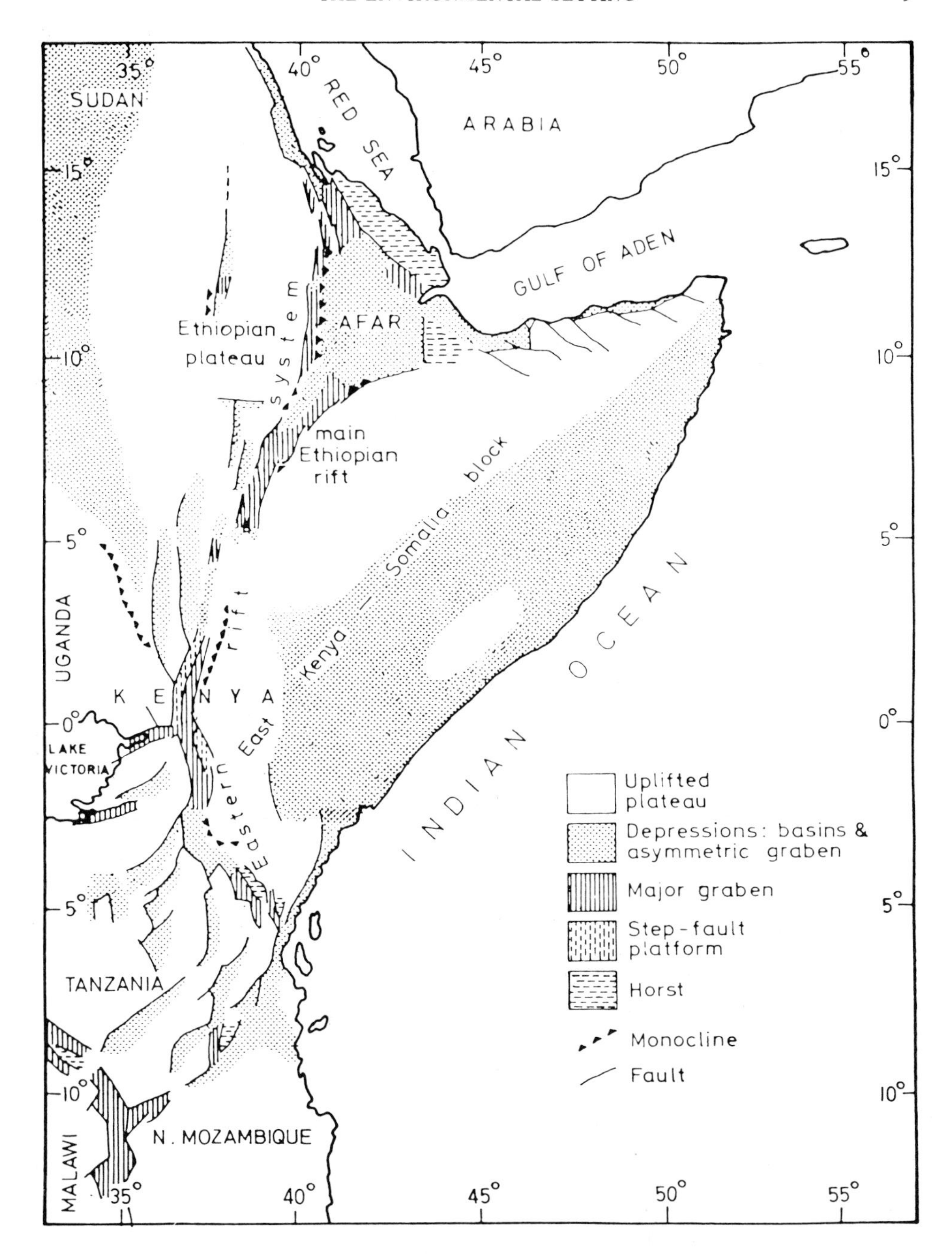

Map 2.1:2. Structural elements of the Eastern Rift system (from Baker *et al.*, 1972).

Precambrian metamorphic and igneous rocks form the foundation of the continental threshold (Furon, 1963, pp. 317–318; Baker *et al.*, 1972, pp. 6–8; Morgan, 1973, pp. 5–7) and surface in a broad shield extending from Southern Tanzania through Kenya and Uganda to the Southern Sudan and Ethiopia. Lake Victoria is a shallow depression in this shield. Uplift and warping followed by volcanism and faulting of the African basement began in the early Tertiary and continued in several phases through the Pleistocene. Only minor faulting has occurred since the later mid-Pleistocene (Baker *et al.*, 1972, p. 40). Development of the large rift systems, volcanism, and formation of the large lakes are all results of this continued tectonic activity. Upper Palaeozoic and Mesozoic sediments are mostly found preserved in the tectonic troughs (Furon, 1963, p. 311). Tertiary to recent volcanics constitute the uppermost portion of the highlands extending from Northern Tanzania through Central Kenya into Eastern Uganda and Southern Ethiopia (Furon, 1963, pp. 317–327; Baker *et al.*, 1972, pp. 25–31; Morgan, 1973, pp. 4–12).

Baker *et al.* (1972) presented a preliminary history of volcanism in Eastern Africa (Map 2.1:3). The largest major volcanic episode in Eocene-Oligocene produced the Trap Series basalts which cover much of the Ethiopian and Somali plateaus. In East Africa, the Miocene marks the onset of nephelinite volcanoes in the Kenya–Uganda border region. Basaltic eruptions in the Turkana Depression are contemporaneous. These were followed by flood phonolites on the Kenya dome from late Miocene to early Pliocene, and trachytes from Pliocene onward. During the Pliocene, nephelinite centers were characteristic of the Gregory Rift and the Kavirondo trough; basalt-trachyte volcanoes erupted on the rift floor in Central Kenya and North Tanzania; and to the East of the Rift, the phonolitic shield volcano of Mount Kenya was built up. During the Quaternary, nephelinite-carbonatite volcanism occurred in North Tanzania. During the Holocene, Mount Meru produced nephelinites. At Kilimanjaro, late Pleistocene–Holocene flows are phonolitic, following a predominantly basaltic phase.

It is at the margins of the main domal volcanic mass of Kenya that four volcanoes loom to great elevations: Mounts Kilimanjaro and Meru to the South, Mount Kenya to the East, and Mount Elgon in the extreme Northwest. The former two mountains consist mainly of basalt, phonolite, nephelinite, trachyte, and a minor variety of other volcanic rocks (Downie and Wilkinson, 1972, pp. 115–186; Baker *et al.*, 1972, pp. 11–17). The rocks of Mount Kenya include basalt, phonolite, and trachyte (Baker, 1967, pp. 19–57; Baker *et al.*, 1972, pp. 11–16). Mount Elgon consists in part of nephelinite and trachyte (Furon, 1963, p. 325; Baker *et al.*, 1972, pp. 11–17). The Aberdares, likewise of volcanic origin, and made up largely of basalts (Baker *et al.*, 1972, p. 15) form the Eastern shoulder of the great Eastern Rift crossing the Kenya highlands. Volcanism, mainly of basaltic nature (Furon, 1963, p. 283), also led to the buildup of the high Virunga mountains sitting in the great Western Rift, between lakes Kivu and Edward. Finally the old Precambrian basement of gneiss, schists, and quartzite (Whittow, 1966; Morgan, 1973, p. 26) surfaces as a horst in the block of the Ruwenzori emerging in the Western Rift between Lakes Edward and Albert.

With the exception of the Precambrian crystalline block of the Ruwenzori, mountains high enough to allow Pleistocene or recent glaciations were only created through the vigorous volcanism. Volcanic activity continuing to the present may have obliterated traces of earlier glaciations.

The gross morphology and structure varies remarkably between the individual high mountains of East Africa. Only the major characteristics are sketched here, with reference to the topographic maps in Chapter 3.

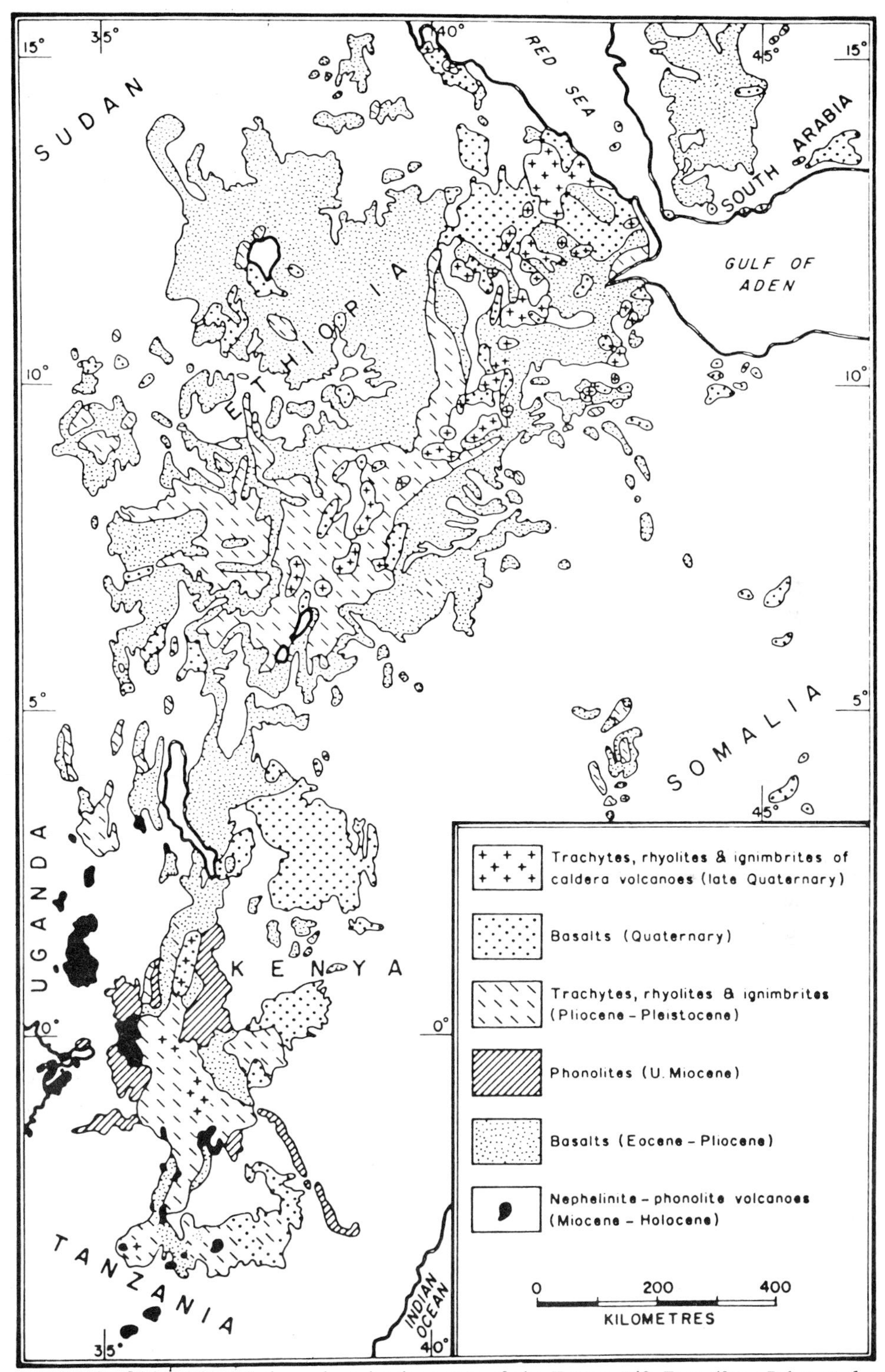

Map 2.1:3. Distribution of the main volcanic groups of the Eastern Rift Zone (from Baker *et al.*, 1972).

The Kilimanjaro massif (Downie and Wilkinson, 1972) rises from a lowland plain to a broadly zonally oriented plateau-like base around 4000 m (Map 3.1:1; Fig. 2.1:2). Perched on this sit the three separate peaks. The easternmost, Mawenzi, is strongly weathered, and is characterized by sharp crests and needles, and precipitous faces. A broad saddle separates Mawenzi from the main peak, Kibo, in the central portion of the massif. Kibo is a volcanic cone of nearly perfect rotational symmetry. The cone and the crater are only broken by the 'Great Breach' in the Western sector. The recent glaciation is essentially confined to this highest peak of the Kilimanjaro massif. To the West of Kibo extends the Shira Plateau which contains a third, very minor peak. To the South of the Kilimanjaro massif, the nearly symmetric volcanic ash and cinder cone of Mount Meru rises directly from the lowland plain.

Mount Kenya (Baker, 1967) is the strongly eroded and dissected central plug of a shield-shaped volcano (Map 3.3:1, Fig. 2:2). The glaciers are embedded in a pronounced rock relief. A meridionally oriented basin separates Mount Kenya from the Aberdares (Map 3.4:1, Fig. 2.1:1) to the West. The Aberdares are a broad meridionally oriented range of volcanics with comparatively gentle topography.

Mount Elgon (Map 3.5:1, Fig. 2.1:1) is a broad volcanic cone with a wide caldera opening towards the Northeast. Abundant vegetation covers all but the higher portions of the mountain. The Virunga are steep volcanic cones mostly covered by lush vegetation.

The Ruwenzori (Whittow, 1966) are the only mountain massif of East Africa consisting of Precambrian crystalline rocks rather than volcanics (Map 3.2:1, Fig. 2.1:1). This is a factor in their distinctly different gross morphology. Whittow (1966) recognizes a strongly fragmented summit planation surface comparable to the 'Gipfelflur' of the Alps, and proposes a sequence of at least three erosion cycles. In terms of extended areas at preferred elevations, that would be suitable for ice accumulation, the planation morphology is believed to be critical in the glaciation history of the Ruwenzori (Whittow, 1966). Erosion and valley systems appear to be at least in part structurally controlled.

The formerly glaciated mountains of Southern Ethiopia are mostly isolated volcanic cones. By contrast, the country's highest mountain massif, the High Semyen in the North, is made up of largely horizontal layers of volcanic rock. Steep escarpments, mountain butresses, and wide valleys dominate the gross morphology.

2.2. Atmospheric Circulation and Climate

The large-scale circulation over equatorial East Africa (Map 2.2:1) is in the following discussed on the basis of surface and upper-air maps and meridional-vertical transects for the months of January, April, July, and October.

The surface climate of the Indian Ocean off the coasts of Eastern Africa has been studied recently on the basis of ship observations during 1911–70. Data processing and analysis have been described in detail elsewhere (Hastenrath and Lamb, 1979). Maps 2.2:2 portray some characteristics of the surface wind field over the Western Indian Ocean. Information of comparable quality is not available for the surface circulation over land, but a rather coarser picture can be assembled for the lower-tropospheric flow pattern (Atkinson and Sadler, 1970, Maps 2.2:3).

During December to February (Maps 2.2:2 and 2.2:3, January) strongest surface heating and a low pressure center (Thompson, 1965, pp. 77–82; Jackson, 1961, maps

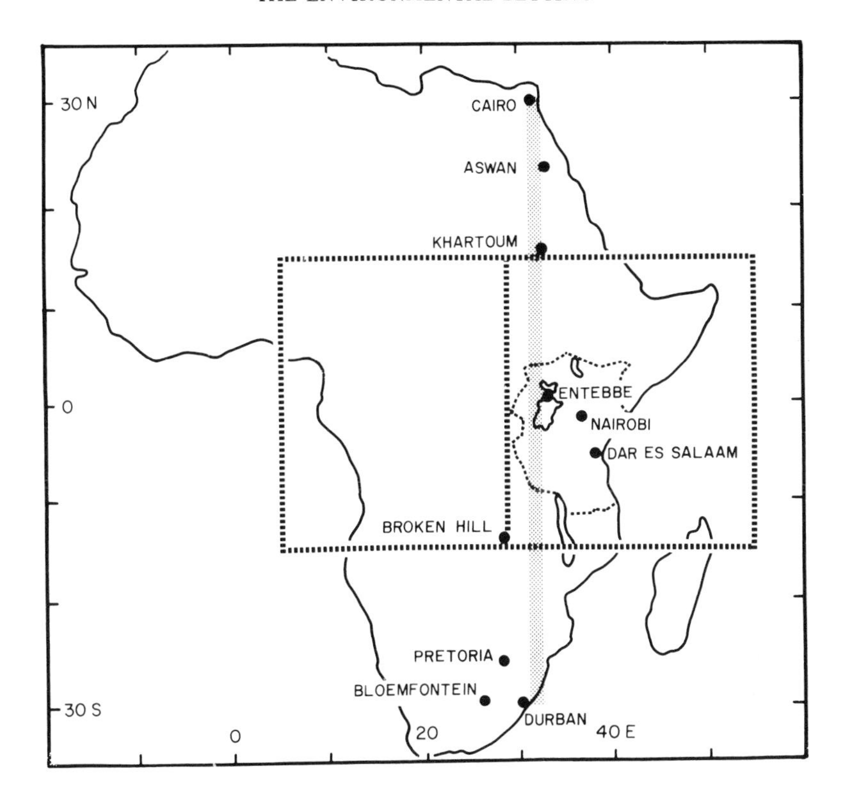

Map 2.2:1. Orientation map. Areas presented in Maps 2.2:2, 2.2:3, 2.2:4, are marked by broken line. Transects in Figs. 2.2:1 and 2.2:2 are indicated by shading.

53–55; Hastenrath and Lamb, 1979, chart 2) are situated in the Southern hemisphere and the Northwest monsoon sweeps the Western Indian Ocean and adjacent East Africa, extending to well beyond the Equator. Cyclonic inflow is apparent in the summertime hot continental region of Southern Africa. The wintertime Northern hemisphere is characterized by clear sky, while largest cloudiness is found over the African continent to the South of the Equator (Maps 2.2:4, January).

Concomitant with the cooling of the Southern hemispheric continent and the northward displacement of the equatorial low pressure trough (Thompson, 1965, pp. 83–88; Jackson, 1961, maps 53–55; Hastenrath and Lamb, 1979, vol. 1, charts 3–6) in the course of March to May is a change in the large-scale flow pattern. In particular, a confluence zone between Southeasterlies from the Southern and Northeasterlies from the Northern hemisphere moves northward (Maps 2.2:2 and 2.2:3, April). Concurrently, skies over Southern hemispheric Africa clear while cloudiness increases to the North of the Equator.

During July and August, strong surface heating and heat lows dominate the lower troposphere over Southern Asia (Thompson, 1965, pp. 89–94; Jackson, 1961, maps 53–55). The Southeasterly trades of the Southern hemisphere recurve near the Equator and form the origin of the Northern hemisphere Southwest monsoon (Maps 2.2:2 and

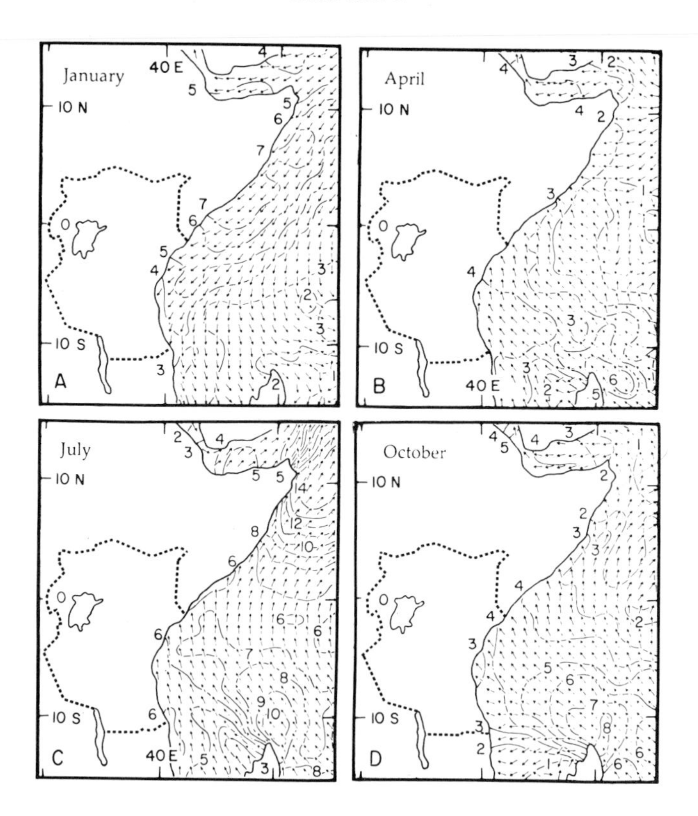

Map 2.2:2. Surface resultant wind direction and speed (m s^{-1}) during January, April, July, and October. Based on ship observations during 1911–70.

2.2:3, July). Associated with the buildup of the Northern summer circulation, cloudiness increases in the Northern hemisphere, but decreases further over Southern Africa (Maps 2.2:4, April, January).

In the course of September to November, temperature contrasts between hemispheres reverse, the lower-tropospheric heat lows over Southern Asia disappear and a belt of minimum pressure is re-established in the equatorial region (Thompson, 1965, pp. 95–100; Jackson, 1961, maps 53–55; Hastenrath and Lamb, 1979, vol. 1, charts 10–12). A confluence zone between currents from the Northern and Southern hemispheres develops again to the North of the Equator (Maps 2.2:2 and 2.2:3, October). This marks the transition to the December to February conditions (Maps 2.2:2 and 2.2:3, January). This development in the circulation pattern is associated with a decrease of cloudiness in the Northern hemisphere, and an increase in the equatorial region and Southern Africa (Maps 2.2:4, July, October, January). Throughout the year, the low-latitude low pressure trough stays to the North of the belt of largest rainfall (Thompson, 1965, pp. 10–14).

The seasonal shift of zonal wind regimes (Serviço Meteorólogico Nacional, 1965) is illustrated in the meridional-vertical cross sections, Fig. 2.2:1 (source: U.S. Weather Bureau, 1958–63). During December to February (Fig. 2.2:1, January), the Northern hemisphere temperate latitude Westerlies extend far equatorward, the maximum being

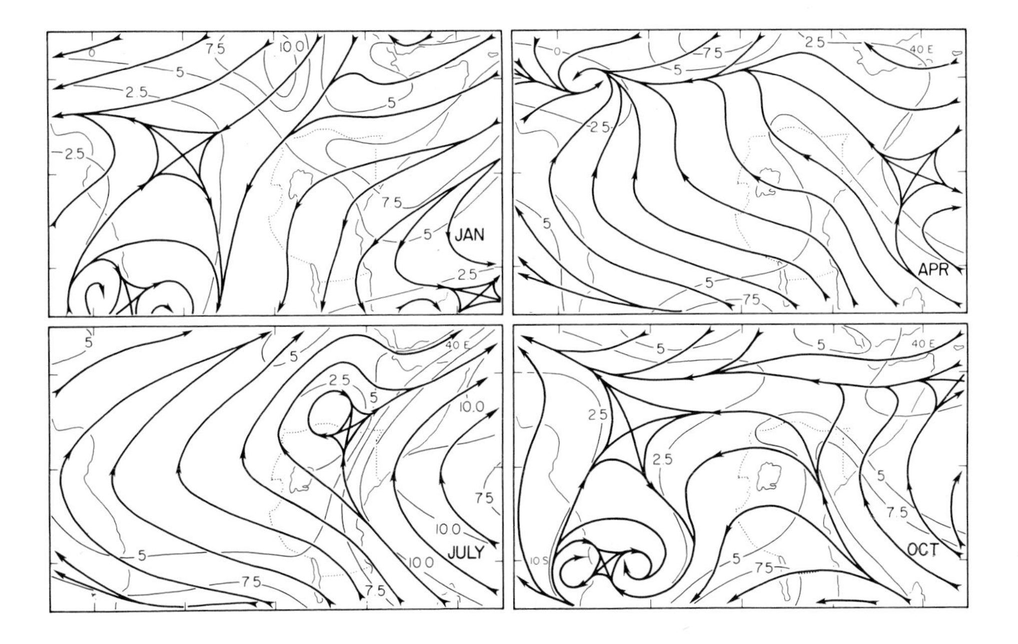

Map 2.2:3. 'Gradient level' streamlines and isotachs (m s^{-1}) during January, April, July, and October. Schematic after Atkinson and Sadler (1970).

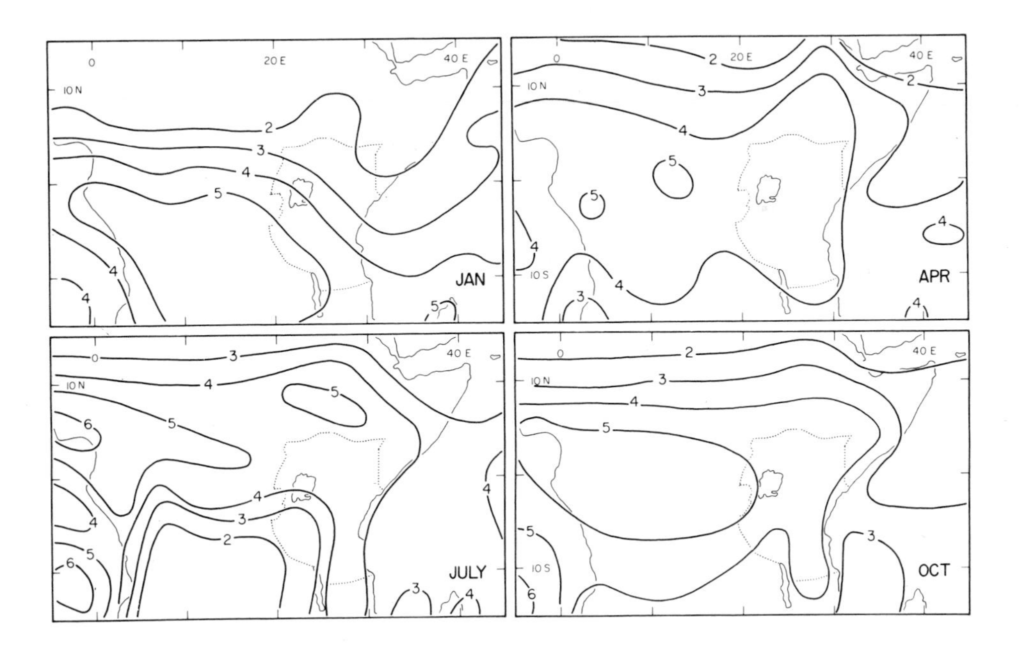

Map. 2.2:4. Total cloudiness in oktas during January, April, July, and October; Schematic after Atkinson and Sadler (1970).

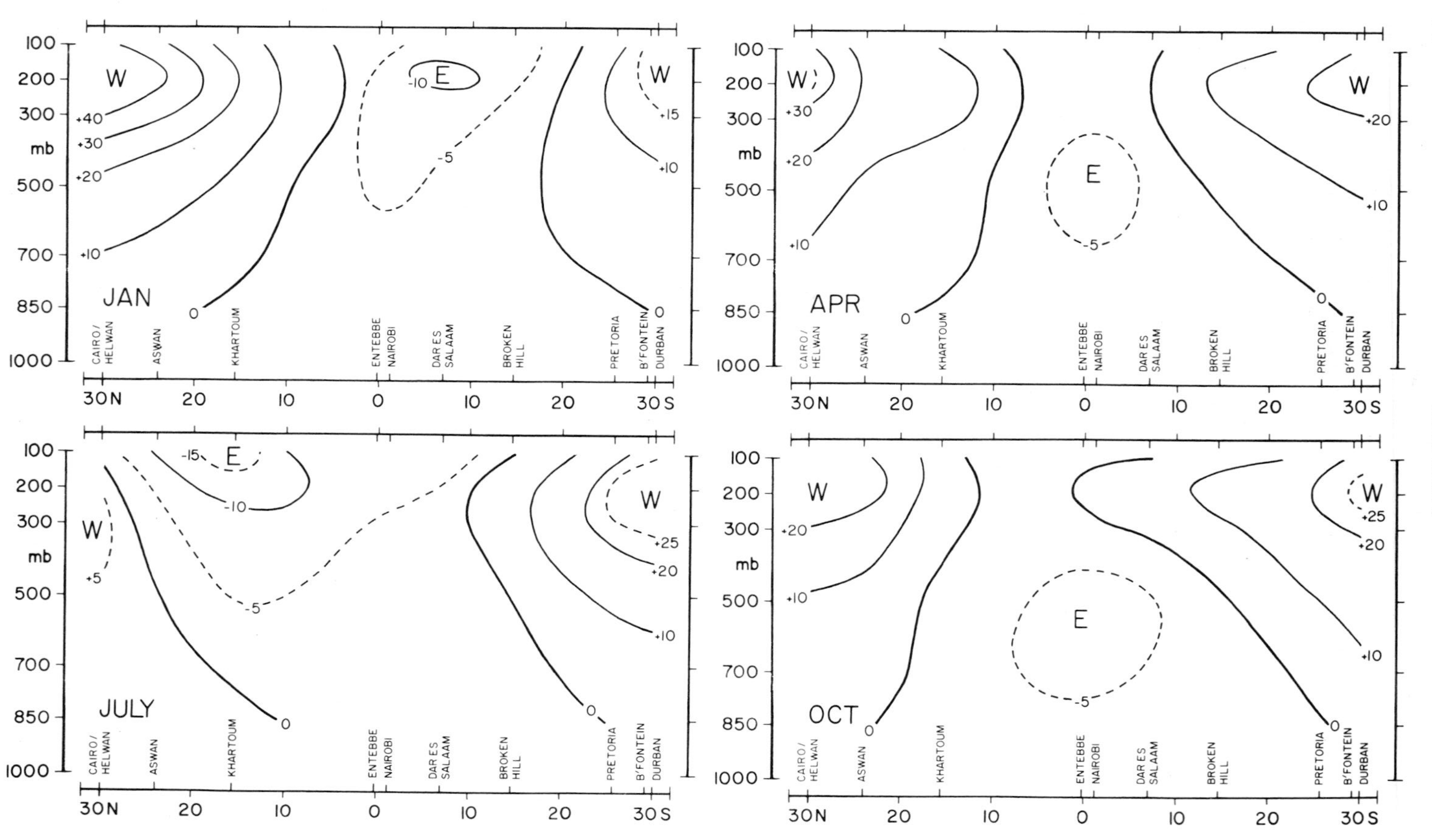

Fig. 2.2:1. Meridional transects of zonal wind component along 30° E during January, April, July, and October. Westerlies positive, isotachs in m s^{-1}. Period 1958–62 (source: U.S. Weather Bureau 1958–63). For approximate location of transects and stations see Map 2.2:1.

found in the upper troposphere near 30°N. The tropical Easterlies occupy a comparatively restricted low-latitude portion of the cross section, while the Southern hemisphere Westerlies are strong throughout the year (Fig. 2.2:1). In the course of March to May, the boundary between the Northern hemisphere temperate latitude Westerlies and tropical Easterlies retracts northward (Fig. 2.2:1, April). In July and August (Fig. 2.2:1, July), the Northern hemisphere temperate latitude Westerlies are weak, while the tropical Easterly wind regime becomes intense. The tropical easterly jet is found in the high troposphere near 15–20°N. The Northern summer monsoon flow reflects itself in Westerly wind components in the lower troposphere in the equatorial region. The Southern hemisphere temperate latitude Westerlies extend furthest equatorward at this time of year. The months September to November (Fig. 2.2:1, October) bring a return to conditions characteristic of the Northern winter circulation. The meridional-vertical cross sections, Fig. 2.2:1, thus complement the lower-tropospheric flow charts, Maps 2.2:2 and 2.2:3.

Characteristics of the regional climates in low latitudes are largely patterned by the precipitation activity. The pronounced local relief (Section 2.1) implies a strong topographic control of rainfall in East Africa (Jackson, 1961, map 3; East African Meterological Department, 1958, 1971; Johnson and Mörth, 1960; Brown and Cochemé, 1969, pp. 7–48; Griffiths, 1972, pp. 313–348; Morgan, 1973, pp. 29–42). Map 2.2:5 shows the spatial pattern of annual rainfall, and Table 2.2:1 summarizes the altitudinal zonation

TABLE 2.2:1

Altitudinal zonation of East African mountains

1. *Precipitation*: belt of maximum rainfall and high regions, elevation (m), aspect, and annual total (mm). Source: Thompson, 1966; Brown and Cochemé, 1969; Coutts, 1969; Mörth, 1970; East African Meteorological Department, 1971; Osmaston and Pasteur, 1972; Water Department, Republic of Kenya, unpublished records.

2. *Vegetation belts*: altitudinal limits (m). Source: Hedberg, 1951.

	Kilimanjaro	Mount Elgon	Aberdares	Mount Kenya		Ruwenzori
Precipitation						
high regions						
elevation (m)	>4500			4500–4800		3500–4500
annual total (mm)	<200			<900		2000–3000
belt of max rainfall						
elevation (m)	S 1800		E	S, W, E 2500–3000 m	N	
annual total (mm)	2000	1200–1400	2000	>2000	1000	
Vegetation belts						
Alpine belt						
	4100	3550	3650	3600		3800
Ericaceous B. (m)						
	2200	2900	3450	N 2900, S 3400		W 2700, E 2850
Montane Forest B. (m)						

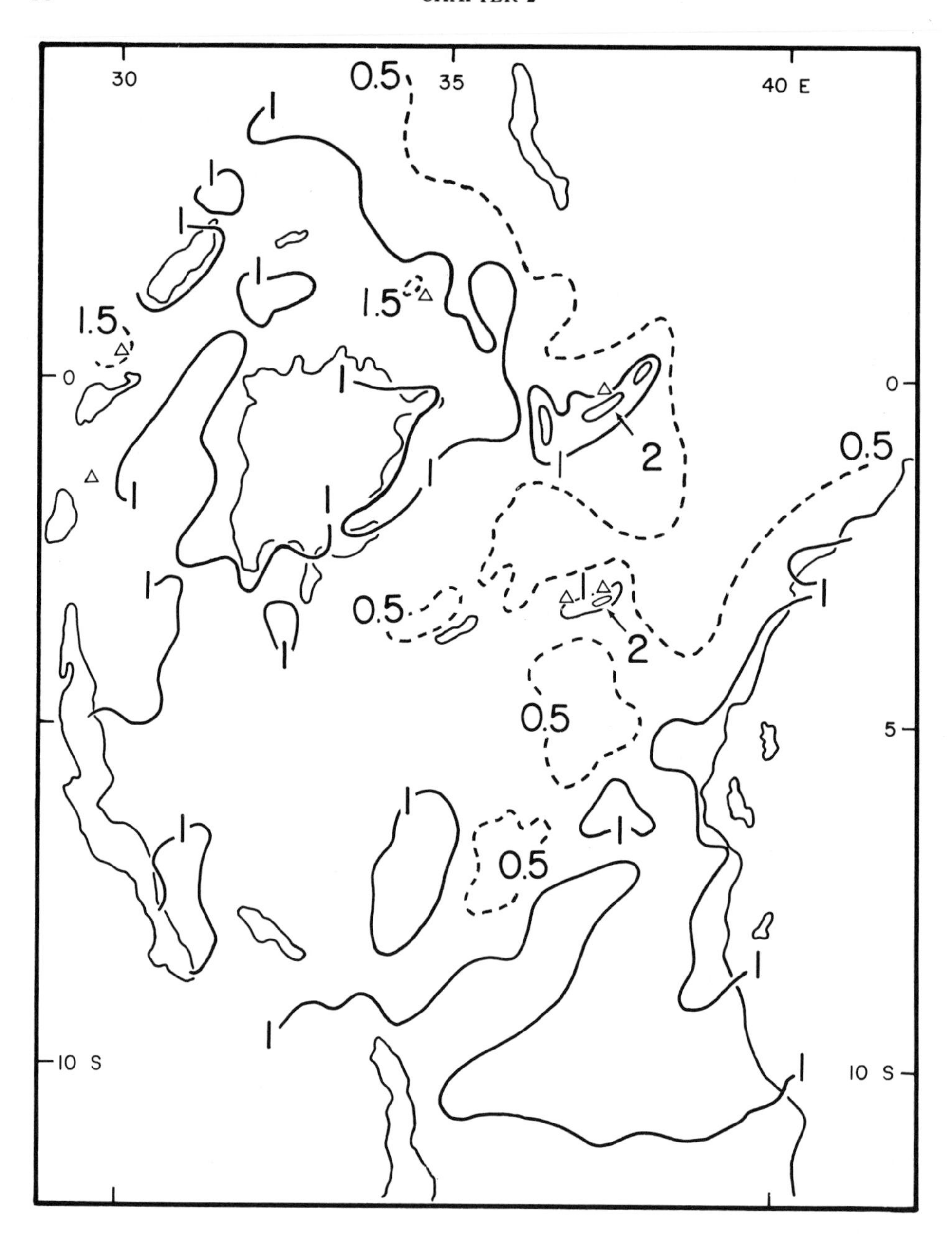

Map. 2.2:5. Annual rainfall in East Africa; in m. Schematic after East African Meteorological Department (1971) and Griffiths (1972).

of precipitation and vegetation on the East African mountains. The arrangement in Table 2.2:1 is from 'dry' (Kilimanjaro) towards 'wet' (Ruwenzori) according to vegetation information and the limited precipitation data for the high regions. The annual rainfall distribution is characterized by a broad decrease towards the semi-arid outer tropics in the North, a tendency for a decrease towards the interior of the highlands, and enhancement of precipitation in the mountains, especially large precipitation being found on the Southern and Eastern flanks of Kilimanjaro and Mount Kenya. A characteristic feature of the rainfall distribution at Kilimanjaro (Coutts, 1967; Mörth, 1970; Salt, 1974), Mount Kenya (Thompson, 1966), other mountains of East Africa – and of the tropics at large for that matter (Hastenrath, 1967) – is the existence of an altitudinal belt of maximum rainfall at intermediate elevations, from where precipitation amounts decrease towards the peak region.

Cloudiness and precipitation conditions differ considerably between the various mountain regions of East Africa (Map 2.2:5, and Table 2.2:1). Kilimanjaro rises out of the dry Masai steppe that has annual rainfall as low as 500 mm. The Kenya highlands harboring Mount Kenya, the Aberdares, and Mount Elgon, are on the whole moister, with annual precipitation totals from around 700 to more than 1300 mm. Western Uganda at the base of the Ruwenzori is likewise rather abundant in rainfall, with annual totals of 1000 to more than 1400 mm (Jackson, 1961; Brown and Cochemé, 1969; East African Meteorological Department, 1971).

The annual march of rainfall differs greatly between various parts of East Africa. A single rainfall maximum mostly between March and May is found in much of the Northern and the coastal regions, broadly coincident in timing with the passage of the aforementioned trough of low pressure and winds of southerly components (Maps 2.2:2 and 3.2:3, April). The slopes of Kilimanjaro participate in this rainfall regime. The wet zone on the Southern and Eastern sides of Mount Kenya, as well as large parts of central Kenya and Uganda share this maximum, in addition to having a second seasonal peak around September to November. The latter is similar in its timing to the southward shift of the aforementioned low pressure trough and concomitant confluence zone (Maps 2.2:2 and 2.2:3, October). As during the maximum in the first half of the calendar year, lower-tropospheric flow over East Africa at this time of year is predominantly from the Southeast. This underlines the dominant topographic control in the origin of the maximum rainfall band on the Southern and Eastern flanks of Mount Kenya and Kilimanjaro. The two precipitation peaks in Central Kenya are popularly referred to as the 'long' and the 'short' rains. Rainfall totals during the former and the latter season are essentially uncorrelated.

The diurnal march of rainfall is in large part of East Africa characterized by a convective maximum in the afternoon and evening (Thompson, 1957; East African Meteorological Department, 1968; Hastenrath, 1970). The local circulation systems in the mountains (Brinkman *et al.*, 1968) are particularly conducive to an afternoon cloud maximum and clear skies at night. The inverse pattern is found at the large inland lakes, where the convergence of mesoscale winds over the warm water produces a nighttime maximum of rainfall (Flohn and Fraedrich, 1966; Fraedrich, 1972).

Within equatorial East Africa, the large-scale thermal pattern is horizontally rather uniform (East African Meteorological Department, 1970), although appreciable contrasts are conceivable on the smaller scale as a consequence of the pronounced topography. As illustrated in the meridional profile, Fig. 2.2:2, isothermal surfaces rise and the annual amplitude decreases as one proceeds from the subtropics of either hemisphere towards the Equator. Over East Africa, the mean annual elevation of the 0°C minimum, mean, and maximum temperature in the free atmosphere is about 3500, 4750, and 6000 m.

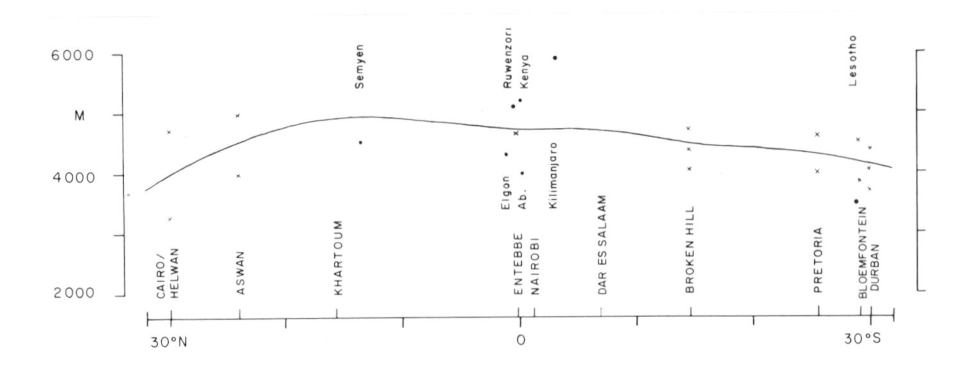

Fig. 2.2:2. Annual mean elevation of 0° C isothermal surface, with annual range indicated by crosses. For approximate location of profile and stations see Map 2.2:1. Period 1958–62 (source: US Weather Bureau, 1958–63).

The annual temperature range is about 2°C, with lowest values in March–April and highest in July–August. As common in low latitudes, the annual is far exceeded by the diurnal temperature range, the latter being of the order of 10–20°C.

Although all of the high mountains of East Africa are contained within a few degrees from the Equator, an insolation asymmetry results between northward and southward facing slopes, which reverses between hemispheres. This radiation asymmetry can be expected to be largest for the mountains furthest away from the Equator and with somewhat reduced cloudiness. Rather less trivial is an East–West asymmetry in radiation, resulting from the effect of the vigorous diurnal mountain circulations. As has been pointed out further above, downslope winds are conducive to clear skies at night and in the morning hours, when the sun is at easterly azimuths, whereas the upslope currents favor cloudiness and precipitation in the afternoon when the sun is to the West. As a consequence the westward facing slopes receive systematically less insolation than those with eastward exposure. Such a zonal asymmetry in radiation is most pronounced at mountains with a marked diurnal cycle in cloudiness, but vanishes in regions with nearly continuous cloud cover. The North–South contrast, and even more so the East–West asymmetry in insolation, are important factors in the ice distribution on mountains in equatorial East Africa and elsewhere in the tropics.

Proceeding from the mountains of equatorial East Africa poleward, High Semyen in Northern Ethiopia provides a reference to the outer tropics. The seasonality in insolation geometry and thermal regime becomes more noticeable. Northern Ethiopia is on the whole less humid than East Africa (Jackson, 1961), although annual precipitation totals in High Semyen are of the order of 1000–1400 mm (Hurni, 1982, pp. 62–73). The highlands of Lesotho represent the subtropics of the Southern hemisphere. Seasonality in insolation geometry and temperature is appreciable. Annual precipitation totals in the highlands of Lesotho range around 700–1400 mm (Jackson, 1961). High Semyen and the Lesotho highlands, in contrast with the high mountains of East Africa, thus exemplify the latitudinal variation of climatic conditions and altitudinal zonation on the African continent.

2.3. Vegetation

The plant cover is among the more climate-sensitive components of the environment. This section presents first a review of altitudinal belts of climate and vegetation and subsequently a synopsis of vegetation history as related to climatic change.

The modern altitudinal zonation of vegetation on the mountains of East Africa has been comprehensively studied by Hedberg (1951). The present review is largely based on his work, but reference is made to Hedberg (1951, 1964), Salt (1954), Coetzee (1967), Lind and Morrison (1973), and Hamilton (1982) for floristic details. In particular, Hamilton (1982, p. 79) modified Hedberg's (1951) original scheme of vegetation belts, and remarked that boundaries between the belts are not always well defined.

Fig. 2.3:1, reproduced from Hedberg (1951), and Table 2.2:1 illustrate the three-dimensional distribution of plant life at the major mountains of East Africa. Proceeding from the savanna or steppe upward, Hedberg distinguishes successively (a) Montane

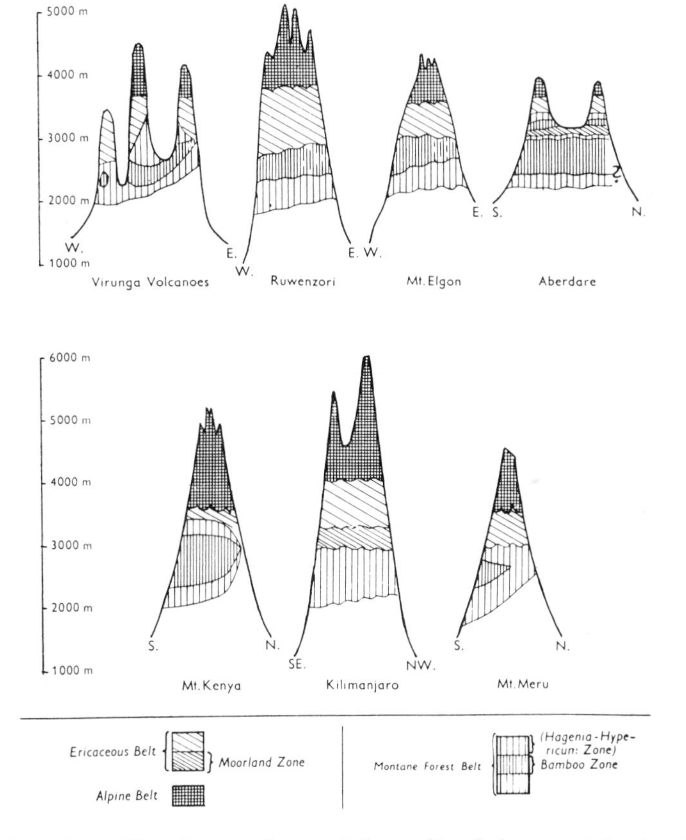

Fig. 2.3:1. Schematic profiles showing the vegetation belts of the mountains investigated. The wettest side of each mountain is turned to the left; the letters at their lower parts denote the quarters (W–E; S–N etc.). Where the Montane Forest Belt is differentiated into distinct zones this is indicated by separate marking for the Bamboo Zone. The zone next below is then a Montane Rain-forest Zone, the one next above a Hagenia-Hypericum Zone. Only the vertical distances are drawn to scale. Reproduced from Hedberg (1951).

Forest Belt, (b) Ericaceous Belt, and (c) Alpine Belt. Characteristics and altitudinal limits of these vegetation belts vary considerably between mountains, and in part even more so between different sides of the same mountain.

The Montane Forest Belt extends from a lower limit around 1700–2300 m to an upper bound at 3000–3300 m, and is characterized by broad-leaved hardwood trees and some conifers. On some mountains three distinct zones can be distinguished within this belt, proceeding upward: the Montane Rain-Forest Zone, consisting mostly of broad-leaved evergreen hardwood and some conifers; the Bamboo Zone; and the Hagenia-Hypericum Zone. All three zones are developed on some mountains, notably on Mount Kenya. At Ruwenzori and Elgon the uppermost zone is lacking, and on Kilimanjaro and the Northern side of Mount Meru the two upper zones are absent.

The Ericaceous Belt has a lower limit around 2600–3400 m, from where it extends upward to 3500–4100 m. The appearance of this belt differs between mountains, varying from dense forest to open scrub. This diversity is believed to be due to climatic and edaphic differences and the effect of fire. Hedberg (1951) points out that the midlatitude concept of upper forest or tree line limit is inappropriate for East Africa, in that giant Senecios, which must be classified as trees, ascend on some of the mountains almost to the upper limit of vascular plants.

The Alpine Belt occupies the altitude domain above the Ericaceous Belt, but again presents a rather diverse appearance on the different mountains. At comparable elevations, the vegetation varies from dense and moist forests of Giant Senecio in the Ruwenzori to open desert-like grass communities on Mount Meru. Hedberg stresses that the term chosen is not meant to imply a direct homology with the Alpine Belt of Europe, both climate and flora being entirely different. The frequent change of freeze and thaw is regarded as characteristic of the Alpine Belt of equatorial mountains. The predominantly diurnal freeze-thaw cycle also leads to a wealth of soil frost phenomena to be reviewed in Section 2.4.

Spatial contrasts (Fig. 2.3:1, Table 2.2:1) are particularly pronounced within the Montane Forest Belt. As a rule, the lower limit of this belt rises from the wetter towards the drier regions and even sectors of the same mountain. The Bamboo Zone sandwiched between the Montane Rain-Forest Zone below and the Hagenia-Hypericum Zone above is best developed in the wettest locations. Thus it is found in the Ruwenzori, at Mount Elgon, in the Aberdares, on the South but not the North side of Mount Kenya and Mount Meru, while it is altogether absent on Kilimanjaro. The Montane Forest Belt as a whole peters out in the 'treeless gaps' in the North sectors of Mounts Kenya and Meru.

In the lower portion of the Ericaceous Belt a Moorland Zone is developed on Kilimanjaro in particular, fire being considered as a major factor. The Ericaceous Belt on Mount Kenya and in the Aberdares is much narrower than on Kilimanjaro and in the Ruwenzori. By comparison with the Montane Forest Belt, however, the Ericaceous and Alpine belts are rather more uniform between mountains. The vegetation of these two highest belts manifests a variation from wetter towards drier conditions, progressing in the sequence from the Ruwenzori over Mount Kenya, the Aberdares, and Mount Elgon, to Kilimanjaro and Mount Meru. Hamilton (1982, p. 220) points out that a large portion of the high-altitude species is endemic and speculates on the problems of long-distance dispersal.

In this environment some very remarkable forms of plant-life occur. Senecio and Lobelia are among the most conspicuous components of the high mountain vegetation of East Africa. Four characteristic form types strike the eye: (a) barrel or column-shaped high trunks with an envelope of dry plant matter; (b) low cabbage-shaped growth with a diurnal opening and closing mechanism; (c) candle shape; and (d) dendritic forms. Particularly intriguing about this morphology is the convergence with the high mountain 'Páramo' vegetation of the South American Andes, where taxonomically altogether different plants (Espeletia, Puya), chose the same life forms, seemingly in response to a similar climatic environment!

Insight into the history of vegetation is provided by palynological work in the mountains and at lower elevations on the highlands of East Africa (Hedberg, 1954; Morrison, 1961, 1968; Van Zinderen Bakker, 1962, 1964; Coetzee, 1964, 1967, 1968; Van Zinderen Bakker and Coetzee, 1972; Livingstone, 1967, 1975, 1976a, b, 1980; Kendall and Livingstone, 1967; Kendall, 1969; Hamilton, 1972, 1973, 1974, 1982; Morrison and Hamilton, 1974; Rzoska, 1976; Flenley, 1979a, b; Perrott, 1982a, b). The present review concentrates on the implications in terms of a climatic history, whereas the original sources should be consulted for floristic details. Flenley (1979a, b), Livingstone (1975, 1976a, 1980), and Hamilton (1982) have discussed much of the palynological work, in part in context with palaeolimnological results to be reviewed in Section 2.5.

Pollen profiles have been evaluated from the Cherangani Hills of Kenya at 2900 m (Van Zinderen Bakker, 1962, 1964; Coetzee, 1967), from Lake Mahoma at 3000 m, and from other elevations in the Ruwenzori (Livingstone, 1967), from Kigezi in Western Uganda at 2500 m (Morrison, 1961, 1968; Morrison and Hamilton, 1974), from Sacred Lake on Mount Kenya at 2400 m (Coetzee, 1964, 1967), and from Pilkington Bay of Lake Victoria at 1100 m (Kendall, 1969). These and other more recent cores are discussed in Hamilton (1982, pp. 111–191). The longest pollen profile, extending over more than 30 000 years has been obtained by Coetzee (1967) from Sacred Lake on Mount Kenya. Among the limitations pointed out by Livingstone (1975) and Flenley (1979a, b) are the shortcomings in the absolute chronology and a stratigraphic change at the base. Nevertheless, this core provides a self-consistent vegetation sequence. Only this profile is reproduced here (Fig. 2.3:2) as a general example.

Livingstone (1975, 1980) discusses the difficulties in the interpretation of pollen profiles. There is a controversy between the opinion that most grass pollen in the profiles at all levels in the highlands originate from the alpine moorlands (Van Zinderen Bakker, 1964; Coetzee, 1967) and the notion that in the modern vegetation grasses are important at all altitudes and that the identification of grass pollen leaves much to be desired (Livingstone, 1980). A useful synopsis is found in Hamilton (1982, pp. 111–191). More extensive study of the modern pollen rain appears desirable. Pollen analysis has concentrated on the highland areas, where the flora is small and includes many genera familiar from elsewhere. Even for this seemingly simple environment interpretations meet with difficulties, in part related to the ambivalent altitude dependence of plants and the deficient ecological understanding of vegetation belts.

Livingstone and Clayton (1980) propose a novel approach to interpreting the tropical grass flora in terms of paleoclimate. While grass pollen is too similar to allow identification below the family level, grass cuticle available in the sediments of African lakes is identifiable to genus. Most grass species at low altitudes use 4-carbon photosynthesis, whereas species using 3-carbon photosynthesis prevail at high elevations; a regular transition is found at intermediate altitudes. Livingstone and Clayton believe that this altitude

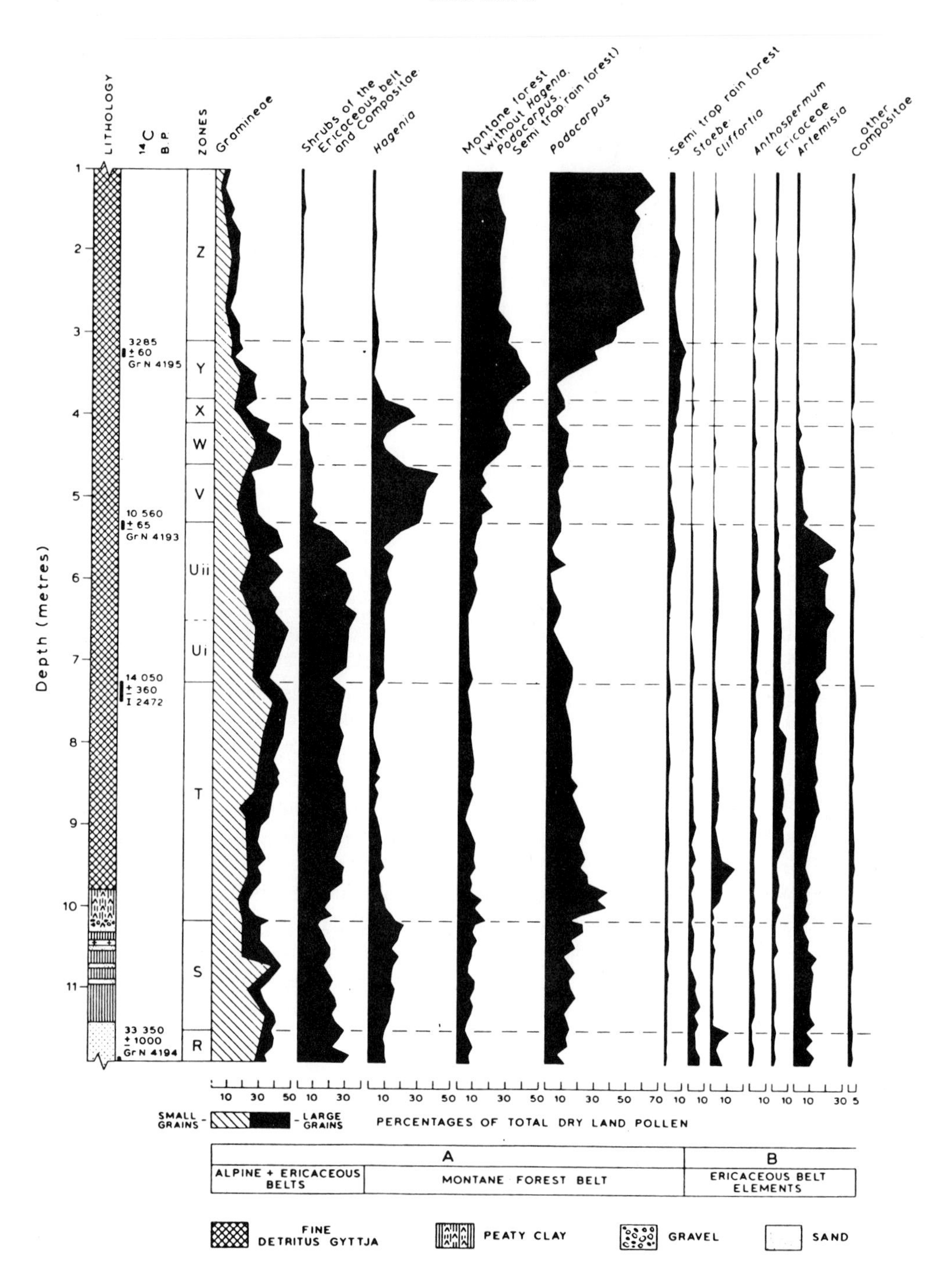

Fig. 2.3:2. Summary pollen diagram from Sacred Lake, Mount Kenya, 2400 m. Values are percentages of total dry land pollen (from Coetzee, 1967).

dependence is related to the cold sensitivity of the 4-carbon photosynthetic pathway, and suggest the possibility of estimating paleotemperatures to within about 1°C. Clearly this method is still in its infancy.

Despite the aforementioned limitations, the major traits of the vegetation history are emerging. The various pollen profiles published for East Africa in context point to conditions drier and cooler than at present during the period extending from 14 700 to between 12 500 to 9500 B.P. A change to drier and more seasonal conditions, or a combination of both is indicated for around 6500 B.P. A marked change in pollen types since about 2000 B.P. is believed to be related to human activity.

Flenley (1979a, b) presents a synopsis of these results in terms of altitudinal zonation, as illustrated in Fig. 2.3:3. A lowering of vegetation limits is indicated from around 30 000 to between about 20 000 to 15 000 years B.P., followed by a rise to sometime after 10 000 years B.P. This pattern is broadly consistent with the altitudinal shifts of vegetation belts in the South American Andes and the mountains of Australasia (Flenley, 1979a, b) and with water level changes of East African lakes to be discussed in Section 2.5.

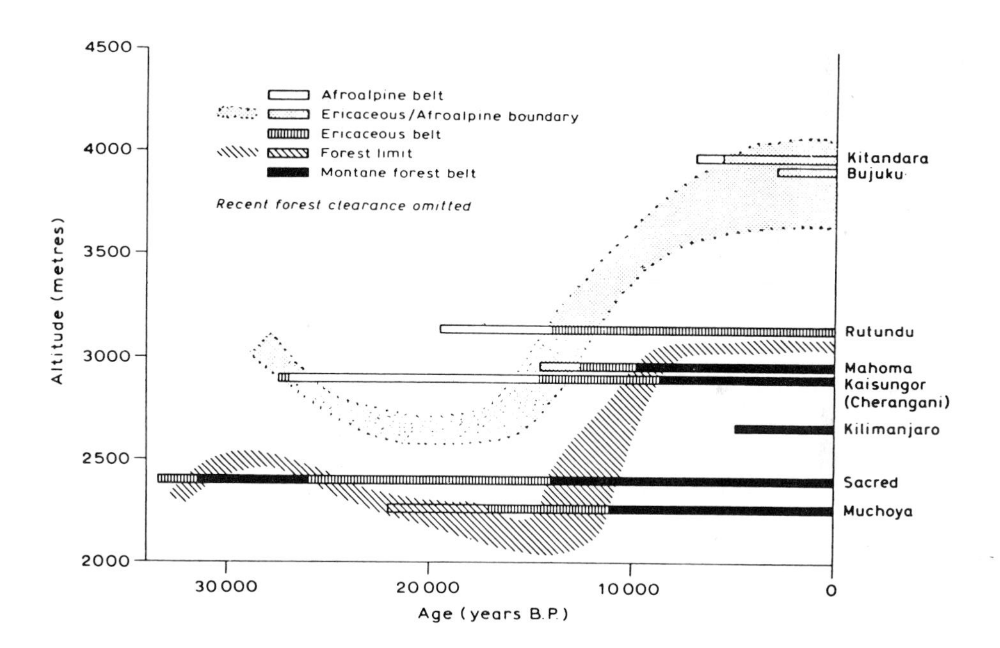

Fig. 2.3:3. Summary diagram of Late Quaternary vegetational changes in the East African mountains (from Flenley, 1979a).

2.4. Subnival Soil Forms

Soil frost processes are an important characteristic of the high mountain environment, but most of the recent studies (Furrer and Freud, 1973; Furrer and Graf, 1978; Hastenrath, 1973, 1974b, c, 1976, 1978; Winiger, 1979, 1981) are not readily accessible. Therefore, a comprehensive synopsis is presented here, based primarily on Hastenrath (1978).

Much attention has been paid to the large soil patterns prevalent in higher latitudes, both in field observations and laboratory experiments (see Washburn, 1973). Concerning the small frost features indigenous to the high mountains of the tropics, however, work since Troll's (1944) classical treatise has been more limited. Within the frame of lithological-edaphic and vegetational conditions, development, size, and spatial arrangement of individual form types would essentially reflect the climatic control. For the recognition of large-scale atmospheric factors governing the behaviour of soil frost phenomena, East Africa is of particular interest: high mountains extend in immediate vicinity of the Equator, with precipitation, glaciation, and vegetation conditions in the peak regions contrasting and ranging from semi-arid to hyper-humid; the highlands of Northern Ethiopia and of Lesotho, respectively, may span the bridges for a comparison towards the outer tropics and the subtropics of the African continent.

2.4.1. NOMENCLATURE

Partial nomenclatures have been published by Troll (1944, 1958), Nangeroni (1952), and Washburn (1973). For purposes of the present synopsis terms shall be proposed for various form types, along with a brief descriptive account and reference to the pertinent literature.

Needle ice

Needle ice (Troll, 1944, figs. 11, 13, 16; Hastenrath, 1973, photo 1; Hastenrath and Wilkinson, 1973, pl. 1; Furrer and Freund, 1973, photo 9) is among the more widely observed soil frost phenomena. With nighttime frost, ice needles grow in bundles of an asbestos-like structure perpendicular to the cooling surface. Fine earth material with some moisture seems to be a prerequisite. Sub-zero surface temperatures are evidently needed, but needles do not seem to form at excessively low temperatures. In the diurnal freeze-thaw cycle of tropical mountain regions, needles reach a few cm, whereas in temperate latitudes they have been observed to grow to 20 cm and more. Under conditions of daily change of frost the ice needles lift crumbs of fine earth off the subsoil; with the morning thaw these crumbs sink back to the surface and arrange themselves in stripes oriented in the direction of local sunrise.

Turf exfoliation

Needle ice is a powerful agent in destroying the vegetation cover in mountain regions. Its effect in causing vegetation scars and turf exfoliation (Troll, 1944, figs. 7, 19, 33; Hastenrath, 1973, photo 2; Hastenrath and Wilkinson, 1973, pl. 4), can be enhanced by human activity and excessive grazing.

Vegetation terracettes

Micro-terracettes with typical dimensions of tens of cm occur on occasionally very gentle slopes (Hastenrath, 1973, photo 3; Hastenrath and Wilkinson, 1973, pl. 3). Circular and crescent-shaped growth of plants is frequently encountered in the mountains of lower latitudes, (Furrer and Freund, 1973, photo 3; Hastenrath and Wilkinson, 1973, pl. 2; present Photo 2.4.2:1), and may be a related phenomenon.

Thufur

Vegetation covered mounds with dimensions of tens of cm (Troll, 1944, fig. 8; Hastenrath and Wilkinson, 1973, pl. 5) have long been known from subpolar regions, whence the Icelandic name 'thufur'. Similar forms have been observed in the mountains of Southern Africa.

Stone polygons

Miniature stone nets (Troll, 1944, figs. 35, 36; Hastenrath and Wilkinson, 1973, pls. 8–10; present Photos 2.4.2:2, 2.4.2:6), occur in a variety of appearances and dimensions ranging around 5 to 20 cm; they are only a few cm deep. With pebbles contained in a fine material matrix of great water holding capacity, stones tend to develop an arrangement around an occasionally watch-glass shaped core of fine earth nearly devoid of large pebbles. The sorting extends to a few cm depth. With a more continuous spectrum of grain size in the basic material, net patterns tend to be less distinct. There are indications for the dimensions of the net pattern to increase with the prevalent grain size.

Stone stripes

In sloping terrain, stone polygons gradually deform, and pass into stripe patterns following the gradient (Troll, 1944, figs. 30, 31; Hastenrath, 1973, photos 7, 10, 11, 13; present Photos 2.4.2:3, 2.4.3:7, 2.4.2:8), overall dimensions being similar to the polygons. There are indications for either polygons or stripes to predominate in certain areas, despite the seeming availability of both horizontal and sloping surfaces. A variety of sorting arrangements and dimensions occur, apparently in response to grain size distribution and depth of penetration of the freeze-thaw cycle. Large particle size and deeper-reaching freezing and thawing seem to contribute towards larger overall dimensions of the pattern.

Fine material (mud) polygons and bands

In contrast to the stone polygons and stripes, these patterns (Hastenrath, 1973, photos 4–6; Furrer and Freund, 1973, photos 7, 8, 11; Löffler, 1970, photo 48; present Photo 2.4.2:4) consist predominantly of fine material and contain very few larger pebbles. Overall horizontal and vertical dimensions are similar though not identical to the stone polygons/stripes. Dimensions can vary remarkably between different regions. Again, a gradual transition can be observed in the field from near-symmetric polygon nets on horizontal terrain to elongated polygons and bands on slopes. Embryonal forms display a sort of crack pattern with a near-absence of stones in the crack, or 'net' portion. Needle ice has apparently not been observed under this pattern, but macroscopically amorphous ice can be found within the mud cake. Some ideas of Bremer (1965) may be pertinent here, namely that processes active in gilgai-phenomena and a conventional frost mechanism may overlap in the mountains of the tropics and subtropics.

Fine earth buds and ribbons

Bulges of moist fine earth protrude from within a body of much coarser and seemingly drier material (Troll, 1944, fig. 42; Hastenrath, 1973, photos 8, 14, 15; present Photo 2.4.2:5). These protuberances take the form of 'buds' on horizontal surfaces, and of irregular-shaped 'ribbons' on slightly sloping terrain. Buds have a typical diameter and height of a few cm, ribbons are about one cm wide and high and some tens of cm long. Invariably, abundant moisture supply is available, often derived from patches of melting snow. A relative deficiency of fine material versus abundance of coarse particles has been suggested as an essential factor in the origin of this feature (Troll, 1944).

Cake polygons

Fine material plates of high water holding capacity and devoid of coarse particles are raised above a net of stones; horizontal dimensions are of the order of 20 cm (Troll, 1944, fig. 38; Hastenrath, 1971, photo 6; present Photo 2.4.2:9). Sites are characterized by an abundance of soil moisture. Furthermore, abundance of fine material and selective scarcity of large particles, with lack of intermediate grain sizes seems essential.

Mud grooves

On mountain slopes with appreciable restriction of the horizon, the morning thaw may begin at the higher elevations. With abundance of soil moisture, mud may start flowing from the higher portions of the slope down to ground portions that thaw out only later. The melting mud tracks are highest at their rims, providing channels for mud flow in subsequent daily cycles (Hastenrath, 1973, photo 16; present Photo 2.4.2:10). Abundant soil moisture is apparently a prerequisite of this phenomenon.

Needle ice occurs in a wide altitudinal range; turf exfoliation, vegetation terracettes, and the latitudinally restricted thufur belong to altitudinal belts with more or less continuous vegetation cover; towards higher elevations, they give way to a variety of sorting patterns and other soil frost forms.

2.4.2. MOUNTAINS OF EAST AFRICA

Three field trips were undertaken to the Kibo cone of Kilimanjaro (Map 3.1:1). For earlier observations reference is made to Meyer (1900), Jaeger (1909), Lange (1912), Klute (1914, 1920), and Flückinger (1934). In June 1971 the conventional route from Marangu on the South side over the Saddle Plateau between Kibo and Mawenzi to Gillman's Point was chosen. The excursion in August 1973 extended from Loitokitok on the North flank up the eastern slope of Kibo to Uhuru Peak; the South side was explored subsequently. In April 1974 the approach was over the Shira Plateau up the western slope of Kibo. Thus observations could be gathered in the various sectors of the mountain massif.

Lowest elevations of observed sites were around 3500 m for needle ice, 3500–3600 m for turf exfoliation as a result of frost action, and 3600–3700 m for micro-terracettes in the vegetation cover. Crescent-shaped and circular growth of plants was found near 3600 m on the Shira Plateau, and above 3800 m on the southern and northern flanks of the massif (Photo 2.4.2:1). The various features just mentioned become on the whole most prominent at 3800–4200 m. Differences in the lowest elevation of observed sites between the various sectors of the massif are not regarded as significant; possibly altitudinal limits are somewhat higher on the South side of Kilimanjaro.

Photo 2.4.2:1. Circular growth of plants at 3850 m along Marangu route of Kilimanjaro. Pocket knife as scale is 9 cm long. (Photo S.H., June 1971).

Photo 2.4.2:2. Polygonal stone sorting, 4300 m, tarn on Saddle Plateau of Kilimanjaro. Pocket knife as scale is 9 cm long. (Photo S.H., June 1971).

Stone polygons are remarkably rare on the Saddle Plateau and the South flank of Kibo. During the 1971 field trip (Hastenrath, 1973) they were observed in the vicinity of a tarn on the Saddle Plateau at 4300–4350 m at localities with abundant soil moisture and also partly under water (Photo 2.4.2:2). They do occur more abundantly on the Shira side of Kibo.

Stone stripes can be encountered on Kilimanjaro in varying dimensions. Lowest observed sites were at 4250 m on the South side and around 4200 m on the North and West sides of the mountain. On the Saddle Plateau, stone stripes become abundant above 4400 m, and they are of small dimensions. Stone stripes are made up of rock fragments with a typical size around 1 cm; they are 1–2 cm wide and 1–2 m higher than the bands of fine material separating them; the latter have a width of 2–5 cm. These dimensions are only meant to indicate the characteristic magnitude. The stripe width appears to increase with the coarseness of the material. Cuts across this stripe pattern show that it is at most a few cm deep. The stone stripes follow closely the irregularities in the terrain, and bend smoothly around obstacles such as larger boulders (Photo 2.4.2:3). Near the lowest elevations of occurrence stone stripes were found only on slopes of more than 20°, whereas on the Saddle Plateau between 4450 to 4460 m they are common on large surfaces of only 1–2° slope.

Stone stripes of comparable size occur also on the Shira side of Kibo at sites with fine material. However, stone stripes of somewhat larger dimensions prevail. Well-developed features abound from about 4200 m upward. Stone stripes in the crater of Kibo at about 5880 m on a slope with northerly aspect, are also of large dimensions. It is recalled that insolation on the West side of Kilimanjaro is reduced as a consequence of diurnal circulations. Thawing may be a less frequent and deeper-reaching event; this would also hold for the much higher elevations of Kibo crater, which is situated well above the mean annual 0°C isothermal surface. A similar suggestion of pattern size to increase with more episodic change of frost in locations of lower overall temperatures arises from observations at Mount Kenya and in High Semyen.

Fine material polygons and bands (Photo 2.4.2:4) were encountered in the ascent towards the Saddle Plateau at lowest elevations of 4100 m on the Loitokitok and at 4250 m on the Marangu route. Upward of 4400 m they cover vast areas of the plateau, alternating with the miniature stone stripe patterns discussed above. Fine material polygons have been reported by Klute (1914) and Flückinger (1934). Elevated flat plates consisting of fine material, with a thickness of about 2–4 cm and a diameter of 20 cm or more are separated by wide cracks; these are somewhat lower and are filled with small stones. On sloping terrain the polygons become more and more elongated, and the pattern degenerates into parallel mud and stone stripes with relative proportions similar to the polygon net. Cuts across such a pattern identify it as a surface phenomenon of less than 5 cm depth. Below that depth the plates of fine material and the stone-filled cracks are underlain by unsorted debris material. In contrast to the saddle plateau, fine material polygons and bands do not belong to the more conspicuous features on the Shira side and in the crater of Kibo.

Fine earth buds and ribbons were found at 5200 m on the eastern slope (Photo 2.4.2:5), at 4800 m on the southern flank, and above 4450 m on the Shira side of Kibo. Abundant

Photo 2.4.2:3. Stone stripes bending around obstacle; 4250 m, along Marangu route of Kilimanjaro. Pocket knife as scale is 9 cm long. (Photo S.H., June 1971).

Photo 2.4.2:4. Fine material polygons at 4250 m, along Marangu route of Kilimanjaro. Box as scale has diameter of 7 cm. (Photo S.H., June 1971).

soil moisture, often derived from melting snow, is characteristic. A limitation to higher altitudes is also apparent in other mountain regions to be discussed later.

Of all regions, Mount Kenya (Map 3.3:1) could be studied most extensively. For earlier observations reference is made to Troll (1944) and Zeuner (1949). In July 1971 the Naro Moru track from the West was used. An excursion in July 1973 covered the Eastern and Northern sectors of the massif, approximately along the Chogoria and Sirimon tracks. In the course of 1973–74, all sectors of the peak region were visited repeatedly, and the approach from the West was commonly used on the frequent field trips in connection with the study of mass budget and secular behaviour of Lewis Glacier. The various sectors of the massif display appreciable contrasts in precipitation and vegetation conditions (Table 2.2:1). Since all three aforementioned approach routes were covered within a few weeks' span in June–July 1973, an immediate spatial comparison of periglacial forms is possible.

Needle ice, turf exfoliation, and micro vegetation terracettes are again the soil frost phenomena occurring at the lowest elevations. On the Western side of the mountain massif, the lower limit is around 3900 m and the preferred distribution domain between 3950 and more than 4200 m. On the Eastern flank (Chogoria route) these phenomena could be observed in good developement down to 3600 m, and on the North side (Sirimon track) even another 100 m lower. However, the preferred altitudinal domain was found to be around 3900–4200 m in all sectors of the massif. The preferred altitudinal domain may warrant a more reliable spatial comparison than singular lowest occurrences of a certain form type, the latter being more easily dependent on accidental circumstances along the travel route. Accordingly, differences in vertical distribution between the various sectors of the massif are considered to be small. However, the form types just discussed are unmistakably best developed along the Naro Moru track in the West, where needle ice plots and crescent-shaped and circular vegetation patterns occupy large surfaces. The Naro Moru track is presently the most frequented route for tourists and mountaineers. Human interference very conspicuously combines with soil frost in a progressive destruction of the vegetation cover. In the high regions of Kenya, grazing is precluded by nature conservation regulations.

Other subnival soil forms, such as stone polygons (Photo 2.4.2:6) and stripes (Photo 2.4.2:7), fine material (mud) polygons (Photo 2.4.2:8) and bands, fine earth buds and ribbons, cake polygons, occur with preference upward of 4300–4400 m. No systematic differences in terms of development and altitudinal domain of these form types could be detected between the different sectors of the mountain. Climatic differentiation between the various quadrants presumably may fade out in the high regions.

Some vertical differentiation is apparent in that stone stripes and polygons and fine earth polygons and bands begin to appear at elevations as low as about 4200 m, whereas fine earth buds (Photo 2.4.2:9) and ribbons seem to stay essentially above 4700 m, although Troll (1944; fig. 42) has described such features from the vicinity of Hall Tarn on the eastern side of Mount Kenya at elevations of 4400 m.

In the prevalent stone stripe patterns, the typical overall band width is about 10 cm, the bands of stone fragments being mostly about three times as wide, but at least about equally wide as the bands of fine material separating them. Stripes tend to bend around obstacles such as boulders, and flags of fine earth appear in their lee. Sorting extends down to a depth of 2–3 cm. The fine material has a grain size of 2–4 mm compared to

Photo 2.4.2:5. Fine earth ribbons, 5200 m, East slope of Kibo. Pocket knife as scale is 9 cm long. (Photo S.H., June 1971).

Photo 2.4.2:6. Stone polygons on shore of Naro Moru Tarn in the upper Teleki Valley, Mount Kenya, at 4200 m. Box as scale has a diameter of 7 cm. (Photo S.H., July 1971).

5–20 mm for the light and coarse material. The fine material mostly has a fresh moist appearance, and is darker than the stone stripes (Photo 2.4.2:7). Then, the bands of fine material are higher than the stone stripes, in great contrast to the features described for Kilimanjaro. In some instances, the fine material is dry and has a lighter color; then the stone stripe portion of the pattern is generally higher. Stone stripes of substantially large dimensions with overall band width of up to about 30 cm and proportionally larger depth were repeatedly observed.

Mud polygons in the vicinity of 4200–4400 m were less abundant and conspicuous than on Kilimanjaro. Stone polygons resulting from a radial sorting were found at several sites between 4200 and 4300 m, with diameters between about 5 and 15 cm. The largest and best developed polygons were found in places with abundant soil moisture around ponds, and also under water.

Mud grooves were first described for Mount Kenya on slopes near Lewis Glacier at elevations between about 4500 and 4700 m. The phenomenon is illustrated in Photo 2.4.2:10.

Seasonal variations could be studied on the West side of the massif, since the Naro Moru track was frequented as an approach route for field work on Lewis Glacier in the course of 1973–74 and again during 1977–82. Most conspicuous changes were in the lower limit and preferred altitudinal domain of needle ice stripes and active turf exfoliation. The relatively low elevations during June and July have been described above. From January to April needle ice is not of common occurrence below 4200 m, and plots with strictly linear fine earth stripes decay to dust. However, upward of 4300 m it is commonly encountered at all seasons. During the January–February dry season, soil moisture and atmospheric humidity are comparatively small. At the same time temperature at these altitudes increases by a few degrees, so that nighttime frost does not occur regularly below 4200 m. Stone polygons and stripes and fine material (mud) polygons can be found all year round, although climatic conditions may not be equally favorable for development during all months.

Wide areas of the massif were visited in July 1971 and again in the course of 1973–74, and 1977–83. Furthermore, Troll's (1944) observations from 1934 are available. This allows some idea on longer-term variations in characteristic soil frost phenomena. Most form types were encountered in good development both in July 1971 and during 1973–74, but there were some surprising exceptions. In July 1971, beautiful stone stripes covered large surfaces on the steep northwestward facing inner slope of the innermost large moraine of Lewis Glacier above the Teleki Valley (Photo 2.4.2:7). The site was easily located two years later, but the large and exceedingly regular stripe pattern had disappeared, and until January 1983, it has not yet re-formed. Despite extensive search, no such pattern could be detected in the area in June 1973 and later years. Mud grooves observed on the large Lewis Glacier moraine in June 1974 had likewise disappeared by July 1973, and have not re-formed since. Soil moisture conditions are considered essential in the origin of both form types. It is suggested that the relative dryness of the years since the early 1970s has caused the disappearance of certain soil frost phenomena.

A seasonal development of material sorting was observed at the steep slopes between Thomson's Flake and Lewis Glacier. The slopes above the Lewis Glacier carry a snow

Photo 2.4.2:7. Stone stripes on moraine of Lewis Glacier, Mount Kenya, at 4650 m. Box as scale has diameter of 7 cm. (Photo S.H., July 1971).

Photo 2.4.2:8. Incipient mud polygon formations on moist soil at about 4170 m along Chogoria route, East side of Mount Kenya. Pocket knife as scale is 9 cm long. (Photo S.H., June 1973).

Photo 2.4.2:9. Fine earth mounds, 4700 m, Mount Kenya. Box as scale has diameter of 7 cm. (Photo S.H., July 1971).

cover during the season of abundant precipitation through January, which is later ablated or buried by scree. From January to March a distinct material sorting is observed, with nearly one-m-wide bands of coarse rock fragments. It is believed that change of frost may play a role in the sorting process.

The Aberdares (Map 3.4:1) were visited in May 1973 with an approach from the East to the high areas of Oldoinyo Lesatima. Remnants of needle ice and incipient terracette formation in the vegetation cover were observed upward of about 3500 m, and circular growth of plants at about 3950 m. Cryogenic sorting of material was not detected.

An excursion to the Kenyan side of Mount Elgon (Map 3.5:1) was undertaken in June 1973. Weak remnants of needle ice were observed at lowest elevations of around 3800 m and incipient micro-terracettes in the vegetation cover upward of 4000 m. Incipient sorting of small stone material is limited to the highest regions above 4150 m.

A field trip to the Ruwenzori (Map 3.7:1) finally materialized in January 1974, under optimal weather conditions. Periglacial phenomena have been mentioned for the Ruwenzori by de Heinzelin (1952), but without details on specific form types and altitude.

Needle ice remnants and frost scars in the vegetation cover were observed at lowest elevations of 4150 m and up to more than 4300 m in the Stuhlmann and Freshfield Pass areas (Map 3.2:1). There is no stripe arrangement of needle ice remnants, presumably because of a distinct diurnal cycle of direct radiation lacking as a consequence of excessive cloudiness. Sorting of stone material was found in the region the the Scott-Elliot Pass around 4350 m.

Most characteristic of the Ruwenzori is the vertical contraction of altitudinal belts. In this hyper-humid mountain region, ice fields and glaciers clash with exuberant plant

Photo 2.4.2:10. Mud grooves, 4500 m, below Lewis Glacier, Mount Kenya. Box as scale has diameter of 7 cm. (Photo S.H., July 1971).

life over a narrow vertical distance. Polished rock surfaces vacated by ice in the not too distant past are occupied by a shallow moss cover, in absence of any soil formation. In contrast to the other high mountains of East Africa, periglacial processes are poor in the variety of form types and they show little vertical differentiation. Thus frost scars in the plant cover and vegetation terracettes – form types characteristically occupying a low altitudinal domain – are confined to comparatively high elevations. Material sorting embryonal to stone polygons and stripes is found in a narrow altitudinal range between vegetated surfaces and bare rock in vicinity of the glaciers, and is poor in development. Fine material polygons and bands – as well as apparently fine earth buds and ribbons – lack altogether. In perspective with the complex of subnival soil forms and altitudinal zonation in the other high mountains of East Africa, the extreme end of a spectrum is reached in the Ruwenzori.

2.4.3. ETHIOPIA

Mapping of Pleistocene glacial morphology in High Semyen in December 1973 (Hastenrath, 1974) gave the opportunity for observation of subnival soil forms.

Weak remnants of needle ice were found at lowest elevations around 3700 m. Turf exfoliation and micro vegetation terracettes occur upward of about 3750 m, and can be found up to more than 4350 m. This indicates the preferred altitudinal domain of these soil frost forms. Stone stripes, fine material polygons and bands, and cake polygons, are the most conspicuous form types upward of 4250 m. Fine material polygons and bands are similar in overall appearance to their counterparts at Kilimanjaro, for example, but they are considerably smaller, typical diameters being 3–5 cm as opposed to 10–20 cm.

In contrast to the mountains of equatorial East Africa, periglacial forms in High Semyen in the outer tropics of the Northern hemisphere, display a pronounced dependence on aspect. Thus, turf exfoliation, stone stripes, and fine material polygons and bands are better developed and of rather larger dimensions on northward as compared to southward facing slopes. A preference of the westerly over the easterly quadrant is less important, and may be related to insolation asymmetries associated with diurnal circulations.

2.4.4. SOUTHERN AFRICA

Field work in the highlands of Lesotho and the Natal Drakensbergs was carried out during April–May 1971 (Hastenrath and Wilkinson, 1973). Needle ice was found in better development and more commonly at elevations above 2800 m. Circular growth of plants could be observed on gentle slopes above 3000 m. Terracette formation in the vegetation cover occurs from elevations of less than 2000 m to more than 3300 m. Especially at lower elevations, one has the impression that grazing cattle and sheep play a major role in the origin of these terracettes. From about 3000 m upward, needle ice has been observed to be an important agent in breaking up the vegetation cover, thus contributing to the origin of scars and terracettes. However, even at higher elevations, the destructive frost action is seemingly aided by the intensive grazing, which is known to have reached extreme proportions since the turn of the century, in wide areas of Lesotho.

Little vegetation-covered mounds of about 20–50 cm height (thufur) are characteristic of sites with abundant soil moisture, at elevations between about 2900 and more than 3100 m. A clear tendency was observed for the vegetation cover to break up on the northward facing side of the mounds (Hastenrath and Wilkinson, 1973, pl. 5). On steep slopes, the mounds show some asymmetry, with a deformation following the gradient. A cross-section was cut through some of these mounds. The vegetation is limited to a relatively thin superficial coat, and the interior of the mound is made up of a dark, heavy clay-like material.

Stone polygons were observed at elevations above about 3200 m. Two somewhat different patterns were distinguished. In situations with a discontinuous grain size distribution, say with pebbles of up to about 2 cm embedded in a matrix of fine grained clay-like material, the mesh width was of the order of 20 cm, and depth of sorting extended to a few cm. By contrast, with a continuous grain size distribution varying from 1 to 10 mm, and absence of soil in the proper sense, a net-like sorting was commonly observed with a typical mesh width of only about 5–10 cm. The sorting appeared to extend downward only to a few mm.

A more unusual arrangement of rocks presumably also due to frost action was found in the area of Letseng-La-Draai on a slightly sloping plateau at 3100 m. The basalt rock in this location has a plate-like structure. Basalt plates were found strongly tilted, sticking out of the ground. Excavation showed that the strongly tilted plates apparent at the surface could be traced to nearly horizontal plate-like layers in the ground, from where they had been dislocated. Distribution and horizontal displacement at this site was found to have taken place in a counterclockwise turning. A similar configuration has been described by Mohaupt (1932, fig. 5) from the Alps and interpreted as a soil frost phenomenon.

2.4.5. FOSSIL FORMS

Distinct climatic conditions are prerequisite for a variety of periglacial processes. Accordingly, fossil forms are of interest as indicators of past climate. The delicate micro-patterns discussed in the preceding Sections 2.4.1 to 2.4.4 are not easily preserved, but certain macro-solifluidal features are less prone to destruction. Williams *et al.* (1978), Messerli *et al.* (1980), and Hurni (1982, pp. 84–139) have identified such geomorphic evidence in the Semyen mountains of Ethiopia, with the aim of climate reconstruction.

2.4.6. SYNTHESIS

East Africa offers a wide spectrum of high mountain environments. The complex of soil frost forms and its altitudinal zonation is summarized in Fig. 2.4.6:1. For the vegetation and precipitation conditions refer to Fig. 2.3:1 and Table 2.2:1.

In summary, needle ice, turf exfoliation, and vegetation terracettes occur commonly above 3500 m on most East African mountains, with the preferred altitudinal domain being rather higher in the Ruwenzori. Stone polygons and stripes and fine material polygons and bands generally appear upward of 4200 m, becoming more abundant above 4300–4400 m; fine earth buds and ribbons occupy a somewhat higher altitudinal domain. Most of the latter form types seem to be absent in the Ruwenzori, where only stone polygons and stripes in incomplete development were sighted upward of 4350 m.

Altitudinal domains for the various subnival forms narrow from the drier towards the wetter regions – in a progression from Kilimanjaro over Mount Elgon, Aberdares, Mount Kenya, to the Ruwenzori. This spatial pattern has its corollary in the vertical separation between the upper limit of plant cover (Section 2.3) and region of perennial ice (Chapter 4), being largest on Kilimanjaro and smallest in the Ruwenzori. It is also noted that the diurnal temperature range tends to be reduced towards the more cloudy regions. As a consequence, the altitudinal belt with frequent daily change of frost – extending vertically from somewhere below to somewhere above the average 0° C isothermal surface – narrows.

Differences in the periglacial morphology between various sectors of the same mountain are in general not pronounced. Precipitation contrasts between different sides of a massif also tend to fade out towards the high regions, to which soil frost forms are

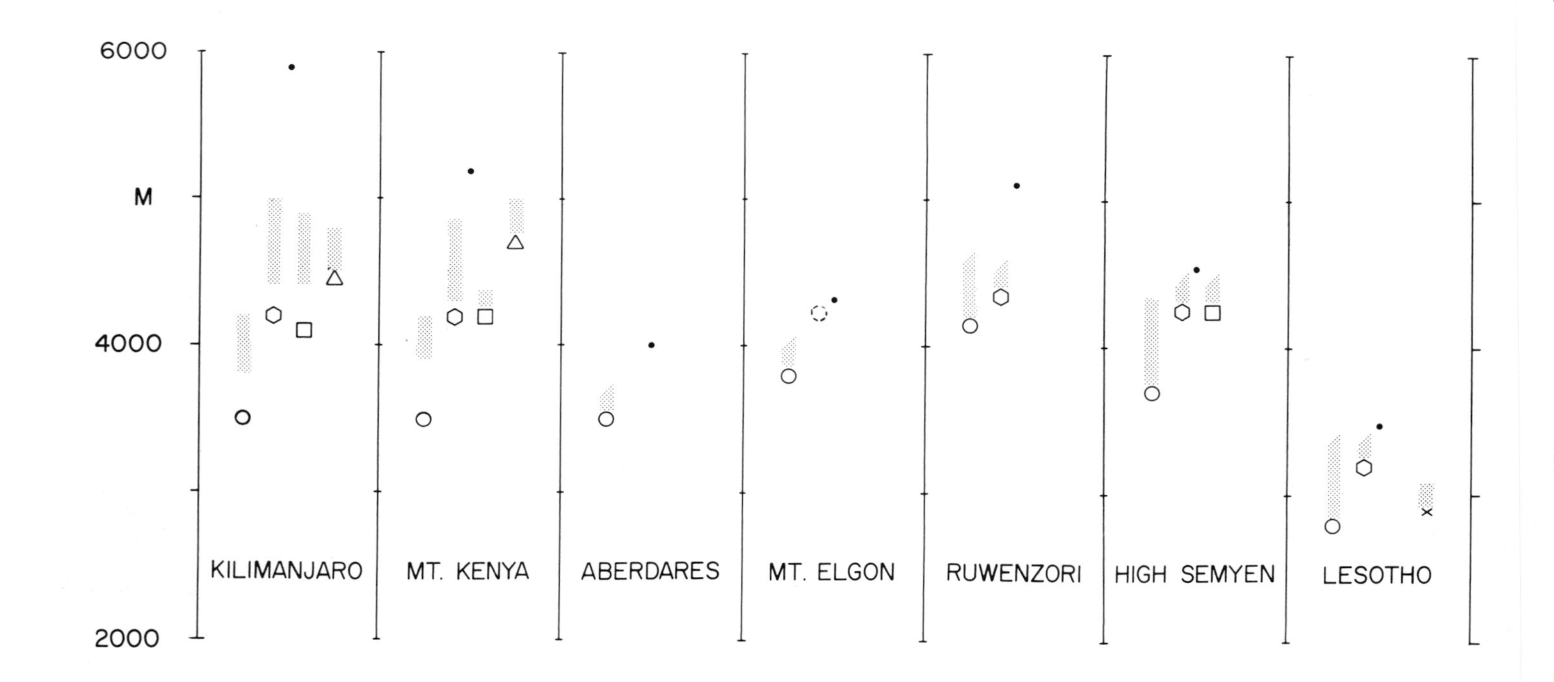

Fig. 2.4.6:1. Altitudinal zonation of subnival soil forms: Kilimanjaro, Mount Kenya, Aberdares, Mount Elgon, Ruwenzori, High Semyen, Lesotho. Complexes of form types: 1. needle ice, turf exfoliation, vegetation terracettes, circular or crescentic growth of plants, circle; 2. stone polygons and stripes, hexagon; 3. fine material polygons and bands, square; 4. fine earth buts and ribbons, triangle; 5. thufur, cross. Lower distribution limit is denoted by center of symbol, and preferred altitudinal domain by shading; tapering signifies an open upper bound of altitudinal domain. Elevation of highest peak is indicated by dot.

typically confined. There are weak indications for altitudinal domains of some form types to be lower on the Northern than on the Southern side of Kilimanjaro. More significant is the prevalence of stone stripe patterns of large dimensions on the Shira side of Kibo. Afternoon cloudiness associated with powerful diurnal circulations reduce insolation on the West side of the mountain. As a consequence thawing may be a more episodic deep-reaching event, in contrast to the shallow diurnal frost change cycle conventionally considered in the tropics. It is recalled that stone stripe patterns of large dimensions also prevail in the crater of Kibo, which is situated well above the 0°C mean annual isothermal surface. Observations at Mount Kenya also point to the importance of episodic processes at particular locations, as opposed to the conventional concepts of a 'diurnal tropical' versus a 'seasonal polar' forcing. It seems that polygon patterns are scarce in some regions, such as in wide areas on Kilimanjaro, despite the seeming availability of horizontal surfaces. By contrast in the highlands of Lesotho stripe patterns are rare, whereas polygons are found more frequently. This leads to the speculation that factors other than slope may play a role in the origin of polygon versus stripe patterns.

In the Ruwenzori, needle ice remnants do not display the regular approximately East–West oriented stripe-like arrangement common in other mountain regions. It is known that this pattern is determined by the direction of solar rays at local sunrise, when the melting of ice needles sets in. The continuously cloudy skies in the Ruwenzori may not allow for this effect of direct radiation in the early morning hours. Fine earth buds and ribbons are regarded as high altitude features, although they are found on a range of elevations in the various sectors of Kilimanjaro and Mount Kenya. They were not observed in the Ruwenzori.

The distributional pattern of fine material polygons and bands deserves particular attention. These form types determine the aspect of wide areas on the saddle plateau, but not on the Shira side of Kibo. They occur on Mount Kenya, and in smaller dimensions also in High Semyen. However, they are conspicuously absent in the hyper-humid Ruwenzori. The embryonal stage of these form types shows a crack pattern suggestive of processes related to desiccation. The picture by Löffler (1970; fig. 48) from Northeastern Anatolia, referred to in Section 2.4.1, almost certainly represents essentially the same feature. A photograph of Schenk (1955; figs. 7, 2) from Spitzbergen is reminiscent of the fine material polygons described here, but it should remain open, whether it is the same phenomenon. Concerning the origin and climatic significance of these features, a suggestion by Bremer (1965) may be pertinent, namely that gilgai phenomena and frost structure soils may overlap genetically in the mountain regions of lower latitudes. The possibility of episodic dryness playing a role in the development of fine material polygons and stripes may deserve attention in further studies.

For Kilimanjaro, Furrer and Freund (1973) and Furrer and Graf (1978) produced a summary diagram in design similar to Fig. 2.4:1. Differences in terminology, routes, and season, may contribute to apparent inconsistencies between the two surveys.

Height limits of form types in High Semyen are similar to equatorial East African, but a marked preference for the poleward and Western quadrants appears in the outer tropics. Consistent with the thermal regime of the subtropics, altitudinal domains in the highlands of Lesotho are displaced to lower elevations, and specific form types believed to be related to the more marked temperature seasonality of the extratropical caps appear, such as thufur and the large arrangements of tilted rock plates described by Mohaupt (1932) for the Alps.

The mountains of New Guinea, likewise near the Equator, are of interest for comparison with East Africa. Löffler (1975) has given the first comprehensive account of soil frost phenomena based on an extensive survey in all high regions of Papua New Guinea. A field trip to the two highest mountains of Papua New Guinea, Mounts Wilhelm and Giluwe, materialized in 1975. Comparing the climate of the New Guinea highlands with the mountains of East Africa, the closest corollary is found in the Ruwenzori, although conditions in the high regions of New Guinea are even more extreme in terms of abundant precipitation and cloudiness, and presumably the small diurnal temperature range. As shown by Löffler, the altitudinal belt with frequent diurnal change of frost – as resulting from the elevation of the average 0°C isothermal surface around 4730 m and the small diurnal temperature range – has a rather small vertical extent. Only small areas of the highest mountains attain this geomorphologically interesting altitudinal zone, and recent subnival soil forms are accordingly rare and poorly developed. Both climatically and in terms of periglacial morphology, the peaks of New Guinea seem to exhibit conditions similar to, but even more extreme than the Ruwenzori.

The High Andes of Ecuador are the third region of the World with high mountains in immediate vicinity of the Equator. During field work in 1974–9, a wide range of altitudes was covered in various parts of the Ecuadorian Andes. The three-dimensional distribution of periglacial morphology is somewhat complicated by edaphic conditions, namely the ubiquitous cover of fine volcanic ashes up to all but the younger moraine stages.

Within equatorial East Africa subnival soil forms thus show a greater diversity of development and altitudinal zonation than in the other high mountains under the Equator, namely the Ecuadorian Andes and New Guinea. Moreover, the contrasts in the altitudinal zonation of soil frost features between mountains within equatorial East Africa appear rather more dominant than the systematic variation with latitude.

2.5. Lake Level Variations

Efforts at understanding the water level variations of East African Lakes as manifestations of climatic change and in the context of glacial episodes date back to Nilsson's (1931, 1935, 1940, 1949) expeditions in the 1920s and 1930s. These and later paleo-climatological endeavors have been hampered by the inability to establish a direct stratigraphic linkage between limnic and glacial events, and by the lack of an absolute chronology. While the correlation of lake and glacier variations is still open to speculation, a coherent lake chronology is beginning to emerge from palaeo-limnological work in the course of the past two decades (Richardson, 1966; Washbourn-Kamau, 1967, 1971; Kendall and Livingstone, 1967; Kendall, 1969; Livingstone and Kendall, 1969; Degens and Hecky, 1974; Butzer, 1971, 1980; Butzer *et al.*, 1972; Richardson and Richardson, 1972; Grove *et al.*, 1975; Livingstone, 1975, 1976a, b, 1980; Street and Grove, 1976, 1979; Gasse, 1977, 1980; Gasse and Street, 1978; Hecky, 1978; Adamson *et al.*, 1980; Street, 1980, 1981; Gasse *et al.*, 1980; Tiercelin *et al.*, 1981; Cohen, 1981; Owen *et al.*, 1982). Fig. 2.5:1 illustrates the major water level variations of East African lakes during the late Pleistocene and Holocene.

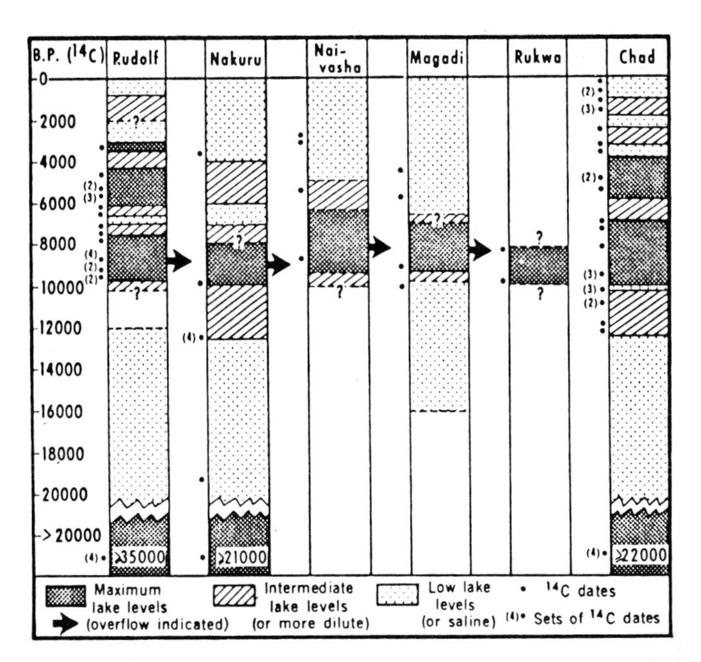

Fig. 2.5:1. Fluctuations of lakes that presently lack outlets (from Butzer *et al.*, 1972, Copyright 1972 by the American Association for the Advancement of Science).

Lake Victoria (Kendall, 1969; p. 164) lay in a closed basin for at least 2000 years before 12 500 years B.P., and actually fell during part of this period. After 12 500 years B.P. the lake rose further and likely achieved an outlet. Vegetation changes are regarded as consistent with an increase in rainfall. Around 10 000 years B.P. the lake became closed again, falling some 12 m below its present outlet. After 9500 years B.P. the lake became open again and retained its outlet thereafter.

Lake Naivasha (Richardson and Richardson, 1972; Butzer *et al.*, 1972) was greatly enlarged from before 9200 to about 5650 years B.P. The lake discharged to the South prior to 5650 and 3040 years B.P. In the course of the past 3000 years the lake has frequently decreased to smaller than its present size.

Concerning the Nakuru-Elmenteita basin (Washbourn-Kamau, 1971; Butzer *et al.*, 1972), there are indications for a high stand before 20 000 years B.P. Between 10 000 and 8000 years B.P. the water stood some 180 m above the modern level. The lake basins were filled to their highest possible level at about 9000 years B.P. Elmenteita and Nakuru then merged and had a common outlet northward into the Menengai Crater.

For Lake Bogoria (Hannington) in the Northern Kenya Rift Valley a high stand is dated at about 4000 years B.P. (Tiercelin *et al.*, 1981).

Lake Rudolf (Butzer, 1971, 1980; Owen *et al.*, 1982) stood high at 10 000–7000, 6500–4000, and about 3250 years B.P., while levels fluctuated around that of the present during 7000–6500 and since 2500 years B.P. During the 10 000–7000 years B.P. episode a periodic overflow to the Nile may have occurred.

Lake Kivu (Degens and Hecky, 1974; Hecky, 1978) was closed prior to 9500 and during 3000–1200 years B.P. A rapid rise of the water level started at about 10 000 and continued to about 4000 years B.P. After that date the climate became drier, reaching the most pronounced aridity around 2500 years B.P.

Lake Albert (Mobutu Sese Seko) was open from at least 28 000 to 25 000 years B.P., but seems to have lacked an outlet thereafter until 18 000 years B.P., and again for a short period prior to 12 500 years B.P. (Livingstone, 1980).

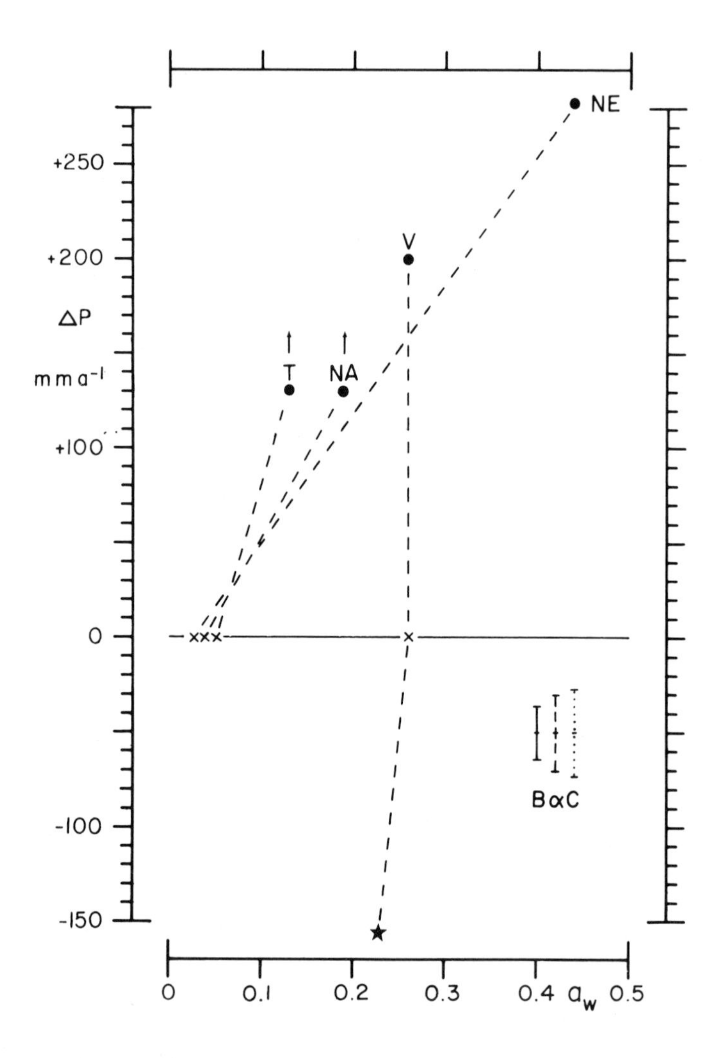

Fig. 2.5:2. Precipitation estimates (P) for Lakes Victoria (V), Turkana (T), Nakuru-Elmenteita (NE), and Naivasha (Na), given as departures from modern values, in mm a^{-1}. a_w is ratio of water to total catchment area. Crosses refer to modern conditions, dots to past wet episodes (10 000–6500 years B.P.), and star to endoreic dry eras (< 12 500 and ~10 000 years B.P.) at Lake Victoria. Vertical arrows at Lakes Turkana and Naivasha symbolize requirement for upward adjustment of estimates, as plots do not account for overflows during the 10 000–6500 years B.P. wet episodes. Vertical solid, broken, and dotted lines in the lower right-hand portion of graph indicate error tolerances corresponding to fractional changes of the order of ±5 percent in land Bowen ratio B_l, albedo α_l, and cloudiness C, respectively.

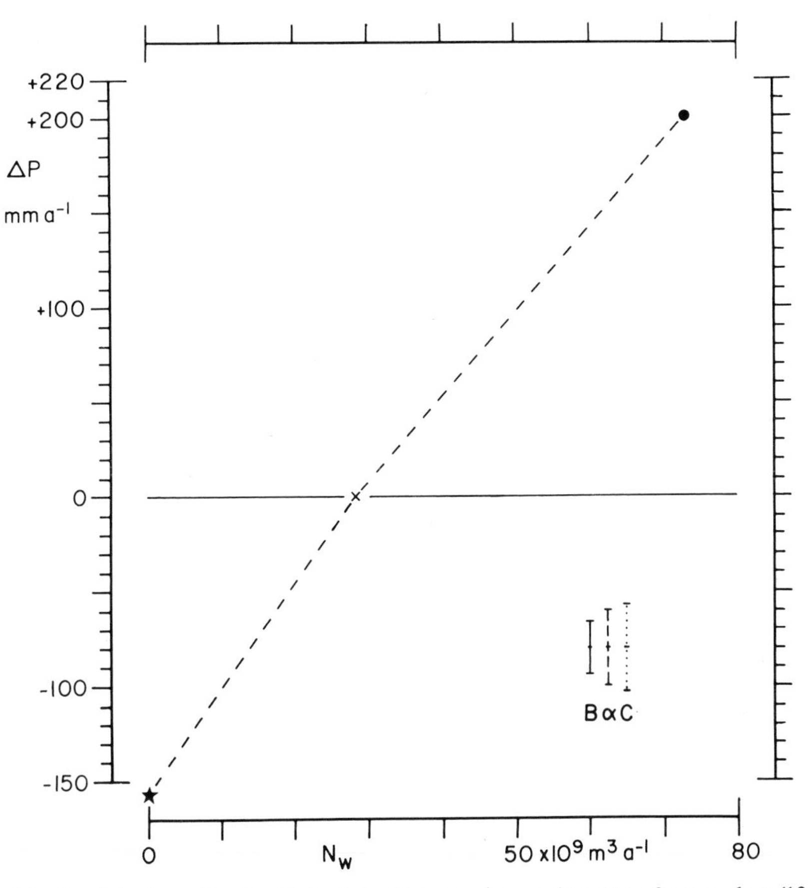

Fig. 2.5:3. Precipitation estimates (P) for Lake Victoria, given as departures from modern (1946–70) value, in mm a^{-1}. N_W is discharge in 10^9 m^3 a^{-1}. Cross refers to modern (1946–70) conditions, dot to past wet episodes (<9500 years B.P., 1880 A.D.), star to endoreic dry eras (>12 500 and ~10 000 years B.P.). Vertical solid, broken, and dotted lines in the lower right-hand portion of diagram have the same meaning as in Fig. 2.5:3.

For Afar and the Ethiopian Rift lakes (Gasse *et al.*, 1980), humid conditions are indicated from at least 27 000 to 20 000 years B.P. and aridity from 17 000 to 12 000 years B.P.

Complementing the evidence from the various East African lakes, the Nile was a banded and highly seasonal river during 20 000–12 500 years B.P., and thereafter became an incised, sinuous, suspended load river (Adamson *et al.*, 1980).

In summary, there are indications of humid conditions during 27 000–20 000 years B.P. in Afar and the Ethiopian Rift. East African lakes stood low prior to a marked rise around 13 000–11 000 years B.P. Between 10 000 and 8000 years B.P. lakes in East Africa and in other parts of the continent were greatly enlarged. After this maximum an apparently concordant positive oscillation between 6000 and 4000 years B.P. is indicated for various basins. This paleo-limnological evidence should be viewed in context with the lowering of vegetation limits from around 30 000 to 20 000–15 000 years B.P. and the subsequent rise to after 10 000 years B.P., as described in Section 2.3.

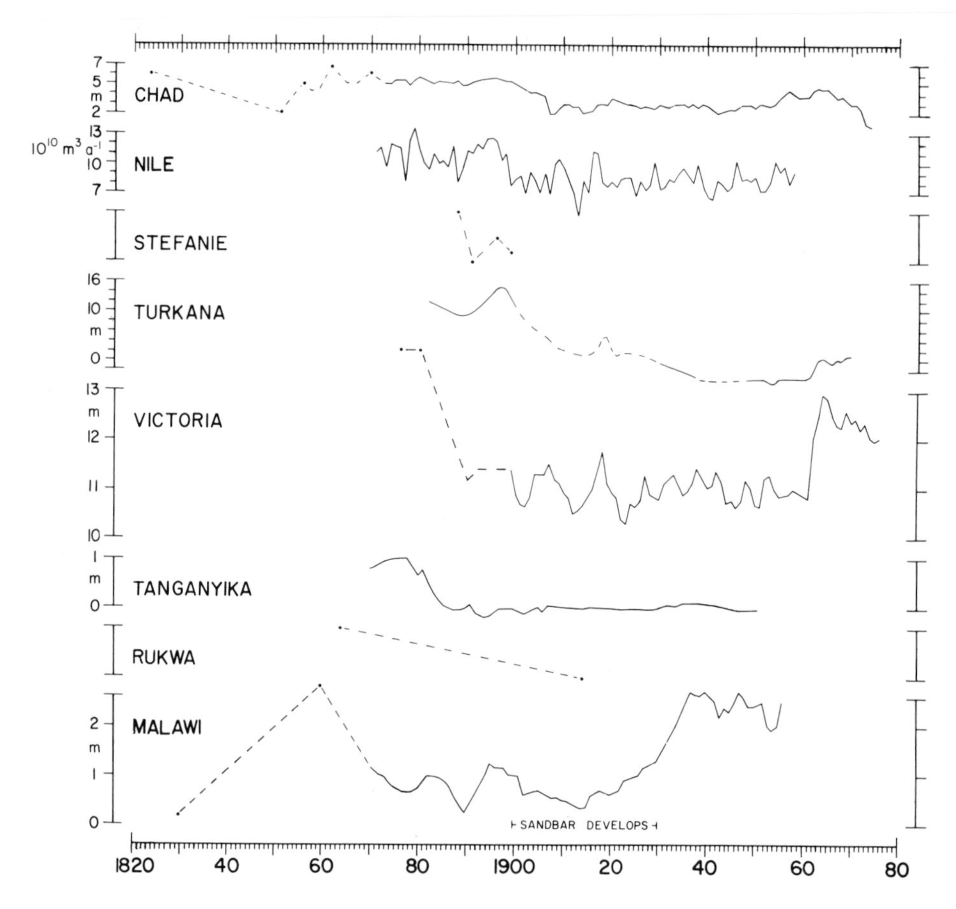

Fig. 2.5:4. Long-term variations of African lakes. (a) Lake Chad (source: Toucheboeuf de Lussigny, 1969); (b) Nile discharge at Aswan in 10^{10} m^3 (source: US Weather Bureau, 1967); (c) Lake Stefanie (source: Grove *et al.*, 1975); (d) Lake Turkana (Rudolf) (source: Butzer, 1971); (e) Lake Victoria (source: Ravenstein, 1901; Lamb, 1966; Mörth, 1967; and unpublished data); (f) Lake Rukwa (source: Hobley, 1914); (g) Lake Tanganyika (source: Academy of Sciences, USSR, 1964); (h) Lake Malawi (Nyasa) (source: Academy of Sciences, USSR, 1964, Lamb, 1966). Lake levels are in m, except for dots and broken lines which indicate that only qualitative information is available. Note that scale for Chad and Turkana differs from the other lakes.

Quantitative paleo-climatic inference has been attempted from former lake levels under consideration of energetic constraints (Hastenrath and Kutzbach, 1983). Analyses include Lake Victoria, and the Turkana, Nakuru-Elmenteita, and Naivasha basins in the Eastern Rift. Results are summarized in Figs. 2.5:2 and 2.5:3. It is found that endoreic equilibrium conditions at Lake Victoria prior to 12500 and around 10000 years B.P. would require rainfall about 150 mm a^{-1} below the modern average. Discharge about three times as large as the modern average could be maintained with a precipitation increment of about 200 mm a^{-1} above the modern average. It further appears that the greatly increased lakes in the Eastern Rift around 10000–8000 years B.P. could be kept in steady-state with rainfall of about 250 mm a^{-1} above the modern average.

Pollen profiles as reviewed in Section 2.3, especially the changes reported to set in after 6500 and 3000 B.P. are intriguing in that they may support the palaeo-limnological evidence. However, considerable further cooperation between palynology, palaeo-limnology, and glacial geology is called for.

Water level variations since the latter part of the 19th century are known in greater detail for various East African lakes (Ravenstein, 1901; Hobley, 1914; Academy of Sciences, USSR, 1964; Lamb, 1966; Temple, 1966; Mörth, 1967; Toucheboeuf de Lussigny, 1969; Butzer, 1971; US Weather Bureau, 1967; Grove, 1973; World Meteorological Organization – UNDP, 1974; Grove *et al.*, 1975). Selected hydrometeorological records are illustrated in Figs. 2.5:4 and 2.5:5.

In Fig. 2.5:4, the water level variations of Lake Chad and the Nile discharge at Aswan are included for comparison. Plots for the East African lakes are arranged proceeding from North to South. Only qualitative information is available for Lakes Stefanie and Rukwa. The most prominent feature most records have in common is a marked drop in the latter part of the past century. This event occurred earliest in the South with a progressive lag northward towards Ethiopia. Thus the highest water level is found around 1860 at Lake Malawi (Nyasa), around 1880 at Lake Victoria, but in the 1890s at Lake Turkana. By way of comparison with other parts of tropical Africa, the Chad and Nile records also show a marked decrease in the latter part of the past century. Irregular fluctuations occurred throughout the 20th century, most conspicuous being relative maxima in the 1910s and in the early 1960s, in particular. The rise of Lake Malawi from the 1920s onward seems to be related to the development of a sandbar (Lamb, 1966). Fig. 2.5:5 illustrates that the water level changes of Lake Victoria are broadly paralleled by the rainfall variations in Central Kenya.

Of particular interest is the drastic drop of the water level of Lake Victoria from around 1880 to the turn of the century (Fig. 2.5:4). Based on the detailed hydrometerological documentation (Ravenstein, 1901; Lamb, 1966; Mörth, 1967; World Meterological Organization – UNDP, 1974), rainfall in the Lake Victoria basin around 1880 is estimated (Hastenrath and Kutzbach, 1983) at about 150–200 mm a^{-1} above the average for the first half of the 20th century.

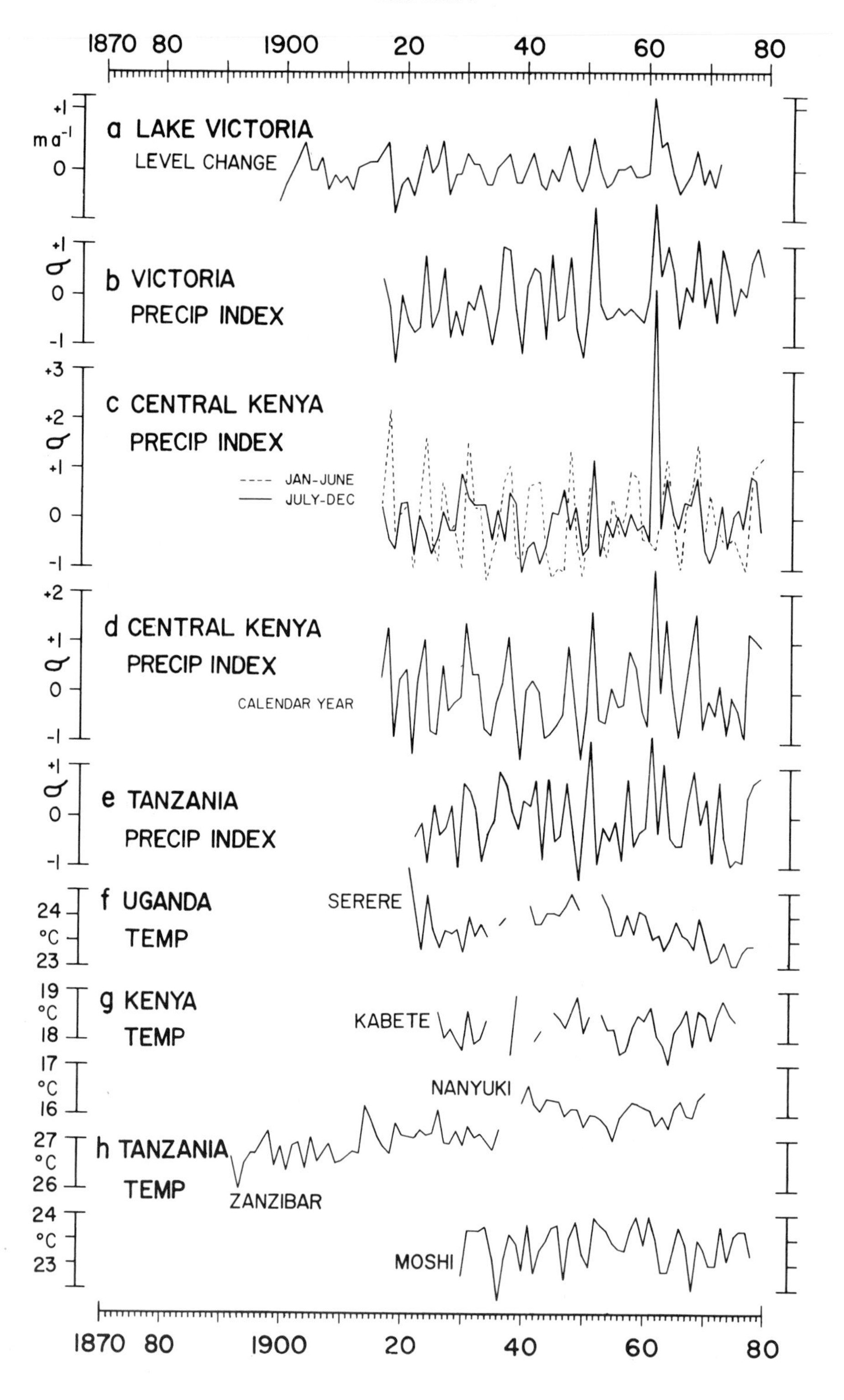
1870 80 1900 20 40 60 80
a LAKE VICTORIA
LEVEL CHANGE
+1
m a⁻¹
0
b VICTORIA
PRECIP INDEX
+1
q
0
-1
c CENTRAL KENYA
PRECIP INDEX
JAN-JUNE
JULY-DEC
+3
+2
+1
0
-1
d CENTRAL KENYA
PRECIP INDEX
CALENDAR YEAR
+2
+1
0
-1
e TANZANIA
PRECIP INDEX
f UGANDA
TEMP
SERERE
24
°C
23
g KENYA
TEMP
KABETE
19
18
17
16
NANYUKI
h TANZANIA
TEMP
ZANZIBAR
27
26
MOSHI
24
23
1870 80 1900 20 40 60 80

Fig. 2.5:5. Long-term variations of selected annual hydrometeorological indices for East Africa. (a) Annual water level change of Lake Victoria, counted from January of indicated to January of the following year, in m a^{-1} (source: Mörth, 1967; and unpublished data); (b) Index of annual rainfall in the catchment area of Lake Victoria. Index is the all-station average of normalized departure (σ) for the following station group: Entebbe, Bukoba, Kalangala, Kisumu, Musoma, Mwanza (source: East African Meteorological Department, 1966c; and unpublished data); (c) Index of January–June (broken), July–December (dash-dotted), and annual (solid) rainfall in Central Kenya. Computed as in (b), for station group: Embu, Kiambu, Kabete, Machakos, Meru, Muranga, Nakuru, Ngong, Nyeri (source: East African Meteorological Department, 1966a; and unpublished data). (d) Index of annual rainfall in Northern Tanzania. Computed as in (b) and (c) for station group: Arusha, Bagamoyo, Dar es Salaam, Kurio Mission, Mbulu, Moshi, Singida, Tabora, Tanga (source: East African Meteorological Department, 1966b; and unpublished data). (e) Annual mean temperature (°C) at Serere, Eastern Uganda (source: East African Meteorological Department, 1970, and unpublished data); (f) Annual mean temperature (°C) at Nanyuki and Kabete in Central Kenya (source: East African Meteorological Department, 1970; and unpublished data). (g) Annual mean temperature (°C) at Zanzibar (broken) and Moshi (solid), Tanzania (source: East African Meteorological Department, 1970; and unpublished data).

CHAPTER 3

PLEISTOCENE AND EARLY HOLOCENE GLACIATIONS

"That the glaciation was due to the former greater elevation of Mount Kenya, which has been reduced by subsidence and denudation. The theory of an universal glaciation is unnecessary, and is opposed by many facts in African geology."

(John W. Gregory, *Glacial geology*, 1894)

" . . . dass also der Grund für seine Hypothese lokaler Ursachen der Kenia-Eiszeit hinfällig ist, und dass gerade die Annahme der Universalität der Eiszeiten unabweisbar ist. Meine diesmaligen Beobachtungen am Kilimandjaro haben diese Annahme endgültig befestigt."

(Hans Meyer, *Kilimandjaro*, 1900)

"Per me il materiale roccioso disseminato in blocchi verso lo sbocco della valle Mobuku o rappresenta veri massi erratici depositati dagli antichi ghiacciai oppure, e questo forse piu verosimilmente, proviene dal graduale disfacimento delle morene, che dovettere la valle a monte del piano di Ibanda."

(Alessandro Roccati, *Ruwenzori*, 1909)

An overview of the geomorphic evidence of former glacial events is presented here based on a survey of the pertinent literature, evaluation of air photographs and topographic maps (Appendix 1), and my own field observations in the course of the 1970s. The separate discussion of conditions in the various mountain massifs will be complemented by an attempt at a synopsis. Correlation of moraine complexes at different mountains is suggested, as seems plausible from their elevation, spatial arrangement, and appearance. The paucity of absolute dating forestalls an ultimate chronology and a coherent large-scale picture. Interpretation in terms of spatial patterns of climate and the relation to the modern glaciation are of particular interest here.

3.1. Kilimanjaro

Traces of formerly much larger ice extent were first noted by H. Meyer (1900, pp. 343–408) during his expeditions in the 1880s and 1890s. Jaeger (1909) in 1906–7 expanded the survey of fossil glacial geomorphology. Klute (1920) mapped the moraines of Kilimanjaro in his stereophotogrammetric map at scale 1:50000 based on the 1912 field work. Nilsson (1931) in his 1927–8 visit confirmed Klute's survey, and complemented it especially on the West side of Kibo and the South side of Mawenzi.

A comprehensive study of former glaciations in the Kilimanjaro massif has been undertaken by the University of Sheffield expeditions in 1953 and 1957 (Downie and Wilkinson, 1972, pp. 43–71). In addition to maps of the various moraine complexes, the report proposes a relative chronology of glacial events (Table 3:1), although absolute

TABLE 3:1

Glaciation phases and moraine stages of Kilimanjaro (Downie and Wilkinson, 1972), Ruwenzori (source: Table 3.2:1), Mount Kenya (Baker, 1967), Aberdares (Perrott, 1982a), Mount Elgon (Hamilton and Perrott, 1978, 1979), Ethiopia (Hastenrath, 1974, 1977; Gasse and Descourtieux, 1979), Southern Africa (Harper, 1969; Hastenrath, 1972; Hastenrath and Wilkinson, 1973). Numbers indicate typical lower limit of moraines (m), and age estimates (years B.P.)

Kilimanjaro peak 5895 m lat 3°04′S	Ruwenzori 5111 m 0°23′N	Mount Kenya 5199 m 0°09′S	Aberdares 4002 m 0°19′S	Mount Elgon 4322 m 1°07′N	Ethiopia 4543 m 13°14′N	Southern Africa 3485 m 29°30′S	code of moraines in Table 3:2 and 3.2:1
5400–4700 m (?) 30–70 yrs	IV 4400 m 30–70 yrs	4500 m (?) 30–70 yrs					IV
Recent Glaciation, 2 phases 5200–4600 m	III 3–5 phases 4600–4400 m (100–300 yrs?)	Little Ice Age VI A, B ⩾4600 m (100–300 yrs?)					III
Little Gl. (Upper) Little Gl. (Lower) 4 phases	II 3500–4300 m (5000–10 000 yrs?)	Retreat II–V 4000–4300 m (> 8000 yrs?)			High Semyen, highest moraines in peak regions 4100–4200 m (→ ?)		II
	KB 3060 m				?		KB
Fourth (Main) Glaciation, 2 phases 3400 m	I 2300 > 15 000 yrs	Younger Maxima I A–D 3400 m (Würm?)	I 3800–3200 m > 12 200 yrs	I 3600–3300 m > 11 000 yrs	High Semyen lowest moraines in peak regions 4000–4750 m (+ Badda, Cacca, Cilalo, Enguolo) Badda > 11 500 yrs		I
Third Glaciation 100 000 yrs	Rwimi Basin 1890 m (100 000 yrs?)	Older Glaciation			Older Glaciation < 2600 m?	Older Glaciation	??
Second Glaciation 300 000 yrs	Katabarua 3000 m (500 000 yrs?)						
First Glaciation 500 000 yrs.							

dates are lacking for all but the oldest (the 'First' through 'Third') glaciations. Independently from the Sheffield study, the moraine systems at Kilimanjaro were mapped on the basis of air photographs and topographic maps (Appendix 1), and own field observations in the course of the 1970s (Map 3.1:1*). Table 3:2 lists the approximate ice-covered area

TABLE 3:2

Approximate ice-covered area [km^2] corresponding to moraine stages I, II, III, and present (1970s, P). Parentheses denote very crude estimates. Refer to Table 8:1 for code of moraine stages

	I	II	III	P
Kilimanjaro				
Kibo	137	41	16.5	4.9
Mawenzi	63	25	–	–
total, Kilimanjaro	200	66	16.5	4.9
Ruwenzori				
Emin		5	0.5	0.1
Gessi		5	0.8	0.2
Speke		9	3.6	1.1
Stanley		9	4.9	1.8
Baker		5	1.9	0.6
Luigi di Savoia		7	0.6	0.1
total, Ruwenzori	(200)	40	12.3	3.9
Mount Kenya	240	81	2.1	0.7
Aberdares	23	–	–	–
Mount Elgon	95	–	–	–
Mount Meru	?			
Virunga	?			
Total East Africa	778	187	30.9	9.5

corresponding to the various moraine stages. Map 3.1:1* broadly confirms the results of Downie and Wilkinson (1972, pp. 43–71). A few moraines are shown in Map 3.1:1* that do not appear in their charts and vice-versa. These minor differences may reflect the extent to which details of surface morphology can be identified from aerial photography.

Downie and Wilkinson's (1972) results are summarized in Table 3:1. The Sheffield expedition (Downie and Wilkinson, 1972, pp. 43–71) inferred three early glaciations largely from evidence other than moraine morphology. Subsequent glaciations are associated with large moraine systems. Fossil moraines (Map 3.1:1*) appear particularly prominent on the Western and Southern side of the Kilimanjaro massif, similar to the pattern of the present glaciation (Chapter 4). A lowest complex of large moraines extends down to about 3400 m. On the Southern side, these moraines are in part found in the lower portion of large U-shaped valleys. Towards the high Shira Plateau in the West, this

* An asterisk denotes that the map referred to is to be found in the back pocket of this book.

complex ends at 3800 m, whereas to the North of the Saddle Plateau and Mawenzi, a few moraine ridges descend to around 3000 m. No moraines are found on the Saddle Plateau, where a large ice sheet seems to have built up from the confluence of ice streams from Kibo and Mawenzi. Moraines on the South and North sides indicate that this ice sheet in turn fed large glaciers. These moraines correspond to the 'Fourth' or 'Main' glaciation of Downie and Wilkinson's (1972) scheme (Table 3:1).

A next higher complex of generally smaller moraines is found somewhat inward from and at a similar elevation as the aforementioned large moraine system. Downie and Wilkinson (1972) refer to this complex as the 'Little Glaciation', distinguishing two stages with altogether four separate phases.

At Kibo but not on Mawenzi, a further complex of comparatively small moraines is found at appreciably higher elevations, namely at 5400 m in the East, and around 5000–4800 m on the South and West sides. Downie and Wilkinson (1972) ascribe these moraines to a 'Recent Glaciation' and distinguish two phases. There are indications of weak moraine ridges immediately below the present ice rim, which could belong to the 20th century.

Table 3:2 shows a decrease of total ice extent on Kilimanjaro from about 200 km^2 at stage I to a third at stage II. From stage III onward the ice cover is limited to Kibo. In the lack of absolute dating, the relative chronology by Downie and Wilkinson (1972) provides a useful frame of reference for comparisons with other high mountains in East Africa and elsewhere in the tropics.

3.2. Ruwenzori

Among the numerous visits to the peak regions around the turn of the century, the expedition of the Duke of the Abruzzi (Roccati, 1907, 1909; Almagià, 1908; Filippi, 1908a, b, 1909a, b; Abruzzi, 1907a) marks the first attempt at a comprehensive scientific exploration. In the lower Mubuku Valley on the East side of the massif, Roccati (1907, 1909, pp. 127–145) claims the existence of moraines down to 1500 m, where Scott-Elliot (1895) from observations during his 1894 expedition proposed a former glacier extent to 1700 m. Roccati further describes moraines, roches moutonnées, and glacier striations particularly from the upper Mubuku Valley, and the Kitandara Valley separating the Stanley and Baker massifs.

Based on his 1927–8 field work, Nilsson (1931) rejects Scott-Elliot's (1895) and Roccati's (1907, 1909, pp. 127–145) earlier claims of very lower-reaching glacial evidence in the Mubuku Valley. Mindful of the shortcoming of his barometric height estimates, he gives a lowest moraine limit of about 2000 m (Nilsson, 1931, 1940). This value also appears in Bergström (1955). Nilsson's (1931) figure for Nyabitaba suggests that his height values are more than 300 m too low. Nilsson's observations are thus broadly consistent with Map 3.2:1*.

In contrast to the aforementioned explorations, the Belgian expedition of 1932 (Grunne, 1933, 1937; Michot, 1933, 1937; La Vallée Poussin, 1933; Meersch, 1933) approached the Ruwenzori from the West. Michot (1937, pp. 226–233) notes U-shaped valleys, roches moutonnées, glacier striations, and a sequence of glacial lakes ranging between 3700 and 4300 m on the Western side. For the Lamia Valley in the Northern Ruwenzori he reports glacial morphology down to 3700 m. However, he speculates that glacial evidence may exist at even lower elevations, and he considers that a swampy depression at 3300 m may be of glacial origin. By way of comparison, Map 3.2:1* indeed shows moraines down to 3000 m in the Lamia Valley, and at even lower elevations elsewhere in the Northern Ruwenzori.

Based on his studies on the Western flanks and the Northern part of the Ruwenzori in particular, de Heinzelin (1951, 1952, 1953, 1962) proposes a spatial correlation and tentative chronology of moraine stages, although absolute dates are lacking (Table 3.2:1).

TABLE 3.2:1

Glaciation phases and moraine stages in the Ruwenzori, according to de Heinzelin (1962), Osmaston (1965), and Osmaston and Pasteur (1972), and the present study (S.H.). Numbers indicate typical lower limits of moraines (m), and age estimates (years B.P.)

de Heinzelin			Osmaston			S.H.		
	m	yrs		m	yrs		m	yrs
						IV	4400	30–70
Lac Gris 4 phases		few 100	Lac Gris 4–6 phases	4000	inner 100 outer 200–300	III 3–5 phases	4600–4400	
Lacs Noir et Vert 2 phases		few 1000	Omurubaho 2 phases	3630	5000–10 000	II	3500–4300	
			Kichubu and Bigo	3060		KB		
Butahu		few 1000	Lake Mahoma	1950	>15 000	I	2300	
			Rwimi Basin	1890	100 000			
Crête Ruamya-Haute Ruanoli	3800–4100	few 10 000 –100 000	Katabarua	3000	500 000			

A stage 'Crête Ruamya – Haute Ruanoli' with moraine termini at 3800–4100 m is estimated at tens of thousands to around 100 000 years old. A 'Butahu' stage is estimated at tens of thousands of years. The associated glacial cirques are given an elevation of 3000–3200 m on the Western flanks and 2900–3100 m in the Northern part of the Ruwenzori. The maximum glacier stage in the Bujuku Valley on the Eastern side is considered as contemporaneous with this event. A stage 'Lacs Noir et Vert' at somewhat higher elevations is estimated as a few thousand years of age from the freshness of forms. A stage 'Lac Gris' finally is thought to be only a few centuries old.

Carbon-14 dating of a core retrieved from Lake Mahoma at the lowest reaching large moraines near the juncture of the Mahoma, Mubuku, and Bujuku Valleys, permits Livingstone (1962) to infer the age of deglaciation as 14 700 years B.P. Accepting de Heinzelin's (1962) proposed correlation, this would correspond to the retreat from his 'Butahu' stage.

Osmaston (1965) mapped moraines of the entire Ruwenzori from field study and air photographs. Barring absolute dates, he proposed a spatial correlation and relative chronology. His map sketches are not readily accessible.

Independently, the moraines of the Ruwenzori were mapped (Map 3.2:1*) on the basis of the available air photographs and maps (Appendix 1) and my own observations on a January 1974 field trip. The approximate ice-covered area corresponding to the various moraine stages is given in Table 3:2. Comparison shows that a few of the moraines in Map 3.2:1* are missing in Osmaston's (1965) sketches and vice-versa. As with the Kilimanjaro and Mount Kenya maps (Maps 3.1:1* and 3.3:1*), this may reflect the limitations in the evaluation of ground details from aerial photography.

Map 3.2:1* and Tables 3:1 and 3.2:1 show a sequence of various moraine complexes proceeding upward in the major valleys. A lowest group (I) of large moraines ends at 2230 m at the juncture of the Bujuku, Mubuku, and Mahoma Valleys. The elevation determined from air photographs and topographic maps (Appendix 1) and field observations agrees with the value given by Osmaston (1965) and Osmaston and Pasteur (1972, p. 136), while Whittow *et al.* (1963, p. 583) list a figure of 2134 m. These moraines may have corollaries at the East side of the Portal Peaks at around 2500 m, in the Lamia Valley at 3000–2800 m, and on the West side of the Ruwenzori around 3500 m.

A next higher complex (II) is exemplified by moraines around 3500 m in the upper Bujuku Valley, around 4000–3800 m between Speke and Gessi, and around 4000–4300 m on the East side of the Stanley massif.

The further, still higher moraine complex (III) is found, for example, at 4600–4400 m on the Stanley and Baker slopes of the Kitandara Valley, and at somewhat lower elevations on the East side of the Stanley massif. Three to five separate ridges can be identified in this complex. This complex is not far from the present ice rim, and has a fresh appearance. Weak debris remnants (IV) can furthermore be observed in the field in the immediate vicinity of the present glacier fronts; these features being too weak to be discernible on air photographs. In part they seem to be in locations that were still occupied by ice around the beginning of this century. In the hyper-humid Ruwenzori a shallow vegetation mantle covers surfaces vacated by the ice not long ago.

The contours in Map 3.2:1* are expected to facilitate the comparison with the map sketches of de Heinzelin (1962) and Osmaston (1965) which lack isohypses. In Table 3.2:1 an attempt is made to compare the present survey with glacial stages and estimated ages proposed in the studies of de Heinzelin (1962) and Osmaston (1965). Uncertainties in the correlation of the three schemes remain with respect to the older glaciations and concerning the role of a few smaller moraine groups. In particular, Osmaston (1965) regards the small but quite distinct 'Kichubu' and 'Bigo' moraines at 3060 m as manifestations of retreat phases of the Lake Mahoma glaciation. Corollaries to these features of the Bujuku Valley are not found in other parts of the massif. However, a broadly consistent picture is emerging for the major glacial events. The weak debris ridges (IV) seem to be a feature of the 20th century. Moraine complex (III) consisting of several distinct ridges may from its fresh appearance be a few centuries old. The next lower complex (II) is found at fewer locations. Particularly prominent, however, is the moraine complex (I) near the exit of the large valleys. Still older glaciations as suggested by de Heinzelin (1962) and Osmaston (1965) are not reflected in a prominent moraine morphology.

From inferred ice extent and peak elevations Osmaston (1965) calculated for the Lake

Mahoma glaciation (moraine stage I, Tables 3:1 and 3.2:1) a rise of the firnline from Southeast to West and a temperature drop with respect to the present of 3–6° C. For the Omurubaho glaciation (moraine stage II, Tables 3:1 and 3:2) he obtained a similar rise from Eastsoutheast to Westnorthwest; zonal contrasts being weaker than at present. It is in fact noteworthy that the moraines of stage (I) in particular seem to reach to lower elevations on the East as compared to the West side of the massif. Further field investigations are no doubt in order. It is here suggested that the apparent asymmetry may not solely reflect the climatic pattern, but may be related to the arrangement of the large ice drainage systems.

Reference is made to Table 3:2 for the ice extent corresponding to the various moraine stages. The ice cover during stage I is particularly difficult to estimate. A decrease to less than a fifth is indicated at stage II, when separate ice entities can be distinguished between the different mountains. The rate of decrease to stage II and present is much smaller in absolute, but comparatively pronounced in relative terms.

The moraine stages of the Ruwenzori will be compared, in Section 3.9, with geomorphic evidence from other high mountains in East Africa. It is already to be seen from Map 3.2:1* and Tables 3:1 and 3.2:1 that well-developed moraines attesting to earlier glaciations extend to rather low elevations in the Ruwenzori, broadly consistent with the low ice equilibrium line and glacier termini of the present.

3.3. Mount Kenya

During his 1893 expedition, Gregory (1893, 1894, 1896, 1900, 1921, pp. 149–153) observed a wealth of fossil glacial morphology, including various moraine complexes, roches moutonnées, and glacier striations. However, he rejects the idea of a universal glaciation related to large-scale climatic episodes and interprets the geomorphic evidence of a much larger ice extent as the result of the formerly greater elevation of Mount Kenya, thought to be subsequently reduced by subsidence and denudation (Gregory, 1894; 1921, pp. 150–1). Mackinder (1900, 1930) on his 1899 expedition was able to confirm geomorphic evidence of formerly more extensive glaciation.

During 1927–8 and 1932, Nilsson (1931, 1940) undertook a mapping of some moraines in the Nithi and Gorges Valleys on the East side of Mount Kenya. Nilsson (1931) finds his survey consistent with Gregory's (1894, 1900) observations in the Teleki Valley on the West side of the mountain. He recognizes 5 distinct moraine complexes, at elevations around 3100, 3300–3400, 4400–4500 and 4600–4700 m, respectively, the latter being considered of recent formation. Baker (1967) completed a comprehensive survey of the glacial geology of Mount Kenya including a mapping of moraine complexes in the entire mountain massif. Spatial correlation and relative chronology are proposed in the absence of absolute dates (Baker, 1967, pp. 57–65). The recent work of Perrott (1982b) is aimed at a glacial history. Mahaney (1972, 1976, 1979, 1980) attempts to arrive at age estimates of moraines by methods other than carbon-14 dating.

Independently of Baker's (1967) comprehensive survey, moraine complexes in the entire mountain massif were mapped on the basis of air photo interpretation and my own field observations (Maps 3.3:1* and 3.3:2). As noted by Zienert (1968), a few moraines are apparent on the air photographs that were not identified by Baker; but likewise, a few of the features mapped by Baker could not be confirmed in the aerial photography. The

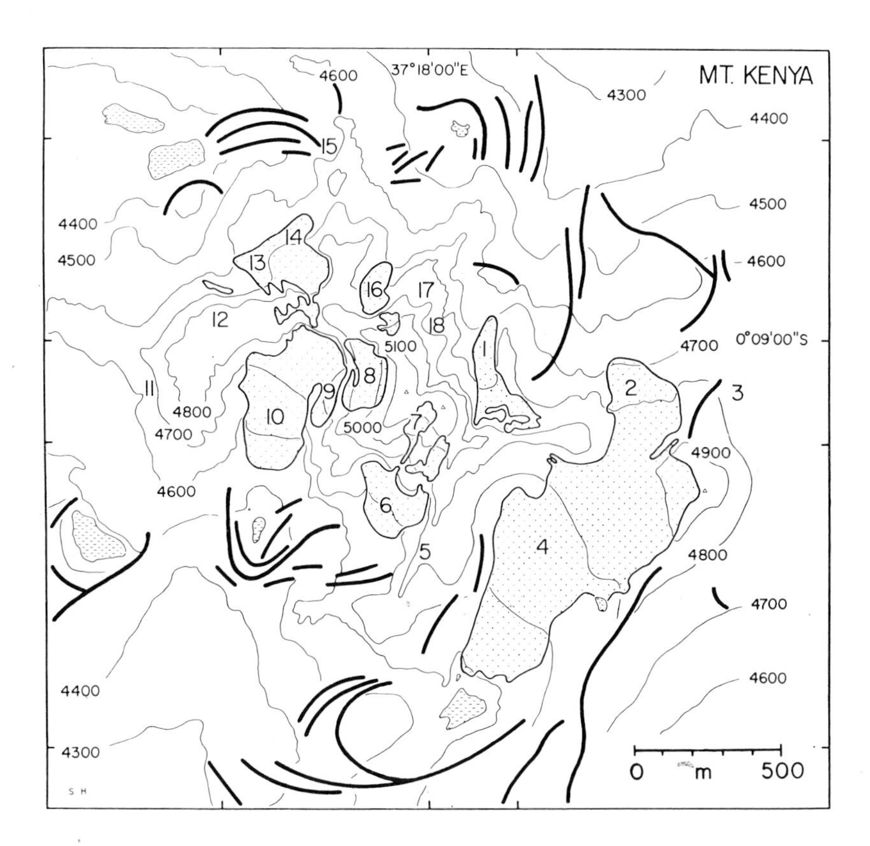

Map 3.3:2. Glaciers and moraine stages in the peak region of Mount Kenya. Scale 1:25,000. Symbols as for Maps 3.1:1, 3.2:1, 3.3:1, and 4.3:1.

limitations of air photo interpretation are thus similar to the state of affairs mentioned for Kilimanjaro and the Ruwenzori. In all important respects Maps 3.3:1* and 3.3:2 are consistent with Baker's (1967) survey. Map 3.3:1* indicates a tendency for lower moraine termini on the Southern and Western, as compared to the Northern and Eastern sides, in similarity to the pattern of the present glaciation.

The moraine stages according to Baker (1967) are included in Table 3:1. Baker (1967, pp. 57–9) considers the evidence of an 'Older' glaciation as fragmentary. However, a sequence of younger glaciations is manifested in the moraine morphology. A lowest complex of moraines extends down to around 3400 m. Baker terms these 'Younger Maxima', distinguishing four stages: I A to D. The next higher moraines are found reaching down to 4000–4300 m, where Baker recognizes the four (retreat) stages II to V. All of the aforementioned complexes are deeply weathered and carry abundant vegetation. A further moraine complex reaching down to about 4400 m, however, is bare and has a fresh appearance. Historical evidence to be discussed in Chapter VI indicates that the ice was almost in contact with these moraines at the end of the last century. Baker refers to this complex as 'Little Ice Age' and distinguishes two stages III A, B. Even further up valley, weak debris ridges are found in locations that have been vacated by the ice since the beginning of this century.

For the ice-covered area corresponding to the various moraine stages, reference is made to Table 3:2. The ice cover diminished from about 240 km^2 at stage I to a third of that, at stage II; the decrease to stage III and the present in relative terms being even more pronounced.

3.4. Aberdares

In the account of his 1932 trip, Nilsson (1935, p. 10) describes small moraines in the highest part of the Aberdares extending down to about 3700 m. It is noteworthy that he gives a summit elevation some 100 m lower than that appearing on the modern topographic maps (Appendix 1).

A visit to the highest part of the Aberdares materialized in 1973. However, the mapping of fossil moraines, Map 3.4:1*, and the entry in Table 3:1 are primarily based on the evaluation of air photographs and topographic maps (Appendix 1). The ice-covered area, estimated from Map 3.4:1*, is listed in Table 3:2. Moraines are found at 3600–3200 m to the East, and at 3800–3600 m to the Northwest of Oldoinyo Lesatima. Small but distinct moraine ridges are also apparent at 3800–3400 m on the Southwest side of the Aberdares. Map 3.4:1 provides the first mapping of moraines in the Aberdares. From a core at about 3800 m, in the summit region of Oldoinyo Lesatima, Perrott (1982a) infers the deglaciation shortly before 12 200 years B.P. For relative chronology and ice extent reference is made to Tables 3:1 and 3:2. This modest fossil glacial morphology will be considered in Section 3.9 in perspective with the evidence from the other high mountains of East Africa.

3.5. Mount Elgon

In the course of his 1927–8 field work, Nilsson (1931) mapped the moraines of Mount Elgon, for which purpose he had to construct a topographic base map in the first place. On the other flanks of the caldera, Nilsson's map shows moraines descending to 3400 m in the North, to 3475 m in the East, to 3800 m in the South, and to 3700 m in the West. Nilsson also recognized moraines at slopes in the interior of the caldera.

During a 1973 field trip to Mount Elgon, fossil moraines were observed in the Eastern quadrant and in the interior of the caldera. The moraine morphology of Mount Elgon was subsequently mapped (Map 3.5:1*) on the basis of air photographs and topographic maps (Appendix 1), and the limited field observations. Results are also entered in Tables 3:1 and 3:2.

On the outer slopes to the South and North of Mount Elgon moraines extend mostly down to about 3600 m, but in one valley to the Northeast to around 3300 m. Moraine morphology is rather less developed on the Western outer slopes of Elgon. Moraines down to 3400 m are also found in the Swam Valley. Very distinct moraine ridges furthermore occupy the northward and eastward facing slopes of the interior of the caldera, descending down to the very bottom of the basin at less than 3800 m. The modern air photographs and topographic maps (Appendix 1) provide a superb basis for geomorphological mapping that was not available to Nilsson. It is therefore not surprising that his survey differs from Map 3.5:1* in moraines identified, elevations, and even the orientation of moraine ridges. Hamilton and Perrott (1978, 1979) obtained two carbon-14 dates from an ice-scooped lake basin at 4150 m in a valley on the Southern slopes of Mount Elgon. The coordinates of the sampling site are given as 1°06′N, 34°34′E. They infer a deglaciation date for Mount Elgon of shortly before 11 000 years B.P., which they regard as concordant with Livingstone's (1962) dates of >14 700 years B.P. for the retreat from the maximum ice advance in the Lake Mahoma area of the Ruwenzori. Alternatively, Hamilton and Perrott (1978) suggest that the ice-scooped rock basins on Elgon may have originated during a standstill or minor ice advance phase and consider as corrolaries the Omurubaho rather than the Kichubu-Bigo (Table 3.2:1) moraines of the Ruwenzori. Hamilton and Perrott (1979) relate the larger moraines and more pronounced glacial erosion on the South side to greater sediment availability, but also note the lower reaching moraines on the North side of Mount Elgon.

For relative chronology and ice extent reference is made to Tables 3:1 and 3:2. It is noted that the ice cover at moraine stage I was substantial, no contribution being indicated for the later stages. The inventory of moraines at Mount Elgon, Map 3.5:1* and Tables 3:1 and 3:2, will be discussed in a larger spatial context in Section 3.9.

3.6. Ethiopia

Early studies of Pleistocene mountain glaciation in Ethiopia are due to Nilsson (1935), Minucci (1938a, b) and Mohr (1963, 1967, 1971). A field survey and mapping of moraines in High Semyen materialized in 1973 (Hastenrath, 1974, 1977). Inference on former glaciation at other mountains of Ethiopia was possible on the basis of air photographs (Hastenrath, 1977; Potter, 1976).

In High Semyen two moraine complexes were mapped, the higher of which at 4100–4200 m, and the lowest at 4000–3750 m. Fossil glacial morphology is best developed in the Northwestern and Western quadrants. Moraines are deeply weathered and carry abundant vegetation. Despite Hurni's (1982) attempt at carbon-14 dating, an absolute chronology of fossil glaciation in High Semyen is still lacking.

At Mounts Badda, Cilalo, Cacca, and Enguolo in Southern Ethiopia, where moraines are identified from aerial photography, there are indications for more pronounced glaciations in the Western quadrant, and moraines seem to extend further down than in High Semyen. This is spatially consistent with Messerli *et al.*'s (1980) report of terminal moraines at 3200–3100 m in the Bale Mountains of Southern Ethiopia, as well as with the low moraine elevations on the Northeast side of Mount Elgon (Section 3.5).

Gasse (1978), and Descourtieux (1979), and Street (1979) obtained a carbon-14 date from the bottom of a small bog embedded in a cirque at 4040 m on the East side of Mount Badda. From this it is inferred that the ice had completely disappeared before 11 500 years B.P.

Results for Ethiopia are also summarized in Table 3:1.

3.7. The Saharan Uplands

Messerli *et al.* (1980) and Messerli (1967) have reviewed the field evidence of former glaciation on the high mountains within and around the Sahara desert. Glaciers existed in the High Atlas of Morocco (31°N, 4165 m). Such evidence is lacking for the Sinai (28°N, 2642 m). For the Hoggar (23°N, 2918 m) there are reports of a nivational zone on East-facing slopes at 2400–2600 m, and of some forms possibly indicative of small glaciers on the summits. In Tibesti (19°N, 3415 m), distinct nivational forms are apparent only above 3000 m, especially around the Northern peaks, but are lacking on the Southern higher summits. The nivation forms are not controlled by aspect and are less clear than in the Hoggar.

For Jebel Marra (13°N, 3042 m) in Western Sudan, Williams *et al.* (1980) report no evidence of former glaciation. Proceeding further around the hydrographic divides of the Chad basin, no reports of former glaciation or the past altitudinal zonation seem to be available for Mount Cameroon (4°N, 4070 m). In fact, recent field observations (Messerli and Baumgartner, 1979, p. 206) indicate the absence of fossil glacial forms, recent volcanic ash cover being suggested as a possible factor.

3.8. Southern Africa

Based on geomorphic evidence, Harper (1969) suggested that the highest peaks of Lesotho were above the snowline during the Pleistocene with a broad periglacial zone extending some 900 m lower. Thabana Ntlenyana, the highest summit of Lesotho and Southern Africa as a whole, reaches only 3485 m. During field work in 1971 (Hastenrath, 1972; Hastenrath and Wilkinson, 1973) land forms suggestive of glacial relief were indeed observed in the highlands of Lesotho, but a clear moraine morphology could not be detected. Geomorphic evidence of Pleistocene mountain glaciation in Southern Africa is thus considered inconclusive (Table 3:1).

3.9. Synthesis

Geomorphic evidence has been reviewed in the preceding Sections 3.1 through 3.8 of former glaciations at the various high mountains of Eastern Africa. Table 3:1 summarizes this information in an attempt at a spatial correlation and relative chronology. The cornerstones of this scheme are the comprehensive studies for Kilimanjaro (Downie and Wilkinson, 1972) including K-Ar datings, and for Mount Kenya (Baker, 1967), and the absolute dates for the deglaciation of the Ruwenzori, Aberdares, Elgon, and Badda, by Livingstone (1962), Perrott (1982a), Hamilton and Perrott (1978, 1979), and Gasse and Descourtieux (1979). Because of the paucity of absolute dates, this synopsis must largely rely on correspondences plausible from elevation, spatial arrangement, and appearance of moraine complexes. In general, lower-reaching glaciers are expected at the mountains with more abundant precipitation and higher summit elevations, or rather larger elevated catchment areas.

Evidence has been advanced by Downie and Wilkinson (1972) for three old glaciations at Kilimanjaro around 500 000, 300 000, and >100 000 years. The second of these may correspond to the old 'Katabarua' (Osmaston, 1965) of the Ruwenzori. The 'Third' glaciation of Kilimanjaro may possess corollaries in the 'Rwimi Basin' (Osmaston, 1965) glaciation of the Ruwenzori, the 'Older' (Baker, 1967) glaciation of Mount Kenya, as well as in the 'Older' glaciations in Ethiopia and possibly even in Southern Africa.

From elevation and spatial arrangement it is here suggested that the 'Fourth' or 'Main' glaciation (Downie and Wilkinson, 1972) at Kilimanjaro may correspond to the 'Younger Maxima I A–D' (Baker, 1967) of Mount Kenya, and to the moraine complex I, or 'Butahu-Lake Mahoma' (de Heinzelin, 1962; Osmaston, 1965; Osmaston and Pasteur, 1972) stage in the Ruwenzori. The ice retreat from these moraines is timed by Livingstone's (1962) Lake Mahoma core at 14 700 years. From the comparable elevations it is here proposed that the rather higher moraines (I) found in the Aberdares and at Mount Elgon also correspond to the 'Younger Maxima, I A–D' (Baker, 1967) at Mount Kenya. As suggested earlier (Hastenrath, 1974), the lowest moraines in the peak region of High Semyen may also pertain to this event. Moraines at the volcanoes in Southern Ethiopia, namely at Badda, Cacca, Cilalo, and Enguelo, may be of similar age. Furthermore, the highest moraines in the peak region of High Semyen may stem from this glaciation although they may belong to the next more recent event.

It is noted from Table 3:1 that ages of >12 200, >11 000, and >11 500 years have been deduced for the disappearance of ice from the summit region of the Aberdares and from cirque basins on Mounts Elgon and Badda, respectively. The 14 700 year date from Lake Mahoma in the Ruwenzori refers to the retreat from a maximum ice advance. Hamilton and Perrott (1978, 1979) in fact regard their Elgon date as 'concordant' with Livingstone's (1962) date for the Ruwenzori. It must remain open whether the dates for Elgon and Badda correspond to a standstill or minor readvance phase (Hamilton and Perrott, 1978, 1979), in which context the Kichubu and Bigo moraines of the Ruwenzori (Table 3.2:1) come to mind.

No absolute dates have been obtained for the more recent glaciations. However, a correspondence is suggested from elevation ranges and spatial arrangement between the 'Little Glaciation' (Downie and Wilkinson, 1972) at Kilimanjaro and stages 'II–V' (Baker, 1967) at Mount Kenya. Complex II in the Ruwenzori, i.e., the 'Lacs Noir et Vert' (de Heinzelin, 1962), or 'Omurubaho' (Osmaston, 1965; Osmaston and Pasteur, 1972) seems a further corollary. Whether the highest moraines in the peak regions of High Semyen belong to the same or the next earlier event remains open. No corollaries can be suggested for the other, lower mountains of Eastern Africa.

Hamilton (1982, p. 31) speculates that the II and I glaciations may be timed at shortly before 11 500 and about 18 000 years B.P., respectively. For the sake of completeness, it should also be mentioned that Hamilton (1982, p. 24) claims to have seen from a long distance, and clearly, moraines on Mount Meru; he gives no elevation or other particulars. I have viewed Mount Meru from various azimuths and have evaluated the available air photographs and topographic maps, and cannot support this contention. Concerning the Virunga volcanoes, air photographs show no moraine morphology. As for Mount Meru, volcanic activity may have obliterated traces of earlier glaciation.

Field evidence has been reviewed in Section 2.5 of greatly enlarged lakes in East Africa around 10 000–8000 years B.P. The inferred precipitation increase as compared to the present, and the associated changes in cloudiness and surface radiation budget during this episode are expected to act towards enhancing the glacier net balance. As there is evidence for the moraine complexes I and III being of a very different age, the moraine complex II (Table 3:1) remains as a possible geomorphic manifestation of a major glacial event related to the episode of enlarged East African lakes. Since a direct stratigraphic linkage between the moraines on and the large lakes outside the mountains does not appear possible, the absolute dating of complex II and other moraines is of particular interest.

A synchroneity is suggested for the multi-phased moraine complexes of the 'Recent Glaciation' (Downie and Wilkinson, 1972) at Kilimanjaro, of the 'Little Ice Age, IV A, B' (Baker, 1967) at Mount Kenya, and the moraines III, i.e., the 'Lac Gris' stage (de Heinzelin, 1962; Osmaston, 1965; Osmaston and Pasteur, 1972) in the Ruwenzori. An age of a few centuries has been suggested for these moraines in the absence of absolute dates.

At even higher elevations and near the present ice rim, small debris ridges are found. For Mount Kenya, these moraines can be shown to be formations of the 20th century (Chapter 4), inasmuch as they are located on ground that has been vacated by the ice since the turn of the century. A corollary exists in the Ecuadorian Andes and the Cordillera Blanca of Peru (Hastenrath, 1981). Table 3:1 thus indicates a broad spatial continuity of glacial events at the high mountains of East Africa.

Table 3:2 suggests a total ice cover on the high mountains of East Africa at moraine stage I as large as about 800 km^2; Kilimanjaro, the Ruwenzori, and Mount Kenya possessing similar shares, but the ice cover of Mount Elgon also being sizeable. From stage II onward ice cover is limited to the three presently glaciated mountain massifs. From stage I to II the ice-covered area has decreased to about a fourth for East Africa as a whole, and to about a third at Kilimanjaro and Mount Kenya. At both stages I and II the ice extent on Mount Kenya still exceeds that of either Kilimanjaro or Ruwenzori. From stage II to III the ice cover diminishes to only about one sixth for East Africa as a whole, but the decrease is particularly drastic for Mount Kenya, in both absolute and relative terms. From stage III onward Mount Kenya contributes only a small fraction to the total ice cover of East African mountains. Large-scale climatic conditions appear less and less conducive to ice cover at the younger moraine stages. Viewed in context, Mount Kenya is less favorable for glaciation than Kilimanjaro in terms of elevation and hence temperature conditions, and less favorable than the Ruwenzori in terms of precipitation. A more detailed account of ice extent and chronology shall be presented in Chapter 4 for the recent glaciation.

The review of field evidence from other high mountains on the African continent suggests large-scale correlations of glacial-climatic events. Development of an absolute chronology may place such inferences on a firm footing.

CHAPTER 4

THE RECENT GLACIATION

"Nach etwa einer Stunde erreichten wir einen etwas erhöhten Ort, von wo aus ich den Schneeberg Kegnia deutlich sehen konnte, da die Luft rein und klar war. Der Berg erstreckte sich von Ost nach Nordwest bei West. Er erschien mir wie eine ungeheure Mauer, auf dessen Spitze ich zwei grosse Thürme oder Hörner erblickte, welche nicht weit von einander stehen, und welche dem Berg ein imposantes Ansehen gaben."

(Ludwig Krapf, *Reisen,* 1858)

"The valley, in fact, ended in a great snowy horseshoe, which was dazzlingly white and beautiful, and from these snowfields I now saw, to my intense surprise and delight, that there descended three superbly green glaciers, the snout of one of which pushed far into the valley. So there were glaciers on these mountains, and there are probably many of them descending from other faces of the great peaks, into numerous valleys which have never yet been entered by any European."

(John E. S. Moore, *Mountains of the Moon*, 1901)

"Wer je den weissen Dom des Kibo gesehen hat, wie er über das satte Grün des Urwaldes in den blauen, klaren Morgenhimmel ragt, wird dieses erhabene Bild nie vergessen."

(Fritz Klute, *Kilimandscharo*, 1920)

In contrast to the formerly much greater ice extent discussed in Chapter 3, the present glaciation is limited to the three highest mountains of East Africa, Kilimanjaro, the Ruwenzori, and Mount Kenya. In all three areas there are indications of a drastic glacier recession since the earliest visits to the peak regions at the end of the last century. The objectives of this chapter are to create an inventory of the present ice extent, and to reconstruct the glacier variations in the course of the past century on the basis of historical documentation.

4.1. Kilimanjaro

The ice cover of Kilimanjaro has decreased greatly since the end of the last century. Quantitative assessment of long-term variations and the inventory of present ice conditions are hampered by the lack of a topographic map at suitable scale (Appendix 1). Before the reconstruction of secular glacier variations, an attempt is made in this chapter to ascertain the present ice extent on Kibo. This effort relies mainly on information from the 1970s. The primary sources include a vertical air photograph shot on 18 March 1972 (Geosurvey Ltd., courtesy of Peter Gollmer, Nairobi); a photograph taken on a hot air balloon flight over the crater of Kibo on 10 March 1974 (frontispiece; courtesy of Alan Root, Nairobi); my own field observations during 1971, 1973, and 1974; a map sketch and other information gathered by John Temple; and a map at scale 1 : 25 000 in Humphries (1953). On this basis, Map 4.1:1 of the present glaciation on Kibo was

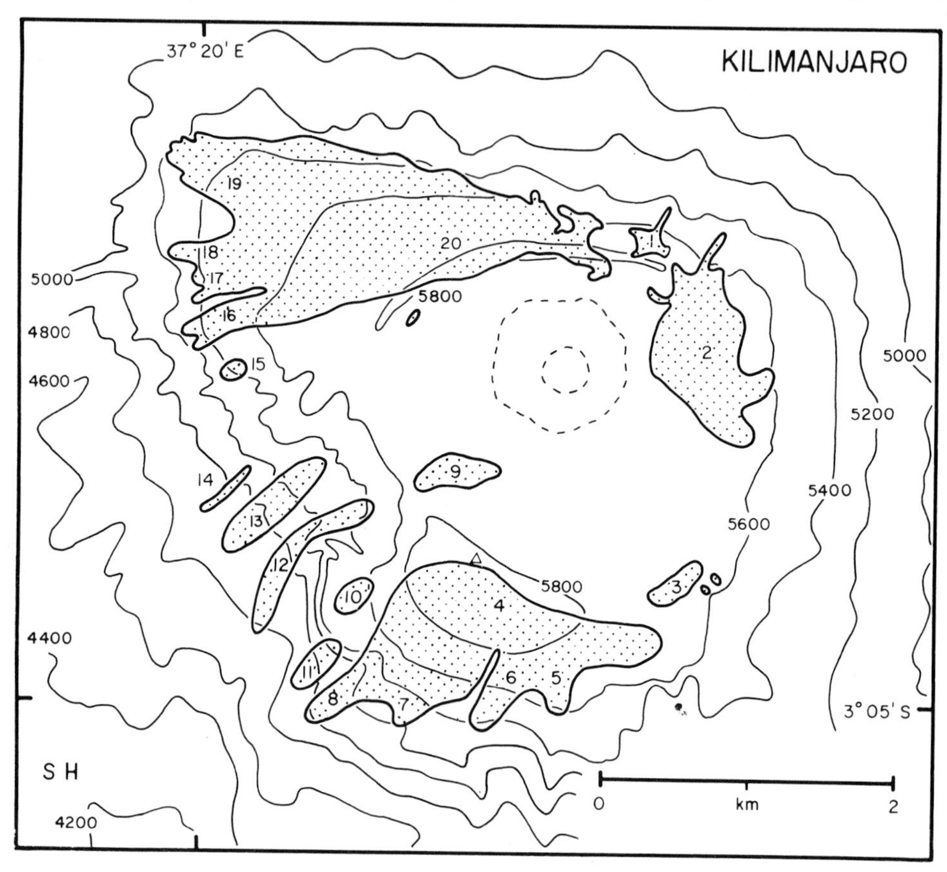

Map 4.1:1. Glaciers of Kilimanjaro. Scale 1 : 50 000. Contours at 200 m intervals. Ice areas stippled. Large numbers denote glaciers listed in Table 4.1:1.

constructed. This is also a source for Table 3:2. Table 4.1:1 relates the nomenclature of glaciers to the numbers entered in Map 4.1:1. Reference is made to Fantin (1968, pp. 46–7), Mountain Club of Kenya (1971, pp. 235–7; 1981, pp. 281–4), and to Table 4.1 for the origin of glacier names. Conventional parameters estimated for the World Glacier Inventory are summarized in Appendix 3.

Kibo is a volcanic cone of approximate rotational symmetry, except for the Great Breach in its Southwest sector. Despite the rather regular shape of the mountain, the ice cover is strongly asymmetric by azimuth. Thus, glaciation is more abundant on the Southern as compared to the Northern side of Kibo. Even more pronounced is the

TABLE 4:1

Origin of names of East African glaciers (sources: Fantin, 1968, pp. 46–47; Mountain Club of Kenya, 1971, pp. 150–152, 235–237; Osmaston and Pasteur, 1972, pp. 168–171)

Kilimanjaro

Glacier no. (cf. Table 4.1:1)	Named after
1	location
3	Friedrich Ratzel, German geographer
4	location
5	Johannes Rebmann, German missionary first to see Kilimanjaro
6	Baron Karl Klaus von der Decken, German explorer on Kibo in 1862
7	Otto Kersten, German explorer on Kibo in 1862
8	Albert Heim, Swiss geologist
9	Giovanni Balletto, Italian medical doctor resident in Tanzania
10	?
11	Walter Furtwängler, German mountaineer on Kibo 1912
12, 13	location
14	shape ?
15	Carl Uhlig, German geographer, on Kibo in 1901–1904
16, 17	A. Penck, German geologist
18	Erich von Drygalski, German geographer
19	W. Credner, German geographer
20	location

Ruwenzori

Glacier no. (cf. Table 4.2:1)	Named after
A 1, 3, 4	Emil Kraepelin, German psychiatrist
A, 2, 5, 6	Emin Pasha, originally Edward Schnitzler, German medical doctor and explorer, Governor of Equatoria
B 1–4, 9	Romolo Gessi, Italian explorer in Africa
B 5–8	Princess Iolanda of Savoia
C 1	Col. James August Grant, British explorer in Uganda
C 2	Vittorio Emanuele III, King of Italy
C 3, 4	Sir Harry Johnston, British Commissioner for Uganda
C 5	John Hanning Speke, British explorer in Uganda
D 1	Queen of Edward VII of Great Britain
D 2	Albert I, King of Belgium
D 3, 4	Queen of Umberto I of Italy
D 5, 14	Sir Henry Morton Stanley, British explorer in Africa
D 6, 11	Elena, Queen of Vittorio Emanuele III of Italy
D 7 ?	coronation of Queen Elizabeth II of Great Britain

Table 4:1, continued

Ruwenzori

Glacier no. (cf. Table 4.2:1)	Named after
D 8, 12	Royal House of Italy during 1906 Italian expedition
D 9	Duke of Edinburgh, consort of Queen Elizabeth II of Great Britain
D 10	Queen Elizabeth II of Great Britain
D 13	August F. Moebius, German astronomer and mathematician
E 1, 4	Sir Samuel W. Baker, English explorer
E 2	shape
E 3, 7	Edward VII, King of Great Britain 1841–1910
E 5	J. E. S. Moore, British zoologist on Ruwenzori in 1900
E 6	Alexander Frederick Richmond Wollaston, British biologist and explorer on Ruwenzori in 1906
E 8	Gottfried Semper, German architect
F 1, 2	Joseph Thomson, Scottish explorer in East Africa
F 3	Vittorio Sella, photographer with 1906 Italian expedition
F 4	Lt. W. G. Stairs, member of Stanley's Emin relief expedition

Mount Kenya

Glacier no. (cf. Table 4.3:1)	Named after
1	Johannes Ludwig Krapf, German missionary first to see Mt. Kenya
2	John W. Gregory, British geologist on Mt. Kenya in 1893
3	Georg Kolbe, German naturalist on Mt. Kenya in 1896
4	Henry Carvell Lewis, American geologist and professor
5	J. D. Melhuish, British dentist and mountaineer resident in Kenya
6	Charles Darwin, English naturalist
7	extreme hardness
8	François Alphose Forel, Swiss geologist
9	Alber Heim, Swiss geologist
10	John Tyndall, British physicist
11	A. R. Barlow, photographer on excursions of J. W. Arthur
12	Point Pigott; J. R. W. Pigott, British colonial official
13	Cesar Ollier, Italian alpine guide, accompanied Mackinder on first climb of Batian
14	Joseph Brocherel, Italian alpine guide, accompanied Mackinder on first climb of Batian
15	Point Peter; the Apostle
16	Sir Edward Northey, British Governor of Kenya in early 1920s
17	J. W. Arthur, Scottish missionary resident in Kenya
18	Sir Halford Mackinder, British geographer, first climbed Batian in 1899

TABLE 4.1:1
Nomenclature of ice entities on Kibo. Numbers refer to Map 4.1:1. Asterisks denote new, and brackets alternative names

1.	Eastern Ice Field*
2.	–
3.	Ratzel Glacier
4.	Southern Ice Field
5.	Rebmann Glacier
6.	Decken
7.	Kersten
8.	Heim
9.	Furtwängler (Western Crater)
10.	Diamond
11.	Balletto*
12.	Great Barranco (Great Breach)
13.	Little Barranco (Little Breach)
14.	Arrow
15.	Uhlig
16.	Little Penck
17.	Great Penck
18.	Drygalski
19.	Credner
20.	Northern Ice Field

contrast between the vigorous glaciers to the West and the rather modest ice extent to the East of the mountain. It is recalled (Chapter 3) that the fossil moraine morphology attests to a similar asymmetry of ice extent during the former glaciations.

Major identifiable ice entities are referred to by numbers in Map 4.1:1. Most of these have been given names by the early explorers. However, some entities have been recognized subsequently and some have in fact formed later.

At the Eastern rim of Kibo several separate ice entities (1, 2, 3) are left, where a continuous ring of ice existed at the end of the last century. The ice-free portions of the Eastern crater include the formerly narrow Hans Meyer, Johannes, and Leopard Notches. Among the aforementioned ice bodies are the remnants of the Ratzel Glacier (3), which has disintegrated into distinct parts.

From the Southern Ice Field (4) four ice streams descend on the Southern slopes of Kibo, namely the Rebmann (5), Decken (6), Kersten (7), and Heim (8) Glaciers. The Decken (6) and Kersten (7) Glaçiers are now separated by a long rock rib, the Wedge, which formerly showed only as an isolated outcrop. Only the Heim Glacier (8) to the Southwest is still of considerable size, whereas the other three form rather short lobes at present.

Proceeding further in a clockwise sense follows the high wall of the Great Breach (Breschenwand). At present, two small ice bodies are found in this area, the Diamond (10) and Balletto (11) Glaciers. Since they were not mentioned by the early explorers (e.g., Meyer, 1890d, 1900a; Jaeger, 1909; Klute, 1920, 1929), it must remain open when these developed into separate entities. The Furtwängler Glacier (9) sits in the Western part of the crater, but barely extends beyond it at present. Two sizeable glaciers, the Great and Little Breach Glaciers (12 and 13) occupy the Western slopes immediately to the North of the Great Breach. These were referred to as the Great and Little Barranco Glaciers by the early explorers (e.g., Meyer, 1890d, 1900a; Jaeger, 1909; Klute, 1920, 1929). To the North of the latter, the remnants of two small glaciers are found, the Arrow (14) and Uhlig (15).

Several large glaciers emanate from the Northern Ice Field (20) in the Northwestern sector of Kibo. There are major changes in ice conditions since the end of the last century, that are interesting in relation to the nomenclature. The southernmost ice streams originating from the Northern Ice Field (20) are the Little and Great Penck Glaciers (16 and 17), but at present the latter is in fact the smaller of the two. They are now separated by a long rock rib, the uppermost portion of which formerly appeared as an isolated rock outcrop, the Ravenstein. The Great Penck Glacier (17) was, during our April 1974 visit, found to possess two distinct tongues. A rock threshold delineates the limit of the more northerly lobe of the Great Penck (17) against the Drygalski (18) Glacier. In the extreme Northwest, the large Credner Glacier (20) descends from the Northern Ice Field (20).

The ice distribution just described is remarkably well depicted in satellite imagery (Appendix 1). Most interesting is the satellite image (180/62:2205–0700) taken in the morning of 15 August 1975, in that snow cover is minimal. This is here reproduced as Photo 4.1:1. The image of 24 January 1976 in the morning is less useful because of abundant snow. Westward facing slopes are shaded on both images. The 15 August 1975 image (Photo 4.1:1) allows us to distinguish most of the ice entities identified in Map 4.1:1 and Table 4.1:1.

In the Northeastern sector of Kibo, two notches in the ice cover are apparent, separating the unnamed small ice entity (1) from both the Northern (20) and Eastern (2) ice fields. The remnant of the Ratzel Glacier (3) in the Southern crater rim can be recognized, set apart from the great Southern Ice field (4). Separate lobes of the four ice streams emanating from the Southern Ice Field (4) can be distinguished, namely the Rebmann (5), Decken (6), Kersten (7), and Heim (8) glaciers. The long rock rib of the 'Wedge' now separating the Rebmann (5) and Decken (6) glaciers right up to their snout portion, can be made out on the satellite image. Identification of ice entities in the Great Breach area to the West is hampered by shading. The ice streams originating from the great Northern Ice Field (20), namely the Little (16) and Great Penck (17), Drygalski (18) and Credner (19) glaciers, are all well depicted. The rock rib of the 'Ravenstein' now separating the Little and Great Penck (16, 17) Glaciers to their snouts can be clearly made out. The satellite image also permits recognition of a rock threshold in the lower portion of the Drygalski (18) Glacier. Embedded in the crater of Kibo one notes the Furtwängler Glacier (9) above the Great Breach to the West, and an extended snow field to the North of the Southern Ice Field (4).

The long-term variations of recent ice extent are to be reconstructed from historical photographs and expedition accounts. Partial compilations of evidence have been presented by Geilinger (1936), Spink (1945), Jaeger (1968), Humphries (1972, pp. 34–6), and Sampson (1974). A drastic ice recession is apparent in the photographs from different

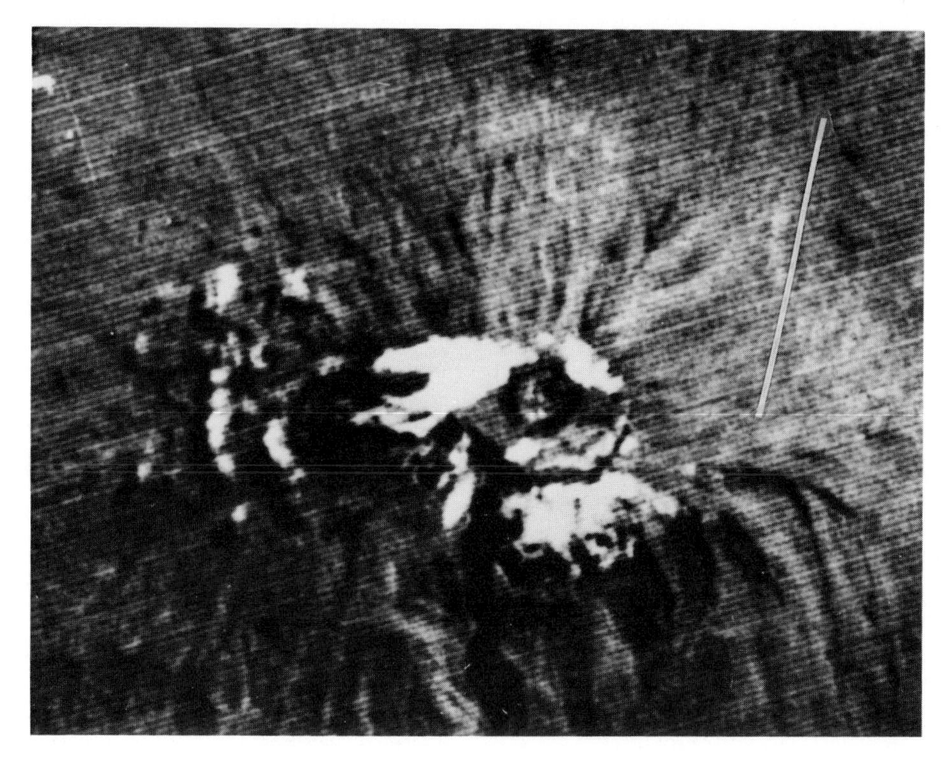

Photo 4.1:1. Part of a LANDSAT 1 (ERTS) satellite image (MSS band 7) of Kilimanjaro region. Date 15 Aug. 1975, 07:00 hrs GMT (NASA).

epochs, although a quantitative reconstruction is precluded by the lack of topographic maps at appropriate scale (Appendix 1). The maps of 1889, 1898, 1912, of around 1930, and the 1960s reproduced as Maps 4.1:2 to 4.1:6, reflect both secular variations of ice extent and phases of mountain exploration. Quantitative evaluation in terms of changes in the position of glacier termini is not warranted. Historical photographs and drawings are listed in Appendix 2 separately for the various portions of the mountain.

The recent glaciation is of limited importance for Mawenzi. During his explorations in 1912, Klute (1920, p. 101) discovered a small cirque glacier on the Southwest side, not visible from below. Apparently the same glacier is still mentioned by Geilinger (1936, p. 7) in the 1930s. Vertical air photographs of 1957 and 1958 (Appendix 1) show snow, but do not allow the confirmation of the existence of a glacier. Repeated visits indicate the lack of perennial ice on Mawenzi in the course of recent decades (R. A. Caukwell, personal communication 1979).

The Eastern rim of Kibo has been the approach route most frequently used since the time of the earliest visitors. The secular disintegration of the ice ring is evidenced by the gradual development of various notches. During his 1887 and 1889 expeditions (Map 4.1:2), Hans Meyer (1890d) observed the progressive development of a depression in the Eastern ice rim, the Hans Meyer Notch (Scharte). By 1898 (Map 4.1:3) another incision, the Johannes Notch (Scharte), had formed somewhat to the South and was free of ice and snow (Meyer, 1900a; Geilinger, 1936). In 1904 two notches were in existence and

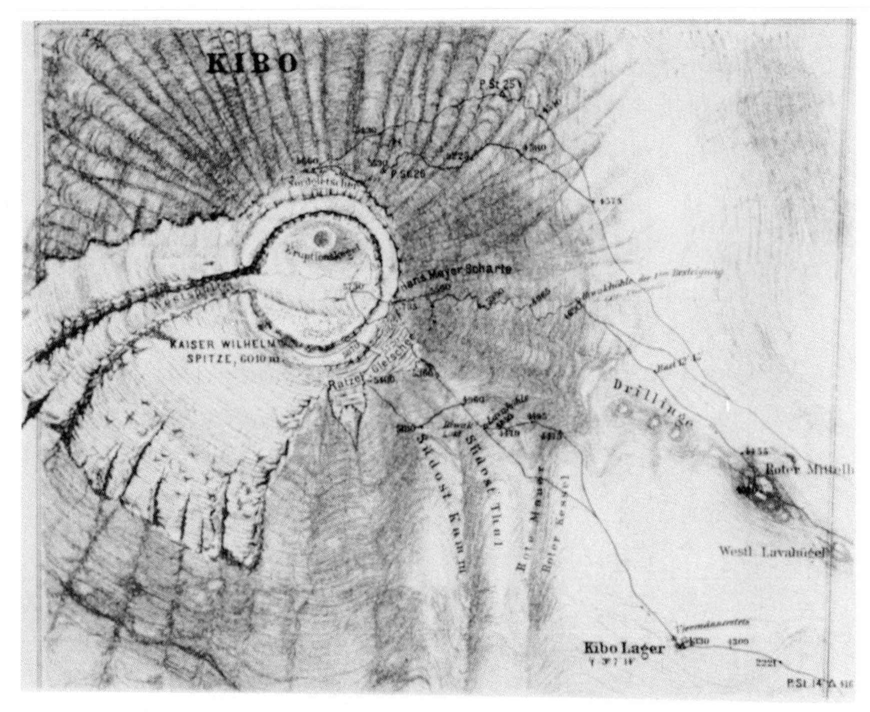

Map 4.1:2. Map of Kilimanjaro by Meyer (1890), date 1889.

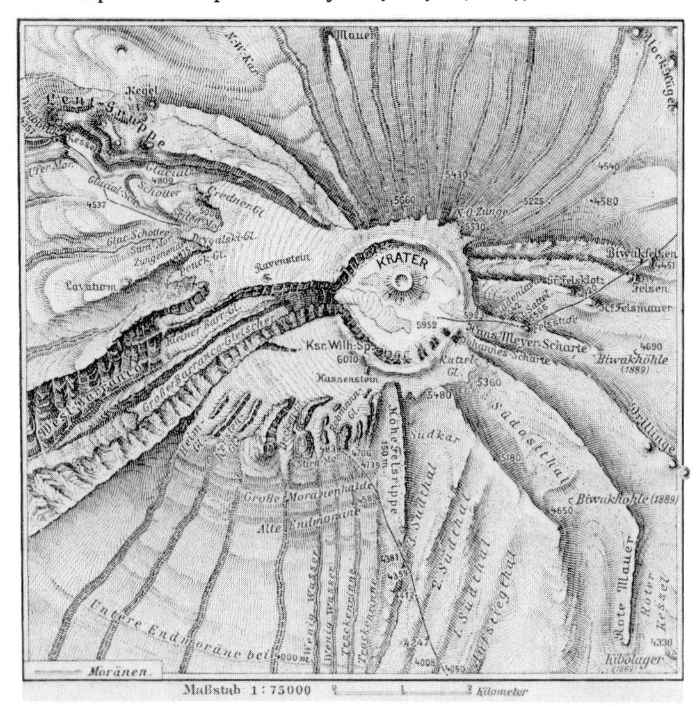

Map 4.1:3. Map of Kilimanjaro by Meyer (1900), date 1898.

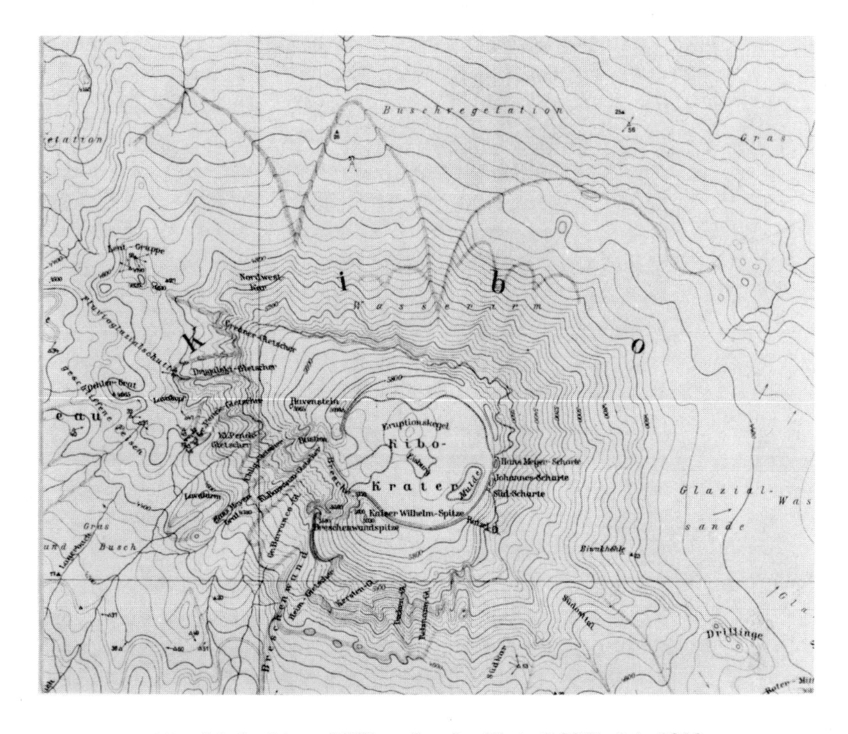

Map 4.1:4. Map of Kilimanjaro by Klute (1920), date 1912.

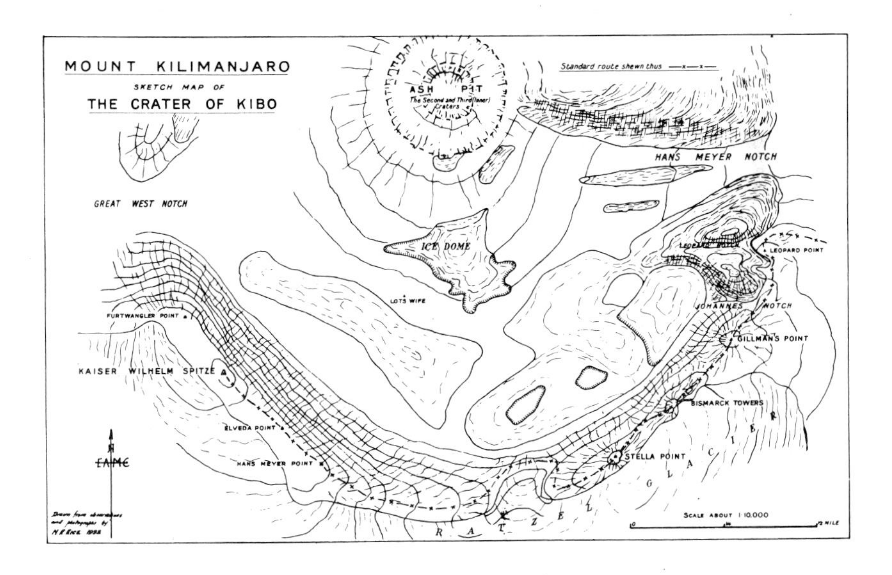

Map 4.1:5. Map of Kilimanjaro by Mountain Club of East Africa (1932), date about 1930.

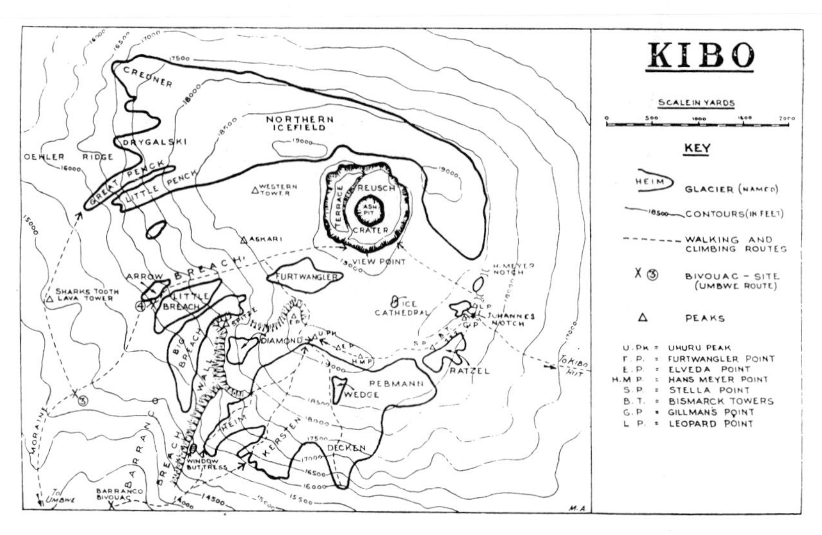

Map 4.1:6. Map of Kilimanjaro by Mountain Club of Kenya (1971), date 1960s.

fully developed (Uhlig, 1908; Geilinger, 1936). By 1912 (Klute, 1920) a third incision, the Leopard Notch, appeared between the Hans Meyer and Johannes Notches (Map 4.1:4). Geilinger (1936) suspects that Klute (1920) may have confused the three notches in the terrain. This may also be true for the sketch map presented by Lange (1912) for July 1909. By 1935 (Geilinger, 1936) a fourth gap in the ice rim, the South Notch, developed South of the Johannes Notch, namely between the Ratzel (3) and Rebmann (5) Glaciers. A depression of the ice cover in this area had already been observed in the preceding years. This notch is not apparent on the slant aerial photographs of Light (1941, photos 170, 171) taken in December 1937/January 1938, and of Spink (1944, fig. facing p. 226, bottom; 1945, fig. facing p. 212, bottom; 1952, plate 35) taken in July 1943 (present Photo 4.1:2). Humphries (1959) describes the development of a gash in the Ratzel Glacier (3) in the course of 1953–7. Two further wide interruptions of the ice ring between the Eastern (1) and Northern Ice Fields (20) were in existence by March 1972 (photo by Geosurvey, courtesy of Peter Gollmer, Nairobi) and 1974 (photo by Alan Root, Nariobi; frontispiece). The ice conditions on the Eastsoutheast side of Kibo in August 1973 are illustrated in Photos 4.1:3a, b.

Photographs of the Eastern side of Kibo taken from lower elevation at various locations on the Saddle Plateau are available from Meyer in 1887 (Meyer, 1888a, photos 8 and 9), in 1889 (Meyer 1890d, photos facing pp. 123 and 158; 1891b, photos facing pp. 141, 183), and 1898 (Meyer 1900a, photos facing pp. 134, 140, 353, 373; 1928, photo facing p. 33), from Uhlig (1904, fig. 44) in 1901, from Klute (1920, plate 4) in 1912, from Latham (1926, photo facing p. 492) in 1926, from Nilsson (1931, fig. 26) in 1928, from Wyss-Dunant (1938, photo facing p. 48, bottom) in the 1930s, from Downie (1964, photo facing p. 7; see also Humphries 1972, plate 4a) around 1953–7. Although seasonally varying snow conditions and differing locations hamper the intercomparison of pictures and reference to the current conditions, these photographs support the notion of a gradual ice recession since the end of the last century, as deduced above from the chronology of notch formation in the Eastern ice rim.

Photo 4.1:2. Kibo from the East in July 1943 (Spink, 1944, 1945).

The South side of Kibo faces a fairly densely populated region with much traffic, yet historical photographs that might contribute towards a documentation of long-term variations in ice conditions are unexpectedly scarce. Meyer (1900a, fig. on p. 200) shows a picture from the Southwest in 1898. Uhlig (1904, fig. 50) presents for 1901 a panoramic view from the South, and *in situ* pictures of various glacier snouts (Uhlig, 1904, figs. 53–58). He proposes a 'Richter' glacier between the Decken (6) and Kersten (7) Glaciers. This seems an entity of smaller dimensions and is not mentioned by other explorers. In fact, Klute (1920, p. 102) specifically notes that he was unable to identify a 'Richter' glacier. Interpretation in terms of secular variations does not appear warranted. A telephoto of the South side of Kibo by Uhlig, probably from 1901, is shown in Jaeger (1909, fig. 19) and is also reproduced here as Photo 4.1:4. A shot by Oehler from the Southsouthwest in 1912 is contained in Klute (1920, fig. 1) and Reck (1921–2, fig. 1), and is here presented as Photo 4.1:5. Klute (1929, plate 4) also shows a photograph of Oehler taken from the Southeast in 1912. Gillman's (1923b, photograph facing p. 3) account for 1921, also contains a photograph from the Southeast. A panoramic picture from the South dating probably from around 1930 is found in Mountain Club of East Africa (1932, photo facing p. 36, top). Mittelholzer (1930, fig. 113) offers an aerial view in 1930 but rather from the Southwest (present Photo 4.1:6). Spink (1944, facing p. 226, top; 1945, facing p. 212, top) shows further aerial photography from the Southwest in 1943. The latter picture is here reproduced as Photo 4.1:7. Geilinger (1936, fig. 1) presents a photograph from the Southeast in 1934.

More recent panoramas from Southwest and South are due to Widmer in 1960 (Jaeger, 1968, fig. 4) and Nicol (1964, fig. 27). Downie (1964, photo facing p. 7) and Humphries

Photo 4.1:3a. Kibo from the Eastsoutheast in Aug. 1973. To the left Rebmann (5), to the right Ratzel (3) Glacier. Panorama continued towards the right in Photo 4.1:3b. (Photo S.H. 15 Aug. 1973)

Photo 4.1:3b. Kibo from the Eastsoutheast in Aug. 1973. To the right remnants of Ratzel (3), to left Rebmann (5) Glacier. Panorama continued towards the left (South) in Photo 9.1:3a. (Photo S.H. 15 Aug. 1973)

Photo 4.1:4. Kibo from the South in 1901 (Jaeger, 1909, fig. 19, by Uhlig).

Photo 4.1:5. Kibo from the Southsouthwest in Dec. 1912 (Klute, 1920, fig. 1, by Oehler).

Photo 4.1:6. Kibo from the Southwest in Jan. 1930. (Mittelholzer, 1930, fig. 113; aerial).

Photo 4.1:7. Kibo from the Southwest in July 1943 (Spink, 1944, 1945).

(1972, plate 4b) show a slant air photograph from the Southeast of unknown date. Comparison of Oehler's photograph of 1912 (Klute, 1920, fig. 1; present Photo 4.1:5) with Nicol's (1964, fig. 27) picture of the early 1960s illustrates a marked ice recession. Other pictures are less readily compared. However, somewhat indicative is the development of a rock outcrop between the Decken (6) and Kersten (7) Glaciers, sometimes referred to as 'the Wedge'. This is apparent in Widmer's (Jaeger 1968, fig. 4) 1960 photograph, on Nicol's (1964, fig. 27) picture, as well as on the air photograph in Downie (1964, photo facing p. 7) and Humphries (1972, plate 4a), but is not readily identified on the earlier photographs. The vertical aerial photograph by Geosurvey (courtesy of Peter Gollmer, Nairobi) in March 1972 illustrates that this isolated rock outcrop has in the meantime developed into a bare rock rib between the Decken (6) and Kersten (7) Glaciers.

For the Western side of Kibo, historical documentation of varying ice extent is somewhat more copious. For 1898 Meyer (1900a, figs. facing pp. 167, 171, 174, 177, and fig. on p. 180) shows pictures from the Westnorthwest, mainly depicting the Credner (19), Drygalski (18), and Penck (17, 16) Glaciers. Uhlig's photo taken from the Meru crater in the Southwest (Jaeger, 1909, fig. 20) dates from 1901 (present photo 4.1:8). For

Photo 4.1:8. Kibo from the West in Oct. 1901. (Jaeger, 1909, fig. 20, by Uhlig).

1906–7, Jaeger (1909, figs. 6, 7, 15) presents photographs covering the West side of Kibo. Jeannel (1950, fig. 13, p. 74) reproduces a picture of the ice margin on the Western side of Kibo taken by Alluaud probably around 1912. A photograph by Oehler in 1912 is shown in Klute (1920, fig. 6). Further photographs of Oehler in 1912 are presented in Klute (1929, plates 1–3) and Reck (1921–2, fig. 4). A shot in 1914 is due to West (1915, fig. facing p. 6). In 1930 Mittelholzer obtained slant aerial photographs of the glaciers in the Northwest (Mittelholzer, 1930, figs. 197, 111, 114; Kurz, 1948, plate 93, bottom) and West (Mittelholzer, 1930, fig. 112; Kurz, 1948, plate 93, top) sectors. The latter picture is here reproduced as Photo 4.1:9. Geilinger (1936, fig. 3) shows a photograph of the West side of Kibo in 1934. Spink (1944, fig. facing p. 226, top; 1945, fig. facing p. 212, top) presents an aerial view from the Southwest in July 1943. The latter picture is here reproduced as Photo 4.1:7. An excellent panoramic view of the glaciers in the Northwest sector, and *in situ* photographs of the snout of the Drygalski Glacier (18) are due to Salt's (1951, fig. 4, and figs. 5–6, respectively) expedition in 1948. The present Photo 4.1:10 is reproduced from Salt (1951, fig. 4). An aerial view around 1953–7 is presented in Downie (1964, photo facing p. 6). Mittelholzer (1930, fig. 114) contains a slant air photograph from the Northnorthwest in 1930 (present Photo 4.1:11). The terminus configuration in the area of the Penck Glaciers (16, 17) during 1953–7 is documented by photographs during the University of Sheffield expeditions (Humphries, 1959, fig. 2; Downie and Wilkinson, 1972, plate 1a). Photo 4.1:12 is a shot of Kibo from the West in March 1974. Photos 4.1:13a–d offer a panorama of the West side of Kibo in April 1974.

The Balletto Glacier (11) may have developed as a separate entity well after the visits of the early explorers. The Diamond Glacier (10) still appears linked with the Southern Ice Field (4) and the upper portion of the Heim Glacier (8) on Uhlig's 1901 photograph (Jaeger, 1909, fig. 20; present Photo 4.1:8). Taken from the crater rim of Mount Meru, this shot lends itself to comparison with later slant air photographs. Unfortunately an undercast of clouds blocks out the lower portions of Kibo. Mittelholzer's (1930, fig. 112; see also Kurz, 1948, plate 93, top) picture (present Photo 4.1:9) from 1930 shows the Diamond Glacier (10) still connected with the large ice mass, but a separation is underway around 1943 as evidenced by Spink's (1944, fig. facing p. 226, top; 1945, fig. facing p. 212, top) pictures (present Photo 4.1:6). On Meyer's (1890a, plate 2) sketch of the crater in 1889, and on Uhlig's (Jaeger, 1909, fig. 20; present Photo 4.1:8) photograph of 1904 an ice stream extends from the crater to barely beyond the Western rim, but does not seem to connect with the glaciers in the Barranco area. This is consistent with Jaeger's (1909, p. 166) description. A secular recession of this glacier is suggested by the aforementioned aerial photographs of Mittelholzer (1930, fig. 112; see also Kurz, 1948, plate 93, top) and Spink (1944, fig. facing p. 226, top; 1945, fig. facing p. 212) for 1930 and July 1943, respectively (present Photos 4.1:9 and 4.1:7). This ice body is commonly called Furtwängler Glacier (9), but it has also been referred to as West Crater Glacier by Geilinger (1936).

The four glaciers in the Barranco sector, namely the Great (12) and Little (13) Breach, the Arrow (14), and the Uhlig (15) Glaciers, have no connection with ice bodies at the crater rim. Speaking of three Barranco glaciers, Geilinger (1936) may mean the former three. The Arrow (14) and the Uhlig (15) are small glaciers, which according to accounts in the 1970s are in the process of disappearing.

The glaciers further to the North (16–19) all connect with the great Northern Ice Field (20) along the crater rim of Kibo. Humphries (1959; 1972, p. 36) infers a 275 m retreat of the Penck Glacier from a 1912 photograph of Oehler (Klute, 1920, presumably fig. 19) and his own field observations in 1957. It appears that the Great Penck Glacier (17) is meant. Humphries (1972, p. 36) further reports a retreat of about 3 m in 3 months during 1957, and a recession of about 45 m in the period 1957–68. By April 1974 the Great (17) was rather smaller than the Little Penck Glacier (16), and the long

Photo 4.1:9. Kibo from the West in Jan. 1930 (Mittelholzer, 1930, fig. 112; aerial).

Photo 4.1:10. Kibo from the West in Nov. 1948 (Salt, 1951, fig. 4, by Swynnerton).

Photo 4.1:11. Kibo from the Northnorthwest in Jan. 1930 (Mittelholzer, 1930, fig. 114; aerial).

Photo 4.1:12. Kibo from the West in March 1974 (by Alan Root, 10 March 1974).

Photo 4.1:13a. Kibo from the West in April 1974. To the right Penck Glaciers (16, 17), to the left Drygalski Glacier (18). Photos 4.1:13a–4.1:13d form panorama from left (North) to right (South) (Photo S.H. 13 April 1974).

Photo 4.1:13b. Kibo from the West in April 1974. To the right Great Breach area, middle Penck Glaciers (16, 17) with Ravenstein rock rib. Panorama continued towards the right (South) in Photo 4.1:13c and towards the left (North) in Photo 4.1:13a (Photo S.H. 13 April 1974).

Photo 4.1.13c. Kibo from the West in April 1974. To the right Great Breach area, to the left Little Penck Glacier (16). Panorama continued towards the right (South) in Photo 4.1:13d, and towards the left (North) in Photo 4.1:13d (Photo S.H. 13 April 1974).

Photo 4.1:13d. Kibo from the West in April 1974. Great Breach area. Panorama continued towards the left (North) in Photo 4.1:13c (Photo S.H. 13 April 1974).

'toothpaste'-like tongue of the Great Penck Glacier (17) shown by Humphries (1959, fig. 2) and Downie and Wilkinson (1972, plate 1a) for 1957 had receded to a prominent rock threshold, over a distance of a few 100 m since 1957 (Photos 4.1:13b, c). Immediately to the South of the Drygalski Glacier (18), a distinct ice lobe forming part of the Great Penck Glacier (17) was encountered during our April 1974 visit. This seems to have developed as a consequence of the changed ice flow conditions in the realm of the Drygalski (18) and Penck (16, 17) Glaciers.

Especially interesting in regard to the long-term glacier recession is the Ravenstein. At the end of the 19th and in the early 20th century this was an isolated rock pinnacle in the upper portion of the Little (16) and Great (17) Penck Glaciers. It appears as such on Meyer's (1900a) and Klute's (1920) maps (Maps 4.1:3 and 4.1:4) for 1898 and 1912, respectively. Uhlig's telephoto of 1904 (Jaeger, 1909, fig. 20; present Photo 4.1:8), Oehler's photographs of 1912 (Klute, 1929, plates 1–3), and West's (1915, fig. facing p. 6) picture of 1914 also show this isolated rock outcrop. The Ravenstein is more conspicuous on Mittelholzer's slant aerial photograph of 1930 (Mittelholzer, 1930, fig. 112; see also Kurz, 1948, plate 13, top; present Photo 4.1:9), developing into an elongated rock wedge on Geilinger's (1936, fig. 3) photograph of 1934. However, on both pictures, the rock outcrop is still surrounded by ice. On Spink's (1944, fig. facing p. 226, top; 1945, fig. facing p. 212) air photograph of July 1943 (present Photo 4.1:7), a long bare rock rib has emerged extending from the rock pinnacle far down the slope. Although its lowermost end is not visible on Spink's photograph, the rock rib seems to separate the Little (16) and Great (17) Penck Glaciers entirely. This state of affairs is illustrated on Salt's (1951, fig. 4) photograph of 1948 (present Photo 4.1:10). The upper part of this rock ridge appears likewise on Widmer's (Jaeger, 1968, fig. 4) photograph of 1960, while its lower portion seems to be hidden from view. A long rock rib separating the Little (16) and Great (17) Penck Glaciers is also seen on Humphries' (1959, fig. 2) photograph of 1957, on a vertical air photograph of March 1972 (Geosurvey, courtesy of Peter Gollmer, Nairobi), and was likewise encountered on our April 1974 visit to this area.

A painting, a sketch, and photographs from Meyer's (1900a, fig. facing p. 167, also on p. 180, figs. facing p. 171, 174, 177) 1898 expedition provide the earliest historical documentation for the Drygalski (18) and Credner (19) Glaciers. Oehler's photograph in 1912 (Klute, 1929, plate 2) as well as Mittelholzer's (1931, fig. 114; see also Kurz, 1948, plate 93, bottom) picture (present Photo 4.1:11) in 1930 show a tripartitioning of the Credner Glacier (19). Although this feature is not apparent on Meyer's painting and sketch, reservations seem in order regarding Geilinger's (1936) interpretation in terms of actual long-term changes. Similarly, it remains open whether Oehler's 1912 photograph (Klute, 1929, plate 3) bears out a change of the Drygalski (18) and Penck (17, 16) Glaciers since 1898. In Mittelholzer's slant aerial photographs of 1930 (Mittelholzer, 1930, fig. 112; see also Kurz, 1948, plate 93, bottom; present Photo 4.1:9) two elongated rock wedges are beginning to show in the upper part of the Drygalski Glacier (18). The more southerly of these rock outcrops is also indicated on Geilinger's (1936, fig. 3) photograph of 1934. The outcrops are located well above the juncture of the Drygalski (18) and Credner (19) Glaciers. This may explain why they do not appear on Salt's (1951, fig. 4) photograph of 1948 (present Photo 4.1:10).

During Meyer's 1898 expedition (Meyer, 1900a, figs. facing pp. 171, 177), the Drygalski

Glacier (18) still reached close to a set of terminal moraines above a rock threshold. Oehler's 1912 photograph (Klute, 1929, plate 3) shows the lower part of the Drygalski Glacier (18), and a rock precipice at somewhat higher elevation is reflected in an irregularity of surface topography of the glacier (Klute, 1929, plates 1–3). Mittelholzer's (1930, fig. 114; see also Kurz, 1948, plate 93, bottom) air photograph in 1930 (present Photo 4.1:11) shows the glacier descending well below this rock step, although the lower portion of the ice tongue appears rather narrower than on Oehler's (Klute, 1929, plates 2–3) photographs in 1912. In Salt's (1951, figs. 4, 5, 6) pictures of 1948 the glacier still extends to beyond the aforementioned rock step, widening somewhat on the less steep terrain below (present Photo 4.1:10). The University of Sheffield expedition (Humphries, 1972, p. 36) found in 1953 that the ice had broken at this rock step, the glacier snout was hanging just over it, and a large patch of dead ice remained in the floor of the valley. By 1957 the ice patch had disintegrated and the snout had retreated well above the step.

No glacier tongues descend to the North side of Kibo, so that secular variations in the configuration of the ice rim are expected to be less conspicuous. Photographic documentation is also somewhat scarce for this sector. Meyer's (1900a, fig. facing p. 126) photograph taken in 1898 from the Northeast shows several small lobes forming the rim of the Northern Ice Field (20). Oehler's (Klute, 1920, fig. 7) 1912 photograph shows the margin of the Northern Ice Field (20) from the Northwest. Slant aerial photographs are available from Mittelholzer (1930, figs. 111, 114, 115; see also Kurz, 1948, plate 23, bottom) from 1930 (present Photo 4.1:11), and from Roy Tuckett (1931, two photographs) probably from 1931. Further photographs from the Northnortheast were taken during the August 1973 field trip (Photos 4.1:14a, b). Intercomparison of the aforementioned sequence of photographs is hampered by the varying azimuths and elevations from which they are taken. However, there is a weak indication of a long-term retreat of the ice rim also on the North side of Kibo. Of interest in this connection is a rock outcrop near the crater rim that is apparent on the March 1972 vertical air photograph (Geosurvey, courtesy of Peter Gollmer, Nairobi), and which was also observed during the August 1973 field trip.

The large crater does not represent a catchment area for the glaciers on the outer slopes of Kibo, despite the topography favoring an outlet to the West. This remarkable state of affairs has already been noted by Jaeger (1909, p. 166). The essentially horizontal rather than inclined surface is expected to be a factor in the radiation budget. A lesser degree of cloudiness over the crater than at the slopes of Kibo, as suggested by Geilinger (1936, p. 16), would also increase the insolation. Finally, the decrease of precipitation with height should be remembered. The role of geothermal effects in the surface heat and hence the mass budget of the crater region remains open, even though Humphries (1972, p. 36) rules out volcanic activity as a cause of the rapid glacier recession. Ice conditions in the outer portions of Kibo crater since the end of the last century are also documented by a sequence of photographs and expedition accounts.

Hans Meyer (1900a, pp. 343–54) experienced a change in the ice ring to the East from 1887 to 1889, that favored the entrance to the crater region. Meyer (1900a, pp. 352–4) encountered a drastic decay of ice cover in the crater from his October 1889 to the August 1898 visits. Whereas the crater floor and the inner walls were largely ice-covered in 1889, most of the floor was ice-free in 1898 although there were still various impressive ice bodies left. Whereas a sizeable ice cascade descended from the Western crater into

Photo 4.1:14a. Kibo from the Northnortheast in Aug. 1973. Rim of Northern Ice Field (20). Panorama continued towards the right (West) in Photo 4.1:14b (Photo S.H. 12 Aug. 1973).

Photo 4.1:14b. Kibo from the Northnortheast in Aug. 1973. Rim of Northern Ice Field (20). Panorama continued towards the left (East) in Photo 4.1:14a (Photo S.H. 12 Aug. 1973).

the upper Barranco in 1889, only a small ice tongue was left in 1898 (Meyer, 1900a, p. 354). A drawing (Meyer, 1900a, fig. facing p. 144) illustrates the ice conditions in the Southwestern part of the Kibo crater in 1898. Uhlig (1904, figs. 47–9; 1908, figs. 16, 17) shows photographs of the crater from his visits in 1901 and 1904.

From his October 1901 visit, Uhlig (1904, pp. 636–7) claims an increase of ice masses since 1898. Unlike 1898, the Johannes Notch was not free of ice and snow in 1901. Furthermore, Uhlig infers an increase of white surfaces in the crater from a comparison of his photographs with Meyer's and Johannes's pictures. A panorama of the crater taken by Uhlig in August 1904 from a location somewhat to the South of the Johannes Notch is shown in Jaeger (1909, plate 13), and is here reproduced as Photos 4.1:15a, b.

Photo 4.1:15a. Kibo crater from South of Johannes Notch in Aug. 1904. Panorama continued to the right (North and East) in Photo 4.1:15b. (Jaeger, 1909, plate 13, by Uhlig).

Photo 4.1:15b. Kibo crater from South of Johannes Notch in Aug. 1904. Panorama continued to the left (West and South) in Photo 4.1:15a. (Jaeger, 1909, plate 13, by Uhlig).

This allows a partial comparison with Meyer's (1900a, fig. facing p. 144) view of the Southwestern portion of the crater. A large 'ice castle' (Eisburg) to the North of the rock face below Uhuru Peak (Kaiser Wilhelm Spitze) is especially conspicuous in both pictures. Uhlig's crater panorama is even more useful a reference in relation to later photographic documentation for the crater.

Mostertz (1929/30, fig. 2) in March 1927 photographed a panorama of the crater from Gillman's Point. Of the 'ice castle' to the North of the rock face below Uhuru Peak

Photo 4.1:16a. Kibo crater from Uhuru Peak in Jan. 1929. Photos 4.1:16a through 4.1:16d form panorama from left (West) to right (Southeast) (Mountain Club of East Africa, 1932, end of issue, by Rice).

Photo 4.1:16b. Kibo crater from Uhuru Peak in Jan. 1929. Panorama continued to the left (West) in Photo 4.1:16a, and to the right (East) in Photo 4.1:16c (Mountain Club of East Africa, 1932, end of issue, by Rice).

Photo 4.1:16c. Kibo crater from Uhuru Peak in Jan. 1929. Panorama continued to the left (North) in Photo 4.1:16b, and to the right (East) in Photo 4.1:16d (Mountain Club of Ease Africa, 1932, end of issue, by Rice).

Photo 4.1:16d. Kibo crater from Uhuru Peak in Jan. 1929. Panorama continued to the left (East) in Photo 4.1:16c (Mountain Club of East Africa, 1932, end of issue, by Rice).

Photo 4.1:17a. Kibo crater from Uhuru Peak in Aug. 1973. Photos 4.1:17a through 4.1:17f form panorama from left (West) to right (Southeast) (Photo S.H. 14 Aug. 1973).

Photo 4.1:17b. Kibo crater from Uhuru Peak in Aug. 1973. Panorama continued to the left (West) in Photo 4.1:17a, and to the right (North) in Photo 4.1:17c (Photo S.H., 14 Aug. 1973).

Photo 4.1:17c. Kibo crater from Uhuru Peak in Aug. 1973. Panorama continued to the left (Northwest) in Photo 4.1:17b, and to the right (Northeast) in Photo 4.1:17d (Photo S.H., 14 Aug. 1973).

Photo 4.1:17d. Kibo crater from Uhuru Peak in Aug. 1973. Panorama continued to left (North) in Photo 4.1:17c, and to the right (East) in Photo 4.1:17e (Photo S.H., 14 Aug. 1973).

Photo 4.1:17e. Kibo crater from Uhuru Peak in Aug. 1973. Panorama continued to left (East) in Photo 4.1:17d, and to the right (Southeast) in Photo 4.1:17f (Photo S.H., 14 Aug. 1973).

Photo 4.1:17f. Kibo crater from Uhuru Peak in Aug. 1973. Panorama continued to left (East) in Photo 4.1:17e (Photo S.H., 14 Aug. 1973).

(Kaiser Wilhelm Spitze), which is so prominent on Uhlig's 1904 (Jaeger, 1909, plate 13; present Photos 4.1:15a, b) and Meyer's 1898 (Meyer, 1900a, fig. facing p. 144) pictures, only a modest remnant is left in 1927. This ice entity is also apparent on a picture by Johannes (1899, fig. bottom of p. 270) of October 1898, and in photographs of the 1902s presented by Geilinger (1930, figs. 147, 149, 150; 1936, figs. 6, 8). These ice cliffs seem to be located in the Eastern portions of the Furtwängler Glacier (11). Nilsson (1931, fig. 3) shot a crater panorama little after Mostertz, namely in January 1928, but from the Southwest. This also illustrates the decay of the 'ice castle'. A panorama taken from a similar vantage point, namely Uhuru Peak (Kaiser Wilhelm Spitze) in January 1929 is included in Mountain Club of East Africa (1932, end of issue) and is here reproduced as Photos 4.1:16a–d. This is similar to Nilsson's 1928 panorama, except that it indicates a rather smaller ice extent. Thus an ice patch near the Breach to the West has separated from the Northern Ice Field, and a gap has arisen between two large ice bodies in the Eastern part of the crater.

Geilinger (1936) notes a strong ice recession in the crater in the course of 1929–35, and suggests that the bare area of the crater doubled during this 6 year period. His photographs of January 1929 and 1935 (Geilinger, 1936, figs. 6, 8) indeed document a remarkable decay of ice structures in the realm of the aforementioned 'ice castle'.

Geilinger (1936, pp. 18–19) infers a gradual deterioration of the 'Ice Dome' in the Northeastern part of the crater from observations of Meyer in 1898, Uhlig in 1901 and 1904, of Lange in 1909, of von Salis in 1914, and of Mostertz in 1927. It did not prove possible to verify these suggestions, which Geilinger is able to offer on the basis of his superb familiarity with the terrain. The entity seems to be referred to also as 'Ice Cathedral'. Geilinger (1936, figs. 9, 10) also presents two photographs of the Ice Dome in February 1935.

Various air photographs of 1930 (Mittelholzer, 1930, figs. 116–19; 1932, one photo; see also Kurz, 1948, plate 94 bottom) of December 1937–January 1938 (Light, 1941, photos 172, 173), and later (Spink 1945, figs. facing pp. 212, 213; 1947, figs. facing p. 328; 1952, plate 36) are of limited use for ascertaining the subsequent changing ice conditions in the crater, because of scale and varying snow cover (present Photos 4.1:7 and 4.1:11). However, Spink's (1945, fig. facing p. 215, top) photograph of July 1943, Humphries (1959, fig. 1) picture of around 1953–7, illustrate the continued decay of the Ice Dome and of other ice features. The more recent ice conditions in the crater are documented by the March 1972 vertical air photograph (Geosurvey, courtesy of Peter Gollmer, Nairobi), my own terrestrial panorama taken from Uhuru Peak (Kaiser Wilhelm Spitze) in August 1973 (Photos 4.1:17a–f), and Root's aerial shot during his hot air balloon flight in March 1974 (frontispiece). The latter two photographic surveys demonstrate a continued ice recession in the Kibo crater, and may serve as an historical reference for future comparisons.

4.2. Ruwenzori

The perennially wet 'Mountains of the Moon' straddling the border of Uganda and Zaire contrast with the other mountains of East Africa in that they carry in close proximity both lush vegetation and a vigorous glaciation. Apart from occasional visits at the end of the last century, the first extensive documentation of ice conditions is available from the expedition of the Duke of the Abruzzi in 1906. Glacier observations have been more abundant since then. Based on the Makerere University expeditions and the evaluation of historical sources, Whittow *et al.* (1963) compiled a glacier inventory and a partial chronology of ice variations through the early 1960s. Temple (1968) and Osmaston and Pasteur (1972) extended this effort. The present account relies on their authority, the evaluation of maps and air photographs (Appendix 1), and own field observations in January 1974. Appendix 3 contains contributions to the World Glacier Inventory.

TABLE 4.2:1
Nomenclature of Ruwenzori glaciers

A. *Mount Emin* 4802 m	
1	Kraepelin 1
	Umberto (disappeared after 1906)
2	Emin 1
3	North Kraepelin
4	Kraepelin 2
5	Emin 2
6	Emin 3
7	one unnamed (disappeared after 1906)
B. *Mount Gessi* 4769 m	
1	Gessi 1
2	Gessi 2
3	Gessi 3
4	Gessi 4
5	Iolanda 1
6	Iolanda 2
7	Iolanda 3
8	Iolanda 4
9	Gessi 0
C. *Mount Speke* 4891 m	
1	Grant
2	Vittorio Emanuele
3	East Johnston
4	Johnston
5	Speke
D. *Mount Stanley* 5111 m	
1	Alexandra
2	Albert
3	Northeast Margherita
4	Margherita
5	East Stanley
6	Elena
7	Coronation
8	Savoia
9	Philip
10	Elizabeth
11	West Elena
12	West Savoia
13	Moebius
14	West Stanley
15	unnamed
E. *Mount Baker* 4873 m	
1	East Baker
2	Y Glacier
3	Edward 1
3	Edward 2
4	West Baker
5	Moore (Mubuku)
6	Wollaston
7	Edward 3
8	Semper (disappeared after 1943)
F. *Mount Luigi di Savoia* 4665 m	
1	Thomson 1
2	Thomson 2
3	Sella (disappeared after 1906)
4	Stairs (disappeared after 1906)

The modern glaciation is described in Map 4.2:1*, Tables 3:2 and 4.2:1, and Appendix 3. Refer to Fantin (1968, pp. 46–7), Osmaston and Pasteur (1972, pp. 168–71), and to Table 4:1 for the origin of glacier names. Six glacier complexes exist on the following mountains: Emin (A), Gessi (B), Speke (C), Stanley (D), Baker (E), Luigi di Savoia (F).

At Mount Emin (A) three glaciers have disappeared since the expedition of the Duke of the Abruzzi in 1906 (Whittow *et al.*, 1963). A glacier to the North of the Kraepelin peak distinguished on air photographs is named North Kraepelin Glacier (A3). The ice entities A4, A5, A6 appear on Osmaston and Pasteur's (1972, map 6) sketch.

Nine separate ice bodies are recognized at Mount Gessi (B). Of these, seven resulted from the division of two larger glaciers in the course of this century. The ice entities B7, B8, B9, are identified on Osmaston and Pasteur's (1972, map 6) sketch.

At Mount Speke (C), the Grant (C1), Vittorio Emanuele (C2) and Johnston (C4) are merely lobes of the summit ice cap, whereas the Speke (C1) and East Johnston (C3) Glaciers are self-contained ice entities. A small cirque-like depression to the North of the Speke (C1) Glacier snout contained ice from 1906 to at least the late 1930s.

Mount Stanley (D), the highest massif of the Ruwenzori also carries the most extensive glaciation. The Stanley ice plateau serves as a feeder to the Alexandra (D1), Margherita (D4), East Stanley (D5), Elena (D6), West Elena (D11), Moebius (D13), and West Stanley (D14) Glaciers. Six glaciers are entirely self-contained, namely the Northeast Margherita (D3), Coronation (D7), Savoia (D8), Philip (D9), Elizabeth (D10), and an unnamed glacier (D15) to the North of the Savoia Glacier (D8). D15 is not mentioned by the earlier surveys, but existed already in 1906.

Mount Baker (E) lacks a large summit ice cap. The East (E1) and West (E4) Baker Glaciers descending towards the Mubuku Valley formed through separation since 1906. The Edward Glaciers (E3, E4, E7) are remnants of the summit ice cap. Separate from these is the Y Glacier (E2) on the West face, named after its distinctive shape. The Semper Glacier (E8) in its vicinity disappeared since 1943. The Wollaston (E6) and Moore (E5) Glaciers in the Eastern part of the massif are self-contained entities. The latter has been referred to as Mubuku Glacier by the early explorers.

On Mount Luigi di Savoia (F), the Sella (F3) and Stairs (F4) Glaciers have disappeared since the beginning of the century. A Thomson 1 Glacier (F1) is described by Whittow *et al.* (1969) as descending from the small ice cap on Weisman Peak in 1960. Osmaston and Pasteur (1972, map 7) further map a Thomson 2 Glacier (F2).

Many of the Ruwenzori glaciers are large enough to be recovered by satellite. However, evaluation of existing imagery (Appendix 1) is precluded by cloudiness.

The reconstruction of long-term glacier variations in the Ruwenzori mostly starts with the documentation of the Duke of the Abruzzi expedition in 1906. Recent topographic maps and air photographs are listed in Appendix 1. Maps of 1906 and 1938 in part reflecting secular glacier variations are reproduced as Maps 4.2:2, 4.2:3, and 4.2:4*. Historical photographs or drawings are listed in Appendix 2 separately for the six mountain massifs. Attention is drawn to the English version of Filippi (1908b) rather than the Italian edition (Filippi, 1908a), because of the better reproduction of photographs.

Mount Emin (A) carried the sizeable Umberto Glacier and two smaller ice entities on its Eastern side in 1906 (Filippi, 1908b, fig. on p. 241, and panorama facing p. 268; 1908a, fig. on p. 221, and panorama facing p. 250; Roccati, 1909, map). A photograph by Hicks (1946–7; fig. on p. 15, bottom) in 1945 shows ice entities on the East side

Map 4.2:2. Map of Ruwenzori by Roccati (1908), date 1906.

of Mount Emin (A). According to Whittow *et al.* (1963, fig. 4) ice remnants had all disappeared by 1959. The upper portion of the Emin Glacier (A2) is seen on the 1906 panorama (Filippi, 1908b, panorama facing p. 268; 1908a, panorama facing p. 250). Part of this glacier also appears in a photograph by D. Pasteur from around 1960, and part of the Kraepelin Glacier (A1) is contained in a photograph of 1959 (Whittow *et al.*, 1963, pp. 591–2). The North Kraepelin Glacier (A3) has first been identified by Whittow *et al.* (1963, p. 592) from air photographs, but not visited.

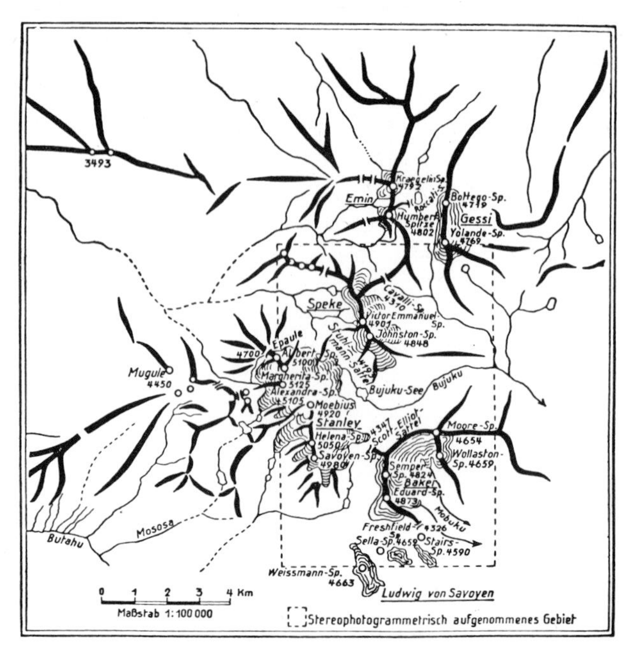

Map 4.2:3. Map sketch of Ruwenzori by Stumpp (Eisenmann, 1939a), date 1938.

The West side of Mount Gessi (B) is covered by 1906 photographs (Filippi, 1908b, fig. on p. 269, panorama facing p. 142; 1908a, fig. on p. 249, panorama facing p. 136). These show three separate glaciers. The major central part later split in two, as shown by photography taken in 1944 (Hicks, 1947, figs. on pp. 14, 16). As a result of this disintegration there are now five glaciers named Gessi 0 through 4 (B0–B4) A bifurcation prior to the 1950s has also given origin to the two glaciers Iolanda 1 and 2 (B5 and B6). A photograph of the former in 1959 is shown by Whittow *et al.*, (1963, fig. 3). Altogether four ice entities, including B7 and B8, are distinguished by Osmaston and Pasteur (1972, map 6).

The conditions on the East side of Mount Speke (C) are documented by photographs in 1906 (Filippi, 1908b, panoramas facing pp. 142, 269; 1908a, panoramas facing pp. 136, 249). At that epoch the Vittorio Emanuele Glacier (C2) was already made up of three main ice streams. Also, the East Johnston Glacier (C3) was already clearly separate from the Vittorio Emanuele Glacier (C2). Subsequently, photographs of these glaciers were taken from approximately the same viewpoint on Mount Gessi (B) around 1932 by Humphreys (Whittow *et al.*, 1963, p. 589), in 1945 (Hicks, 1947, fig. on p. 20, top; Allen and Posnett; Whittow *et al.*, 1963, p. 589), and in July 1959 (Whittow *et al.*, 1963, p. 589). Results have been illustrated in a composite sketch, showing the continued ice retreat (Whittow *et al.*, 1963, fig. 2). A slant aerial photograph exists from December 1937 (Light, 1941, photo 197).

The Speke Glacier (C5) on the South side of the massif appears on a photograph of 1906 (Filippi, 1908b, fig. on p. 235; 1908a, fig. on p. 215). This picture, reproduced here as Photo 4.2:1, also shows an ice stream originating from a small cirque-like depression to the North, and being joined to the Speke Glacier (C5) proper. Photographs of

Mount Speke (C) from the South during the late 1920s are due to Humphreys (1927, fig. facing p. 521, bottom; Osmaston, 1961, fig. 1). In 1934 (Synge, 1937, plate 38; British Museum of Natural History, plate 13) and in 1938 (Stumpp, 1952, map; present Map 4.2:4*) the ice still connected with the Speke (C5) Glacier (Whittow *et al.*, 1963, p. 589). In 1941 (Hodgkin 1944, fig. on p. 304), and in 1943 (Firmin 1944–5, p. 33 top) the small cirque-like depression was still full of ice. Temple (1968, figs. 4, 6) presents a photograph of Speke Glacier presumably from the 1960s, and one of the snout portion taken in January 1967. The latter demonstrates the further ice recession since the time of the Makerere expeditions. This ice patch had disappeared completely by 1955 (Whittow *et al.*, 1963). Osmaston (1961, fig. 2) shows the state of the Speke (C5) and Johnston (C4) Glaciers in 1951. A further picture of the 1950s is given by Whittow and Shepherd (1959, fig. 1). Photos 4.2:2 and 4.2:3 taken in January 1974 offer views of Mount Speke from the Stanley Plateau and the upper Bujuku Valley, respectively. The front of Speke Glacier (C5) in January 1974 is shown in Photo 4.2:4. During the January 1974 visit (Hastenrath, 1975), the frontal retreat of the snout since 1958 was estimated at 30–40 m, the lateral shrinkage at 10–20 m, and the volume loss in the snout portion alone at $5–10 \times 10^4$ m^3. Little historical documentation seems to be available (Whittow *et al.*, 1963) for the Grant Glacier (C1). Temple (1968; fig. 3) presents an excellent aerial photograph of Grant presumably from the 1960s.

Mount Stanley (D) being the highest and most heavily glaciated massif of the Ruwenzori, it has been visited most frequently since the 1906 Abruzzi expedition. A photograph by Stuhlmann (1894 facing p. 288; see also Filippi, 1908b, fig. on p. 206; 1908a, fig. on p. 188) in 1891 shows a panorama of glaciers from the West (present Photo 4.2:5). Some photographs taken from the West in 1926 are shown by Chapin (1927, figs. on pp. 615, 617). Humphreys (1933, fig. facing p. 483) gives an aerial photo from the West in 1931. Further pictures from the West side were obtained by the Belgian expedition in 1932 (Grunne 1933, figs. facing pp. 275, 276, 280, 285; Grunne *et al.*, 1937, fig. 73; de Heinzelin, 1953, plate 1). De Heinzelin (1951, fig. facing p. 205; 1953, plates 1–3) furthermore shows photographs of the West Stanley (D14) and Alexandra (D1) Glaciers in July 1935, July 1939, September 1940, July 1950 and 1952. These pictures document a continued and drastic glacier recession. De Heinzelin (1953) estimates a volume decrease of the West Stanley Glacier (D14) of the order of 10^6 m^3 for the 20 year period from the early 1930s to the early 1950s.

The East side of the Stanley (D) massif is depicted on various photographs of the 1906 Duke of Abruzzi expedition (Filippi, 1908b, panoramas facing pp. 142, 152, 235, figs. on pp. 149, 151, 200, 250; 1908a, panoramas facing pp. 136, 142, 215; figs. on pp. 138, 139, 182; Roccati, 1909, pls. 15, 16, 24, 28). Some of these pictures are reproduced here, namely Photo 4.2:6 (Filippi, 1908, fig. on p. 200, 1908a, fig. on p. 182). Photo 4.2:7 (Filippi, 1908b, panorama facing p. 152, partial), and Photo 4.2:8 (Roccati, 1909, pl. 28). Several of these pictures (Filippi, 1908b, panorama facing p. 142, figs. on pp. 151, 200; 1908a, panorama facing p. 136; figs. on pp. 136, 139, 182; Roccati, 1909, pl. 28) show the steep Coronation (D7) joined to the large Elena (D6) Glacier (present Photos 4.2:6 and 4.2:8). This remained the case at least through 1926 (Whittow *et al.*, 1963, p. 587), as shown by a photograph of Humphreys (1927, fig. facing p. 517) of uncertain date. A slant aerial photograph by Light (1941, photo 197) suggests a similar state of affairs for December 1937. A photograph by Haddow in 1945 (Hicks, 1946, fig. facing

Photo 4.2:1. Mount Speke viewed from Scott-Elliot Pass in 1906 (Filippi, 1908, fig. on p. 235).

Photo 4.2:2. Mount Speke viewed from Stanley Plateau in Jan. 1974. (Photo S.H., 22 Jan. 1974).

Photo 4.2:3. Mount Speke viewed from upper Bujuku Valley in Jan. 1974. (Photo S.H., 20 Jan. 1974).

Photo 4.2:4. Snout of Speke Glacier (C1) in Jan. 1974. (Photo S.H., 21 Jan. 1974).

Photo 4.2:5. Mount Stanley from the West in 1891 (Stuhlmann, 1894, facing p. 288).

Photo 4.2:6. Mount Stanley from the Edward Peak of Baker in 1906 (Filippi, 1908b, fig. on p. 200).

Photo 4.2:7. Mount Stanley from Mount Luigi di Savoia in 1906 (Filippi, 1908b, panorama facing p. 152, partial).

Photo 4.2:8. Mount Stanley from the Southeast in 1906 (Roccati, 1909, pl. 28).

Photo 4.2:9. Coronation (D7) and lower Elena (D6) Glaciers in Jan. 1974 (Photo S.H., 22 Jan. 1974).

Photo 4.2:10. Mount Stanley viewed from the East in Jan. 1974. To the left Savoia Glacier (D8), to the right the ice entity D15 (Photo S.H., 23 Jan. 1974).

Photo 4.2:11. Snout portion of Savoia Glacier (D8) in Jan. 1974 (Photo S.H., 23 Jan. 1974).

Photo 4.2:12. West side of Mount Baker viewed from Stanley Plateau in Jan. 1974. To the right (South) the Y-Glacier (E2), to the left (North) the location of the defunct Semper Glacier (E8). (Photo S.H., 22 Jan. 1974).

Photo 4.2:13. South side of Mount Baker viewed from upper Mubuku Valley in Jan. 1974. From left (West) to right (East) the West Baker (E4), East Baker (E11), and Moore (E5) Glaciers (Photo S.H., 24 Jan. 1974).

p. 215, top) suggests that the Coronation (D7) just barely reaches the Elena Glacier. A photograph in 1953 by Busk (1954, fig. facing p. 145) shows a bare rock threshold between the snout of the Coronation (D7) and the Elena (D6) Glacier. The two glaciers were separated by 60–90 m of sheer rock slabs in 1960 (Whittow *et al.*, 1963). A photograph by Fantin (1968, fig. 115) in 1961 illustrates this state of affairs. The gap had increased further by the time of my January 1974 visit (present Photo 4.2:8). Eisenmann (1939a, plate 11, top) presents a view of the Margherita and Albert peak from the Northeast in 1938. Temple (1968, fig. 3) shows an aerial view of the Stanley massif from the North, presumably from the 1960s.

The Savoia Glacier (D8) on the Eastern side of Stanley (D) is best viewed from Freshfield Pass, and there is a good sequence of photographs taken from this area. Results have been summarized in a panorama sketch by Whittow *et al.* (1963, fig. 18). In 1906 (Filippi, 1908b, fig. on p. 149; panoramas facing pp. 152, 235; 1908a, fig. on p. 138, panoramas facing pp. 142, 215) the snout extended close to an arc of moraines, which can still be appreciated in the terrain at present (present Photos 4.2:7 and 4.2:8). Photographs by Humphreys (1927, fig. facing p. 517; 1933, panorama facing p. 482; Whittow *et al.*, 1963, p. 607; archive of Mountain Club of Uganda) indicate the ice extent in 1927 and 1931. Photographs in 1934 (Synge, 1937, plates 37, 38) indicate only small changes since Humphreys' 1927 visit. A marked shrinkage is borne out by a photograph in January

Photo 4.2:14. Moore Glacier (E5) of Mount Baker in 1906 (Roccati, 1909, pl. 30, fig. 2).

Photo 4.2:15. Moore Glacier (E5) of Mount Baker in 1906 (Filippi, 1980b, on p. 143).

Photo 4.2:16. Moore Glacier (E5) of Mount Baker in June 1958 (Whittow *et al.*, 1963, fig. 19).

Photo 4.2:17. Moore Glacier (E5) of Mount Baker in Dec. 1966 (Temple, 1968, fig. 8b).

1943 (Firmin, 1944–5, p. 25 bottom). The ice retreat until the time of the Makerere expeditions in the late 1950s is conspicuous. Temple (1968, fig. 7) presents a photograph of December 1966 also taken from Freshfield Pass. A further recession was noted on my January 1974 visit (Photos 4.2:10 and 4.2:11). A large ice patch (D15) on the steep rock face to the North of the Savoia Glacier (D8) is evident on the 1906 photographs (Filippi, 1908b, fig. on p. 149, panoramas facing pp. 152, 235; 1908a, fig. on p. 138, panoramas facing pp. 142, 215) and has persisted until at least 1974 (Photo 4.2:10). This ice body is not included in the surveys of Whittow *et al.* (1965) and Osmaston and Pasteur (1972). Little historical documentation seems to exist for the other glaciers of Mount Stanley (Whittow *et al.*, 1963). Slant aerial photographs of Mount Stanley in December 1937 are found in Light (1941, photos 197, 198, 199).

Long-term glacier variations at Mount Baker (E) are fairly well documented. Pictures of 1900 (Moore, 1901, pp. 15, 247) and photographs from 1906 (Filippi, 1908b, fig. on p. 149, panoramas facing pp. 142, 152; 1908a, fig. on p. 138, panoramas facing pp. 136, 142) show that the present East (E1) and West Baker (E4) still formed a single glacier. Whittow (1959) and Whittow *et al.*, (1963) suggest from the absence of lichens on the intervening rock that the division may have taken place since the 1930s. Photographs of 1906 show the Semper Glacier (E8) on the West cliffs of Mount Baker (Filippi, 1908b, fig. on p. 153; 1908a, fig. on p. 141; Roccati, 1909, pl. 18, fig. 2). Photographs by Humphreys (Whittow *et al.*, 1963, p. 590) prove its continued existence in the 1920s. It still appears on the 1938 map (Whittow *et al.*, 1963, p. 590) and on a 1943 photograph (Firmin, 1944–5, p. 33 bottom). It had disappeared by the time of the Makerere expeditions in the late 1950s (Whittow *et al.*, 1963, p. 590). Photo 4.2:12 shows the Y-Glacier (E2) and the location of the defunct Semper Glacier (E8) on the West face of Mount Baker in January 1974. Bere and Hicks (1946, plate 4) present a photograph of the North face of Mount Baker in 1945.

The South side of Mount Baker (E) is depicted on photographs from 1906 (Filippi, 1908b, panoramas facing pp. 142, 152; fig. on p. 149; 1908a, panoramas facing pp. 136, 142; fig. on p. 138). A slant aerial photograph of December 1937 is contained in Light (1941, photo 197). Photo 4.2:13 is a view from the South in January 1974.

Photographs of the Wollaston Glacier (E6) exist from 1906 (Filippi, 1908b, panorama facing p. 152; 1908a, panorama facing p. 142) and the 1920s (Humphreys; Whittow *et al.*, 1963, p. 590). It is also depicted on the 1938 map (Whittow *et al.*, 1963, p. 590). This small glacier was still in existence at the time of the Makerere expeditions in the late 1950s (Whittow *et al.*, 1963, p. 590). Two prominent ice patches on the Western slopes of Cagni Peak are apparent on a photograph in 1906 (Filippi, 1908b, panorama facing p. 152; 1908a, panorama facing p. 142), but had disappeared by 1958 (Whittow *et al.*, 1963, p. 590). The Moore Glacier (E5) has been photographed in 1906 (Wollaston, 1908, figs. facing pp. 68, 96; Filippi, 1908b, figs. on pp. 143, 147, panorama facing p. 152; 1908a, panorama facing p. 142, figs. on pp. 134, 130; Roccati, 1909, pl. 30, fig. 2; Fisher; Whittow *et al.*, 1963, p. 608). The 1906 conditions are illustrated in Photo 4.2:14 (reproduced from Roccati, 1909, pl. 30, fig. 2), and Photo 4.2:15 (reproduced from Filippi, 1908b, fig. on pl. 143). Photography of the Makerere expeditions (Whittow, 1959, fig. facing p. 373; Whittow *et al.*, 1963, fig. 19) shows the snout in June 1958 (present Photo 4.2:16). Whittow *et al.* (1963, pp. 608–9) give a snout retreat of 90 m for the period 1906 to 1958. Based on the study of lichens they suggest a breakdown of about 32 m for the period 1906–30, and approximately 55 m during 1930–58.

Photographs by Temple (1968, figs. 8–10) in June 1958 and December 1966 illustrate the continuing drastic ice recession. The present Photo 4.2:17 (reproduced from Temple, 1968, fig. 8b) illustrates conditions in December 1966.

For Mount Luigi di Savoia the 1906 photographs and map (Filippi, 1908b, figs. on pp. 190, 191, 251, panorama facing pp. 142, 246; 1908a, figs. on pp. 172, 173, 232; panoramas facing pp. 136, 232) document an extensive glaciation. The map sketch of 1938 (Eisenmann, 1939a, p. 46; present Map 4.2:3) still shows three glaciers on this mountain, although they do not appear on the actual 1:25 000 map (Stumpp, 1952; present Map 4.2:4*), being at the margin of the area surveyed. No documentation seems to be available for the intermediate decades, but Whittow *et al.* (1963, p. 592) have compared the 1906 data with conditions in the late 1950s. By that time the Stairs Glacier (F4) had completely disappeared, while the fate of the Sella Glacier (F3) was uncertain. The Thomson 1 Glacier (F1) was still in existence. Osmaston and Pasteur (1972, map 7) further map a Thomson 2 Glacier (F2). Whittow *et al.* (1963) infer an accelerated glacier recession in the Ruwenzori since the 1930s.

4.3. Mount Kenya

The modern ice extent on Mount Kenya is documented by E. Schneider's excellent map of the entire peak region at scale 1:5000 (Forschungsunternehmen Nepal-Himalaya, 1967). The glaciers of Mount Kenya are described in Map 4.3:1 and Tables 3:2 and 4.3:1. Reference is made to Fantin (1968, pp. 46–7), Mountain Club of Kenya (1971, pp. 150–2; 1981, pp. 200–5), and to Table 4:1 for the origin of glacier names. Conventional parameters estimated for the World Glacier Inventory are summarized in Appendix 3.

To the North and East side of the mountain only comparatively small glaciers are found. The Cesar (13) and Joseph (14) glaciers have a common accumulation area in the basin below Firmin Col, but a bifurcation takes place in the lower portion of this complex. The Northey (16) and Krapf (1) glaciers are embedded in steep canyons, thus being largely shielded from insolation. The Gregory (2) connects over a gentle col with the Lewis (4) Glacier, the largest ice body on Mount Kenya. The most extensive glaciation is in fact found to the South and West of the peaks (Batian, Nelion). Remnants of the small Melhuish Glacier (5) to the Southwest of Point Melhuish were still in existence in February 1978, but disappeared before January 1980. The steep Diamond (7) below the twin peaks used to connect with the large Darwin (6) Glacier, but a year-round separation lasted throughout 1978/9. The Tyndall (10), being the second-largest ice body on the mountain, is in part fed by the hanging Heim (9) and Forel (8) Glaciers above. The latter two also separated, and the upper margin of the Forel (8) Glacier receded in the course of 1979.

Concerning the overall stronger glaciation on the Southeastern as compared to the Northwestern side of the mountain, the following factors appear relevant: the location at very low latitude in the Southern hemisphere, the diurnal march of cloudiness reducing the afternoon insolation on the westward facing slopes, and the approximate Northeast-Southwest orientation of the mountain crest. In addition to radiation geometry, details of topographic configuration are to be considered for the location of individual glaciers.

The present ice distribution is in part shown by satellite imagery. An interesting image, (180/60:2367–06573), here reproduced as Photo 4.3:1, is taken in the morning of 24 January 1976. Accordingly, the glaciers on the North side of the mountain tend to be

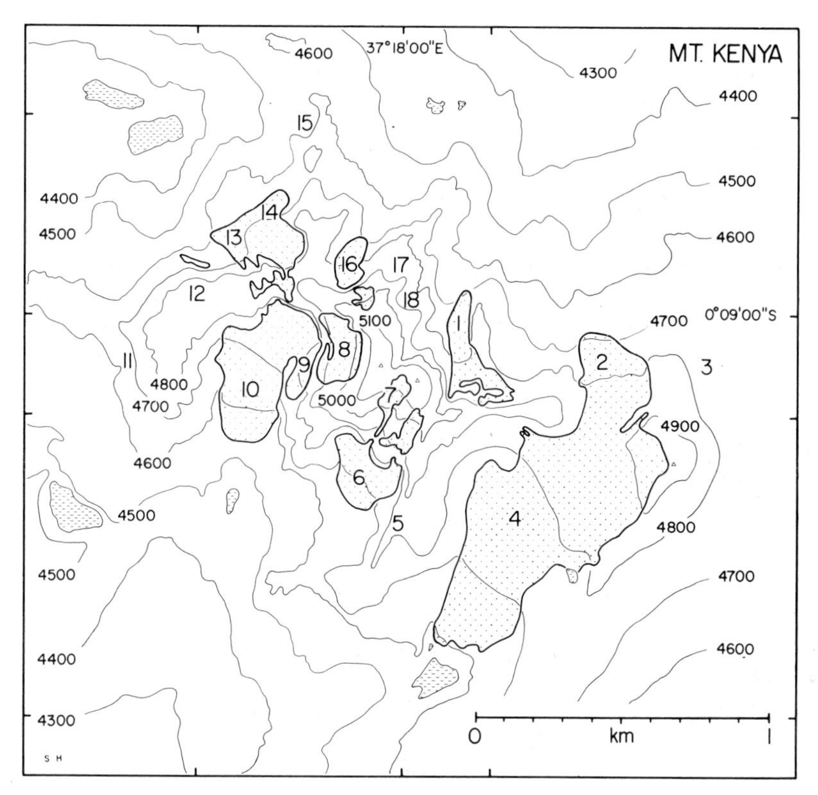

Map 4.3:1. Glaciers of Mount Kenya. Scale 1:25 000. Contours at 100 m intervals. Large numbe denote glaciers listed in Table 4.3:1.

TABLE 4.3:1
Nomenclature of glaciers on Mount Kenya

1	Krapf	
2	Gregory	
(3	Kolbe	disappeared after 1926)
4	Lewis	
(5	Melhuish	disappeared after Feb 1978)
6	Darwin	
7	Diamond	
8	Forel	
9	Heim	
10	Tyndall	
(11	Barlow	disappeared after 1926)
(12	NW Pigott	disappeared)
13	Cesar	
14	Joseph	
(15	Peter	disappeared after 1926)
16	Northey	
(17	Arthur	disappeared)
(18	Mackinder	disappeared)

Photo 4.3:1. Part of a LANDSAT 1 (ERTS) satellite image (MSS band 7) of Mount Kenya region. Date 24 Jan. 1976, 06:57 hr GMT (NASA).

obliterated by snow cover, and westward facing slopes are in shade. Thus, of the glaciers to the North, the Krapf (1) seems to connect with the Lewis (4) Glacier, and the Gregory (2) appears large. The Northey (16), Forel (8), and Heim (9) glaciers are not depicted at all, and the Darwin (6), Tyndall (10), and Cesar-Joseph (13, 14) glaciers appear small. The image is dominated by the Lewis Glacier (4).

The reconstruction of long-term glacier variations is of all East African mountains most nearly satisfactory for Mount Kenya, in part because of the comparatively abundant observations, but especially thanks to the unique 1:5000 map for the peak region (Forschungsunternehmen Nepal-Himalaya, 1967). This excellent topographic reference permits an actual mapping of glacier stages since the turn of the century. To this end, historical photographs (Appendix 2) were evaluated in the terrain in conjunction with the aforementioned topographic map. Map sketches of Mount Kenya in 1893, 1899, around

1920, in 1926, and in the 1960s are reproduced as Maps 4.3:2–4.3:7. Dutton's (1929) map of 1926 is also contained in Mountain Club of East Africa (1932, map no. 3). As for the other mountains, however, it may not always be possible to distinguish unambiguously between ice and merely temporary snow cover, especially in the interpretation of air photographs. The date of potentially informative ground photographs is not always given, and can be inferred only approximately; adret-ubac relationships of differently exposed slopes varying with time of year having been mentioned by Charnley (1959) for Mount Kenya. The quantitative results of glacier reconstruction for Mount Kenya are presented in Map 4.3:8* and Tables 4.3:2–4.3:4.

The Krapf Glacier (1) appears joined with the Lewis (4) over the col West of Thomson's Flake on an 1899 photograph by Hausburg (Mountain Club of Kenya, archive). Baker (1967, p. 66) concludes from a photograph in Dutton (1929, fig. 39) that the two glaciers were still connected in the 1920s. However, photographs in Arthur (1921, fig. facing p. 20, bottom) from about 1919–20, and in Gregory (1921, plate 12a) do not show this juncture. Baker mentions that some of the photographs in Dutton's book in fact date from 1921 and earlier. The lower portion of the Krapf (1) and Gregory (2) Glaciers are shown on slant aerial photographs of Mittelholzer (1930, figs. 87, 88) in 1930. The former picture is here reproduced as Photo 4.3:2. The Krapf Glacier (1) is recognizable on a slant aerial photograph in January 1938 (Light, 1941, photo 182). Hicks (1945–6, fig. facing p. 80) presents a photograph of July 1944. These compare well with Waddington's (Spink 1945, facing p. 216, top) photograph of August 1944, the RAF slant air photograph of 1945 (Spink, 1945, p. 281, top) and the vertical air photography of February 1947 (Appendix 1). The former picture is presented here as Photo 4.3:3. A slant aerial photograph of December 1957 is shown in Charnley (1959, fig. 8), and reproduced here as Photo 4.3:4. The January 1963 map (Forschungsunternehmen Nepal-Himalaya, 1967) indicates some change, but a more drastic shrinkage in ice-covered surface has taken place by 1973. Photos 4.3:5 and 4.3:6 depict the snout conditions in June 1973 and July 1980. Although paint marks established in June 1973 could not be found during the January and July 1980 visits, the snout recession over this seven year interval is estimated at 50–100 m, similar to the other glaciers on the North side of Mount Kenya.

The Gregory Glacier (2) appears on photographs of 1930 (Mittelholzer, 1930, figs. 87, 88), July 1944 (Hicks, 1945–6, facing p. 80), of January 1938 (Light, 1941, photo 182), August 1944 (Spink, 1945, facing p. 216, top; facing p. 217, top), 1945 (Spink, 1945, p. 281, top), December 1957 (Charnley, 1959, fig. 8), and the February 1947 vertical air photography (Appendix 1), in part mentioned before. Reference is made to the present Photos 4.3:2–4.3:5, and 4.3:7. Photo 4.3:7 (Spink, 1945, facing p. 217, top) in particular shows the terminus of Gregory Glacier (2) in August 1944. This sequence confirms a sizeable decrease in ice-covered area from the early 1930s to the 1940s, and to the January 1963 mapping (Forschungsunternehmen Nepal-Himalaya, 1967). A further marked retreat continued to the 1970s. Photos 4.3:8 and 4.3:9 illustrate the drastic decrease in ice extent from June 1973 to July 1980.

A Kolbe (3) Glacier appears on Mackinder's (1900) map for August–September 1899 (Map 4.3:2), and on a map sketch by Melhuish and Arthur (Map 4.3:5) presumably from around 1920 (Royal Geographical Society, London, archive), seemingly extending to below the present-day Harris Tarn. Arthur (1921, p. 17) mentions a 'Kolb' glacier for February 1920. Dutton's (1929) map shows for 1926 (Map 4.3:6) the Kolbe Glacier in

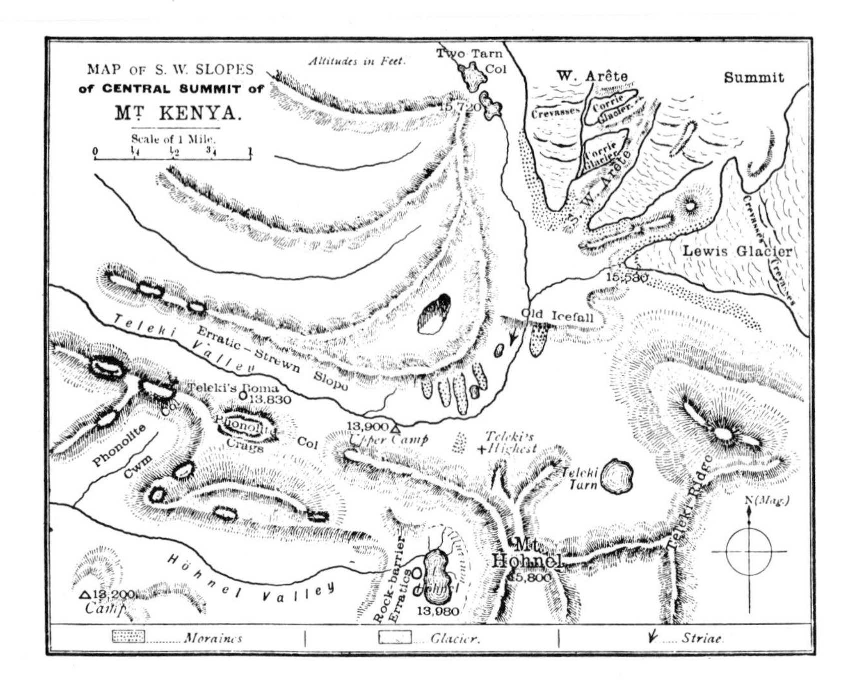

Map 4.3:2. Map of Mount Kenya by Gregory (1894), date 1893.

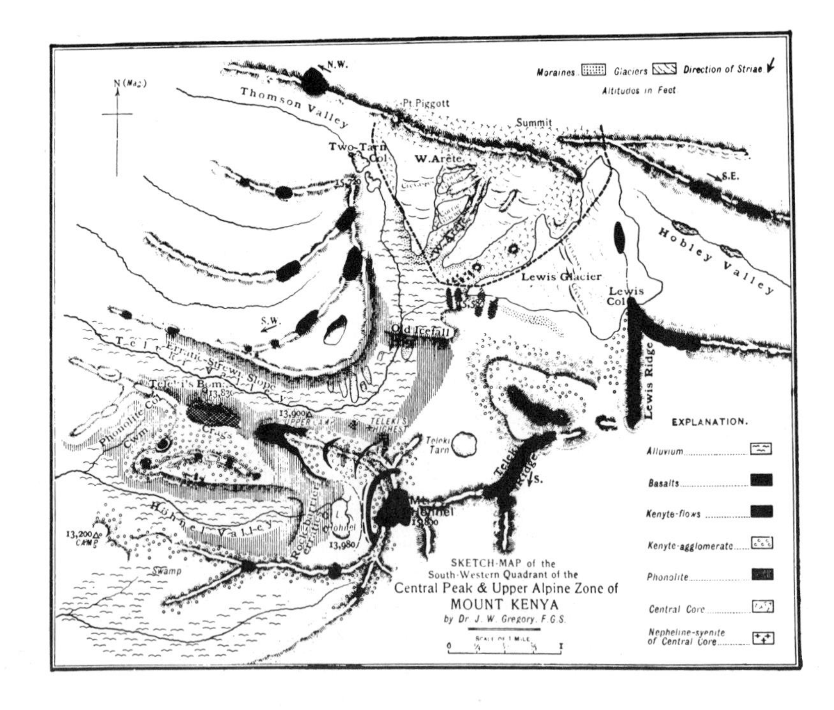

Map 4.3:3. Map of Mount Kenya by Gregory (1900), date 1893.

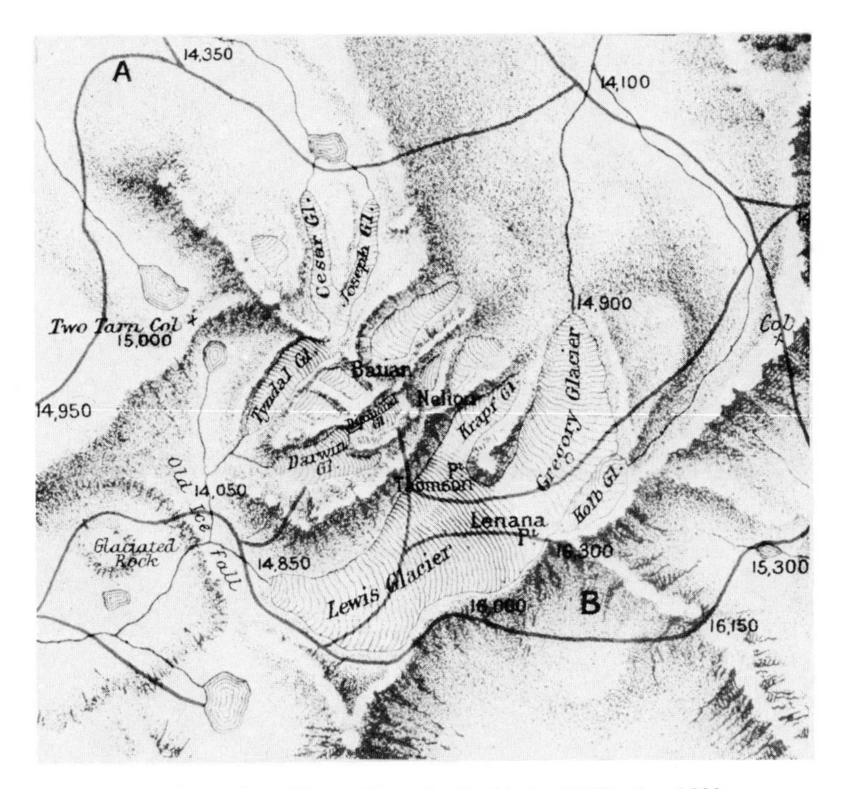

Map 4.3:4. Map of Mount Kenya by Mackinder (1900), date 1899.

contact with a frozen Harris Tarn. Nilsson (1931, p. 269) reports for 1927 a glacier extending down into a little lake at the beginning of the Gorges Valley. A photograph of the mere glacier snout and the lake is also given (Nilsson, 1931, fig. 16). It is believed that this represents Harris Tarn, since reference to Hanging or Simba Tarns in this context does not seem plausible. This state of affairs in 1927 appears consistent with conditions borne out on Mittelholzer's (1930, figs. 87, 88) photographs in 1930 (see present Photo 4.3:2). A slant aerial photograph of January 1938 (Light, 1941, photo 182) lacks resolution in this area. Benuzzi (1947, p. 332) on his escape from a POW camp in February 1943 saw no remnant of the Kolbe Glacier. The 1:25 000 map based on the February 1947 air photographs (Appendix 1) shows a Kolbe 'Snow Field' not quite in contact with the Harris Tarn. By the 1970s this snow field has disappeared altogether.

Lewis Glacier (4), the largest ice body on Mount Kenya, will be discussed comprehensively in Chapter 5. Observations are comparatively abundant and date back to the last century. Gregory (1894, figs. 1–4; 1900, plate 10) presents sketches of the peak area from his January 1893 expedition. In particular, he describes the Lewis Glacier as having burst through the uppermost of apparently four terminal moraines, and suggests that the glacier has been almost stationary for some time (Gregory, 1894, fig. 4 and pp. 521–2). The present Photo 4.3:10 is reproduced from Gregory (1894, fig. 4). Mackinder's (1900) map (present Map 4.3:4), as well as Hausburg's photographs from the same expedition (archive of Mountain Club of Kenya, Nairobi), suggest a snout position not much different

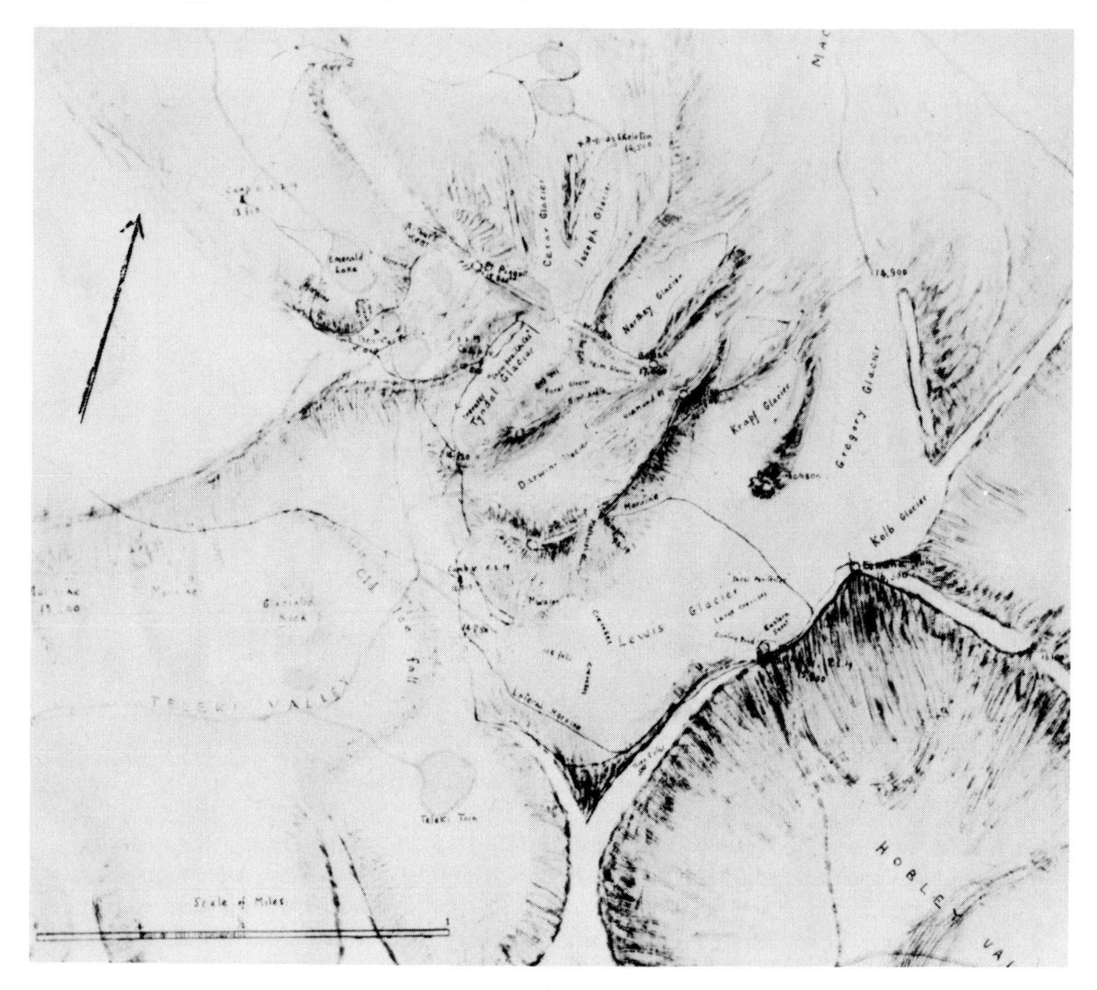

Map 4.3:5. Map of Mount Kenya by Melhuish and Arthur (Royal Geographical Society, archive), date about 1920. Magnetic North direction is indicated in upper left portion of map.

in August–September 1899. Photographs in Dutton (1929, figs. 32, 20) illustrate that by February 1926 the ice tongue had receded from the innermost moraine towards a rock threshold situated just down the valley from the present-day Lewis Tarn. The former picture is here reproduced as Photo 4.3:11. The elevation given in Table 4.3:2 is consistent with Baker's (1967, p. 66) evaluations.

A detailed photogrammetric survey and photographs in April–May 1934 (Troll and Wien, 1949, figs. 3, 4, 6) document the further retreat to the base of this rock precipice. The latter picture is reproduced here as Photo 4.3:12. Slant aerial photographs of January 1938 (Light, 1941, photos 183, 184, 185) do not show the snout region. A photograph in June 1941 (Hodgkin, 1941, second photograph following p. 312), presented here as Photo 4.3:13, shows the ice extending to only the upper edge of the rock face. Air photography (Appendix 1) indicates that the Lewis Tarn was in its incipient stage and

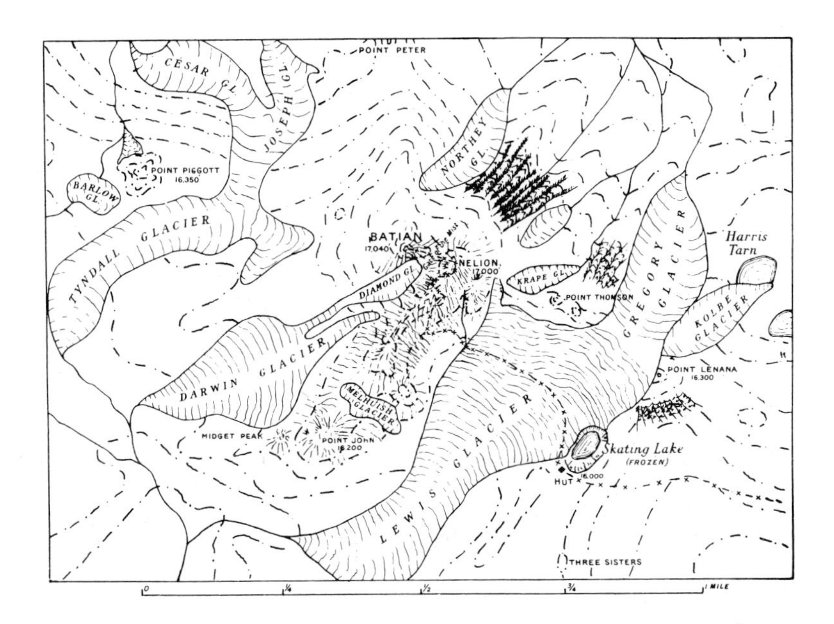

Map 4.3:6. Map of Mount Kenya by Dutton (1929), reprinted in Mountain Club of East Africa (1932), date 1926.

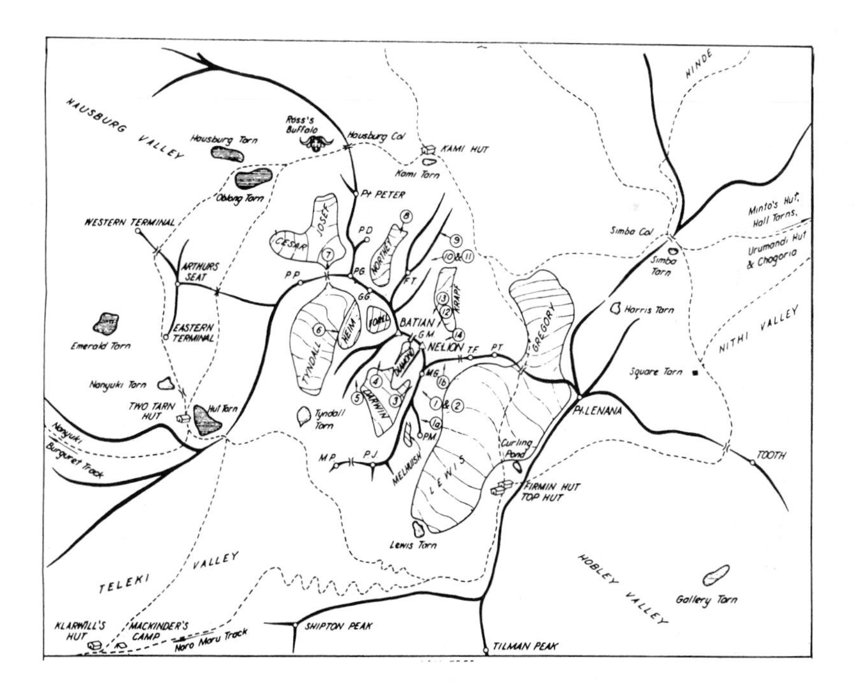

Map 4.3:7. Map of Mount Kenya by Mountain Club of Kenya (1971), date 1960s.

Photo 4.3:2. The Krapf (1) and Gregory (2) Glaciers from the North in Jan. 1930 (Mittelholzer, 1930, fig. 87).

Photo 4.3:3. The Krapf (1), Gregory (2), and Northey (16) Glaciers from the North in Aug. 1944, (Spink, 1945, facing p. 216, top, by Waddington).

Photo 4.3:4. (A) Krapf (1) and (B) Gregory (2) Glacier in Dec. 1957 (Charnley, 1959, Fig. 8; aerial; reproduced from the Journal of Glaciology by permission of the International Glaciological Society).

in contact with the ice in February 1947. An aerial photograph in 1945 (Spink, 1949, p. 281, bottom), and photographs by Cameron and Reade (1950) from February-March 1950 are consistent with this. Charnley (1959, figs. 6, 9) presents aerial photographs from the IGY expedition in December 1957, the former picture being reproduced here as Photo 4.3:14. Conditions in January 1963 are documented by Schneider's map (Forschungsunternehmen Nepal-Himalaya, 1967). During my visits in July 1971 and in 1973–4 large portions of the rock threshold above Lewis Tarn lay bare (Photo 4.3:15), and the ice tongue had become extremely thin, especially on the Eastern flank (Caukwell and Hastenrath, 1977). By February 1978 the snout had receded further, some large caves had formed (Photos 4.3:16 and 4.3:17), as well as another small pond (Hastenrath and Caukwell, 1979). Further retreat is documented to February 1982 (Caukwell and Hastenrath, 1983).

At the Northwestern side of the glacier, in the general vicinity of IGY control point L3 (Maps 4.3:9*, 4.3:10* and 4.3:11*), Helmut Heuberger (personal communication 1977), in January 1969, established paint marks, bearings, and distances to the ice rim. These were re-examined in March 1978. At a bearing of 77 degrees an increase of distance of about 20 m was found, which would correspond to some 10 m in a direction perpendicular to the ice margin.

For Curling Pond, there are some photographs, dating probably from the 1920s and in part from February 1926 (Dutton, 1929, figs. 19, 22, 31; Mountain Club of Kenya, archive), a picture of February 1941 (Douglas-Hamilton, 1941–2, fig. facing p. 217,

Photo 4.3:5. The Krapf Glacier (1) in June 1973. (Photo S.H. 25 June 1973).

Photo 4.3:6. The Krapf Glacier (1) in July 1980 (Photo S.H. 22 July 1980).

Photo 4.3:7. Front of Gregory Glacier (2) in Aug. 1944 (Spink, 1945, facing p. 217, top; by Waddington).

Photo 4.3:8. The Gregory Glacier (2) in June 1973 (Photo S.H. 25 June 1973).

Photo 4.3:9. The Gregory Glacier (2) in July 1980 (Photo S.H. 22 July 1980).

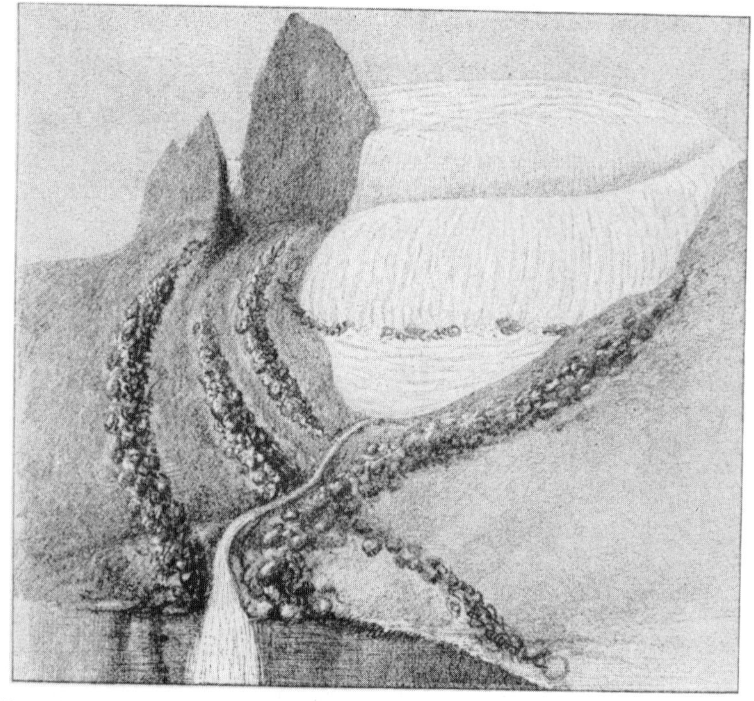

Photo 4.3:10. Gregory's (1894, fig. 4) sketch of Lewis Glacier (4) terminus in 1893.

Photo 4.3:11. The Lewis (4) Glacier in Feb. 1926 (Dutton, 1929, fig. 32).

Photo 4.3:12. The Lewis (4) Glacier in April–May 1934 (Troll and Wien, 1949, fig. 6).

Photo 4.3:13. The Lewis (4) Glacier in June 1941 (Hodgkin, 1941, second fig. following p. 312).

Photo 4.3:14. The Lewis Glacier (4) in Dec. 1958 (Charnley, 1959, fig. 6; aerial; reproduced from the *Journal of Glaciology* by permission of the International Glaciological Society).

Photo 4.3:15. Snout of Lewis Glacier (4) in Sept. 1973 (Photo S.H. 26 Sept. 1973).

Photo 4.3:16. Lower Lewis Glacier (4) in Feb. 1978 (Photo S.H. 27 Feb. 1978).

Photo 4.3:17. View from ice cave at Lewis Glacier (4) terminus to Lewis Tarn. (Photo S.H. 27 Feb. 1978).

Photo 4.3:18. The Curling Pond area of Lewis (4) Glacier in Feb. 1945 (Spink, 1945, facing p. 217, bottom, by David).

bottom), and a photograph of February 1945 (Spink, 1945, facing p. 217, bottom; present Photo 4.3:18). The January 1963 situation is shown on the 1:5000 map of Forschungsunternehmen Nepal-Himalaya (1967). The 1:2500 map based on aerial photography (Caukwell and Hastenrath, 1977; present Map 4.3:9*, at scale 1 : 5000; Table 4.3:5) documents the ice extent in February 1974. Two further maps at the same scale, dated of February 1978 and 1982 based on aerial photography (Hastenrath and Caukwell, 1979; present Maps 4.3:10* and 4.3:11*, at scale 1:5000) bear out the further drastic changes in these two 4 year intervals.

The Melhuish Glacier (5) appears on a photograph of 1899 (Mountain Club of Kenya, archive), a picture of 1919 (archives of Royal Geographical Society, London; and of Mountain Club of Kenya, Nairobi), on a slant aerial photograph of 1930 (Mittelholzer, 1930, fig. 89), and on photographs in April–May 1934 (Troll and Wien, 1949, figs. 3, 4, 6). The latter picture is shown here as Photo 4.3:12. The glacier also appears on slant aerial photographs taken in January 1938 (Light, 1941, photos 183, 184). Cameron and Reade (1950) claim that Melhuish Glacier had disappeared in 1950, but it seems that there may be some confusion. The upper portion of Melhuish is seen on a photograph around January 1958 (Charnley, 1959, fig. 6). The aforementioned photographic sequence in conjunction with the January 1963 mapping (Forschungsunternehmen Nepal-Himalaya, 1967) bear out a progressive ice recession. Visits in 1973–4 and 1978 confirm that the Melhuish had degenerated into an ice or snow field. This disappeared altogether after February 1978.

The Darwin Glacier (6) is shown on photographs in 1899 (Mountain Club of Kenya, archive), in 1913 (Royal Geographical Society, archive), in 1919 (archives of Royal Geographical Society; and of Mountain Club of Kenya, Nairobi), on slant aerial photographs in 1930 (Mittelholzer, 1930, fig. 89) and January 1938 (Light, 1941, photo 183), on photographs of the 1940s (Howard 1955, facing pp. 279, 274; Hodgkin, 1941, third photograph following p. 312; Firmin, 1945–5, facing p. 404), and on the February 1947 aerial photography (Appendix 1). A comparatively large ice retreat is indicated for this interval, whereas the retreat to 1963 (Forschungsunternehmen Nepal-Himalaya, 1967) and the 1970s appears more modest. During a March 1978 visit a paint mark labelled 'Aug 70' was detected, the particulars of which are not known. Distances of 5 and 16 m were taped to the ice rim and to a large ice cave, respectively. Helmut Heuberger (personal communication 1977), in January 1969, measured from IGY point S6 at a bearing of 42 degrees a distance of 61 m ± 2 m to the ice rim. During visits in January 1980 and January 1982, distances of 81 and 80 m, respectively, were taped. Commensurate with the limited accuracy of tape measurements over such large distances, little change is indicated over this two-year interval.

The hanging Diamond (7), Forel (8), and Heim (9) glaciers are depicted on photographs of 1899 (Mackinder, 1900, figs. on p. 465, 467; 1901, fig. facing p. 102) of 1908 (McGregor Ross, 1911, fig. on p. 473), of 1919 (Arthur, 1921, fig. facing p. 17, top) and of the 1940s (Firmin, 1945–6, facing p. 404; Howard, 1955, facing pp. 273, 274). They appear to have essentially maintained their areas until 1963 (Forschungsunternehmen Nepal-Himalaya, 1967), and the 1970s, although a couloir descending to the lower portion of Tyndall Glacier (10) gradually vanished. Since 1978 the connections between the Forel (8), Heim (9), and Tyndall (10) glaciers have broken up, and the ice volume of the Forel (8) and Heim (9) is seemingly decreasing.

Tyndall Glacier (10) is the second largest ice body on the mountain. Changes since the end of the last century are fairly well documented. Gregory's (1894, fig. 2; 1900, plate 10) sketch maps (present Maps 4.3:2 and 4.3:3) and his drawing (Gregory, 1896, facing p. 180; present Photo 4.3:19) of 1893, show the ice tongue close to the large moraine situated below the present-day Tyndall Tarn. A similar state of affairs is indicated on photographs of 1899 (Mackinder, 1900, fig. on p. 467; 1901, fig. facing p. 109; Mountain Club of Kenya, Nairobi, archive) and 1908 (McGregor Ross, 1911, fig. on p. 473). Photographs by Arthur (1921, figs. facing p. 17) in February 1919, and another one of presumably similar date (Gregory, 1921, fig. facing p. 150, top) show the upper portion of Tyndall, but are not conclusive regarding the snout position. By February 1926 (Dutton 1929, fig. 33) Tyndall Tarn had formed, but was still in contact with the ice (present Photo 4.3:20). Slant aerial photographs of January 1938 (Light, 1941, photos 183, 185) lack resolution. Firmin (1945–6, facing p. 404) presents a photograph of Tyndall glacier in January 1946, but the snout portion is not fully visible. Another photograph by Firmin, presumably from the 1940s, is shown in Howard (1955, facing p. 274). Air photographs in February 1947 (Appendix 1) show that the ice tongue had receded to well above Tyndall Tarn. A similar position is suggested for February 1950 (Cameron and Reade, 1950). Further retreat is documented to January 1963 (Forschungsunternehmen Nepal-Himalaya, 1967). A sizeable terminal moraine has formed a little below the present snout position, i.e., well after the ice tongue had separated from Tyndall Tarn. Another small moraine arc was found in immediate vicinity of the ice rim in March 1978. The Tyndall Glacier terminus in March 1978 is illustrated in Photo 4.3:21.

Helmut Heuberger (personal communication 1977), in January 1969, measured from IGY point S8 at a bearing of 11 degrees (Firmin Col) a distance of 116 m to the rim of Tyndall glacier. During the March 1978 visit a distance of 119 m was taped. Heuberger also observed a fresh frontal moraine of 1 m height immediately below the ice rim. In March 1978 a distance of 7.3 m was taped from the innermost (West) end of the moraine to the ice rim, the approximate reference being 'the West end of the East pond'. At the Eastern half of the snout, a mark labelled 'Aug. 70' was found, the particulars of which are not known. From there a distance of 10.6 m was measured to the ice rim.

In January 1982, the distance from IGY point S8 at a bearing of 11 degrees (Firmin Col) to the ice rim was measured to be 131.5 m. Along this line two subpoints were established with red paint at distances from S8 of 51 and 108 m; the distance from the latter point to the rice rim thus being 23.5 m. From the innermost (West) end of the moraine (approximate reference 'West end of East pond') a distance of 17.15 m was taped to the ice rim. From the same mark 'Aug. 70' a distance of 18.10 m was measured to the ice rim. Comparison with the measurements in March 1978 indicates a retreat of the ice front of 7–15 m over the four year interval.

To the North of Tyndall Glacier (10) and on the Southwest side of Point Pigott, a small ice entity existed earlier, the now defunct Barlow Glacier (11). This is evidenced on photographs of 1908 (Royal Geographical Society, London, archive), of 1912 (Jeannel, 1950, pl. 25), on a map sketch by Melhuish and Arthur (present Fig. 4.3:5) around 1920 (Royal Geographical Society, archive), and Dutton's (1929) map (present Map 4.3:6) for 1926. This glacier has since disappeared.

Another small ice body formerly existed to the Northwest of Point Pigott, as is indicated by the aforementioned maps (Maps 4.3:5 and 4.3:6) for 1920 (Royal Geographical

Photo 4.3:19. Gregory's (1896, facing p. 180) sketch of Tyndall Glacier (10) in 1893.

Photo 4.3:20. The Tyndall Glacier (10) in 1926 (Dutton, 1929, fig. 33).

Photo 4.3:21. The Tyndall Glacier (10) in March 1978 (Photo S.H. 2 March 1978).

Photo 4.3:22. The Cesar Glacier (13) in 1908 (McGregor Ross, 1911, p. 469).

Photo 4.3:23. The Cesar Glacier (13) in July 1973 (Photo S.H. 21 July 1973).

Photo 4.3:24. The Cesar Glacier (13) in July 1980 (Photo S.H. 22 July 1980).

Photo 4.3:25. The Joseph Glacier (14) in the early 1930's (Tilman, 1937, facing p. 64).

Photo 4.3:26. The Joseph Glacier (14) in July 1973 (Photo S.H. 21 July 1973).

Photo 4.3:27. The Joseph Glacier (14) in July 1980 (Photo S.H. 22 July 1980).

Photo 4.3:28. The Northey Glacier (16) in the early 1930s (Tilman, 1937, facing p. 66).

Photo 4.3:29. The Northey Glacier (16) in June 1973 (Photo S.H. 25 June 1973).

Photo 4.3:30. The Northey Glacier (16) in July 1980 (Photo S.H. 22 July 1980).

Society, archive) and 1926 (Dutton, 1929). This ice entity which has also since disappeared is here called the Northwest Pigott Glacier (12).

The earliest reference to Cesar Glacier (13) is due to Mackinder (1900, map) in August–September 1899 (Map 4.3:4). His map shows the ice tongue reaching to within 150 m of a lake, presumably Oblong Tarn. Photography by McGregor-Ross (1911, p. 469) in 1908 (present Photo 4.3:22) indicates a snout position at 4400 m, some 150 m from Oblong Tarn. Baker (1967, p. 65) estimates an elevation of 4420 m from this photograph. Arthur (1921, p. 16) describes the snout of the Cesar Glacier as being just 17 m above the upper one of two little lakes in February 1919. Evidently Oblong Tarn is meant. Unless Arthur's estimate is in error, this would indicate an advance of the glacier snout. The later map sketches (present Maps 4.3:5 and 4.3:6) are not informative regarding variations in the extent of Cesar Glacier. The February 1947 air photography (Appendix 1) illustrates a successive and conspicuous recession of the snout. Photographs by Cameron and Reade (1950) for February–March 1950 are compatible with the

February 1947 documentation. A further retreat is documented by the photogrammetric map of January 1963 (Forschungsunternehmen Nepal-Himalaya, 1967). The June–July 1973 field survey revealed a drastic retreat of Cesar Glacier (13) since then. Conditions in July 1973 and July 1980 are illustrated in Photos 4.3:23 and 4.3:24. Retreat of the glacier terminus over this seven-year interval is estimated to be of the order of 50–100 m, similar to the other glaciers on the North side of the mountain.

The earliest reference to Joseph Glacier (14) is again due to Mackinder (1900, map) in August–September 1899 (present Map 4.3:4). His map depicts the glacier tongue as reaching rather close to a lake, presumably Oblong Tarn. The later map sketches (present Maps 4.3:5 and 4.3:6) do not permit us to assess changes of Joseph Glacier. A ground photograph by Tilman (1937, facing p. 64; present Photo 4.3:25), presumably from the early 1930s, suggests a considerably larger ice extent than at present, but does not cover the lower portion of the glacier. A slant aerial photograph of January 1938 (Light, 1941, photo 182) lacks resolution. The Joseph Glacier also appears on a slant aerial photograph in 1945 (Spink 1949, p. 281, top) except for its lower portion. Air photography in February 1947 (Appendix 1) and the January 1963 mapping (Forschungsunternehmen Nepal-Himalaya, 1967) bear out a conspicuous ice retreat. Again, the June–July 1973 field trips revealed a particularly drastic ice decrease since then. Conditions in July 1973 and July 1980 are depicted in Photos 4.3:26 and 4.3:27. During the January and July 1980 and January 1982 visits, no attempt was made to locate the paint marks established in June 1973. Recession of the terminus over this seven year interval is estimated at about 50 m. In January 1982 two bench marks (nail, red paint, mark 'HK', cairns) were established, 16.08 m apart along a line oriented at 172 degrees. The distance to the ice rim was not measured, as it is obliterated by snow at this season of the year.

A small glacier or firn bank on the West face of Point Peter entered on Dutton's (1929) map for 1926 (Map 4.3:6) has since disappeared. This ice entity is here called Peter Glacier (15).

The Northey Glacier (16) appears on the maps of Mackinder (1900) for 1899 (present Map 4.3:4), and on later map sketches (present Maps 4.3:5 and 4.3:6). The glacier is depicted on a ground photograph by Tilman (1937, facing p. 66) from the early 1930s (present Photo 4.3:28), and on a slant aerial photograph by Light (1941, photo 182) in January 1938. The ground photograph in Spink (1945, facing p. 216, top; present Photo 4.3:3) from August 1944, a slant aerial photograph of 1945 (Spink, 1945, p. 281, top), and vertical air photography in February 1947 (Appendix 1) consistently show an ice extent much larger than in January 1963 (Forschungsunternehmen Nepal-Himalaya, 1967). The June–July 1973 field observations revealed a drastic ice retreat since then. The snout conditions in June 1973 and July 1980 are shown in Photos 4.3:29 and 4.3:30. Retreat of the terminus over this seven-year interval is estimated to be of the order of 50–100 m, similar to other glaciers on the North side of Mount Kenya. In contrast to the photographs of the 1930s and 1940s, the uppermost part of the ice body was, in the 1970s, found to be separated by a ledge from the lower, larger part of Northey glacier. The 1963 map (Forschungsunternehmen Nepal-Himalaya, 1967) seems to illustrate the transition.

Two unnamed ice entities are shown on Mackinder's (1900) map for 1899 (Map 4.3:2) between the Northey (16) and Krapf (1) Glaciers. These will here be called the Mackinder (17) and Arthur (18) Glaciers; the latter being the more southerly and larger of the two.

The Arthur (18) Glacier still appears, though unnamed, on Melhuish and Arthur's sketch map (Map 4.3:5) of around 1920 (Royal Geographical Society, archive) and on Dutton's (1929) map of 1926 (present Map 4.3:6). These ice bodies have since disappeared, although the January 1963 map (Forschungsunternehmen Nepal-Himalaya, 1967) shows a snow or firn field in the depression to the North of Point Dutton, the presumed site of the defunct Arthur Glacier (18). The location of the former Mackinder Glacier (17) was probably in a niche to the Southeast of Point Peter.

In conclusion, the large glaciers to the West and South of the massif, Lewis (4) and Tyndall (10), receded essentially continuously from the end of the 19th century to present, with a rather more moderate change of snout positions in recent decades. However, detailed volume assessments for Lewis Glacier, to be presented in Chapter 5, demonstrate that the total ice wastage has been considerable. Records for Darwin Glacier (6) covering a shorter time span also bear out an ice retreat. The Melhuish (5) degenerated into an ice or snow field, and disappeared altogether after February 1978 and prior to January 1980. The Diamond (7), Forel (8), and Heim (9) glaciers seem to have conserved their total surface throughout many decades, presumably thanks to their situation well above the equilibrium line of the mass budget. However, they are showing deterioration in recent years.

Glaciers to the East and North of the massif are generally smaller, so that volume changes are more conspicuously reflected in variations of the snout position. A recession from the turn of the century to present is documented for the twin glaciers Cesar (13) and Joseph (14), with changes in snout position becoming particularly pronounced in recent years. For the Cesar Glacier (13) there are indications for an advance around 1920. If this interpretation is correct, this evidence for the Cesar (13) may parallel the formation of young moraines at Tyndall (10) and even Lewis (4) glacier around a similar time. Reference is here made to Map 4.3:8* of glacier stages and the 1963 map of Forschungsunternehmen Nepal-Himalaya (1967). Observations on the Northey (15), Krapf (1), and Gregory (2) glaciers limited to a more recent time span also display an essentially continuous recession, with rather large changes in recent decades. The fate of the small Kolbe Glacier (3) was to give way to a tarn, to degenerate into an ice field, and to vanish altogether in less than seven decades. Six other glaciers or ice fields disappeared since the beginning of the century.

4.4. Glaciological Studies

Complementing the survey of modern ice extent and recent glacier variations in the preceding Sections 4.1 through 4.3, the present section reviews the limited studies of the physical characteristics of East African glaciers.

At Kilimanjaro, the earliest measurements of surface ice velocity were made. Klute (1920, pp. 107–8) in 1912 surveyed stakes laid out on the lower portion of the Great Penck Glacier (17) within a one-month interval and thus obtained surface velocities of the order of 7 m a^{-1}. Similar attempts by the University of Sheffield expeditions in 1953 and 1957 (Humphries, 1972, pp. 38–41) were not successful. Spink (1943) and Humphries (1972, pp. 39–40) noted the penitentes-like ablation forms on Kilimanjaro, especially in the firn of the caldera. Any comparable development is lacking in the Ruwenzori and on Mount Kenya, which are cloudier, more humid, and are closer to

TABLE 4.3:2

Variation of glacier termini at Mount Kenya. Top number denotes terminus elevation, and bottom number horizontal distance from 1963 position, counted positive for larger ice extent. For Kolbe Glacier bottom numbers refer to 1899 position. Units in m. Refer to Table 4.3:3 for sources

Part 1: 1893 to 1930.

		1893	1899	1908	1912	1913	1919	1920	1920 approx.	1920s	1926	1927	1930
1	Krapf												4520
													+160
2	Gregory												4610
													+90
3	Kolbe		4720					4735	4735		4740	4740	4750
			0					–50	–50		–90	–90	–100
4	Lewis	4470	4470								4450		
		+440	+430								+270		
4a	Lewis-Curling									4790	4790		
										+70	+70		
5	Melhuish												–
6	Darwin					4600	4600						4600
						+100	+100						+100
7	Diamond												
8	Forel												
9	Heim												
10	Tyndall	4450	4450	4450			4450				4450		
		+260	+250	+230			+210				+120		
11	Barlow			4620	–			–			–		
12	NW Pigott							4810			–		
13	Cesar		4400	4420			4360						
			+150	+120			+250						
14	Joseph		4440										
			+220										
15	Peter										4570		
16	Northey												
17	Mackinder		4850										
18	Arthur		4850					–			–		

TABLE 4.3:2, continued

Part 2: 1930s to 1978

		1930s early	1934	1940s	1941	1944	1945	1946	1947	1950	1957	1963	1973	1974	1978
1	Krapf					4580	4580		4600		4600	4600	4690		
						+45	+45		+10		+10	0	−80		
2	Gregory					4660	4660		4630		4630	4620	4690		
						+30	+30		+20		+20	0	−20		
3	Kolbe								(4750)						
									−160						
4	Lewis		4480		4550				4570		4575	4580		4585	4590
			+200		+140				+100			0		−40	−65
4a	Lewis-Curling				4790		4790					4790		4790	4790
					+20		+20					0		−20	−25
5	Melhuish		–								–	4740	4740		4740
												0	0		0
6	Darwin			4610				4610	4610			4620	4640		4640
				+70				+70	+70			0	−30		−30
7	Diamond											4900	4900		
												0	–		
8	Forel											4840	4840		
												0	–		
9	Heim											4700	4700		
												0	–		
10	Tyndall			4480				4480	4480	4480		4505	4515		4530
				+50				+50	+50	+50		0	−10		−30
11	Barlow														
12	NW Pigott														
13	Cesar								4530			4500	4600		
									+10			0	−50		
14	Joseph	<4520							4550			4540	4630		
		+20							+10			0	−140		
15	Peter														
16	Northey	–				4500	4500		4560			4630	4700		
		–				+210	+210		+150			0	−100		
17	Mackinder														
18	Arthur														

TABLE 4.3:3

Sources of data presented in Table 4.3:2. See Table 4.3:4 for code of sources

Part 1: 1893–1930

		1893	1899	1908	1912	1913	1919	1920	1920 approx.	1920s	1926	1927	1930	1930 approx.
1	Krapf												13	
2	Gregory												13	
3	Kolbe		3						9		11	12	13	
4	Lewis	1,2	10								11			
4a	Lewis-Curling									10	11			
5	Melhuish		10				9,10						13	
6	Darwin		10			9	9						13	
7	Diamond		3	5			7							
8	Forel		3	5			7							
9	Heim		3	5			7							
10	Tyndall	1,2	3,4,14	5			7,8				11			
11	Barlow			9	6			9			11			14
12	NW Pigott							9			11			14
13	Cesar		3	5			7							
14	Joseph		3								11			14
15	Peter										11			
16	Northey		3								11			14
17	Mackinder		3											
18	Arthur		3					9			11			14

TABLE 4.3:3, continued

Part 2: 1930s to 1978

	1930s early	1934	1940s	1941	1944	1945	1946	1947	1950	1957	1963	1974	1978
1 Krapf					20, 21	22		24		26	27		
2 Gregory					20, 21	22		24		26	27		
3 Kolbe								24					
4 Lewis		16				22		24		26	27	28	29
4a Lewis-Curling				17		21					27	28	29
5 Melhuish		16								26	27		
6 Darwin			17				23				27		
7 Diamond			23, 17								27		
8 Forel			23, 17								27		
9 Heim			23, 17								27		
10 Tyndall			17				23	24	25		27		
11 Barlow													
12 NW Pigott													
13 Cesar								24	25		27		
14 Joseph	15					22		24			27		
15 Peter													
16 Northey	15				21	22		24	27				
17 Mackinder													
18 Arthur													

TABLE 4.3:4
Code of sources listed in Table 4.3:3

Code	Source	Date	
		Publication	Actual
1	Gregory	1894	1893
2	Gregory	1900	1893
3	Mackinder	1900	1899
4	Mackinder	1901	1899
5	McGregor Ross	1911	1908
6	Jeannel	1950	1912
7	Arthur	1921	1919
8	Gregory	1921	1919
9	Royal Geographical Society, archive		1908, 1913, 1919, 1920, 1920 approx.
10	Mountain Club of Kenya, archive		1899, 1919, 1920s
11	Dutton	1929	1926
12	Nilsson	1931	1927
13	Mittelholzer	1930	1930
14	Mountain Club of East Africa	1932	1930 approx.
15	Tilman	1937	1930s early
16	Troll and Wien	1949	1934
17	Howard	1955	1940s
18	Hodgkin	1941	1941
19	Douglas-Hamilton	1941–42	1941
20	Hicks	1945–46	1944
21	Spink	1945	1944
22	Spink	1949	1945
23	Firmin	1945–46	1946
24	Appendix I		1947
25	Cameron and Reade	1950	1950
26	Charnley	1959	1957
27	Forschungsunternehmen Nepal-Himalaya	1967	1963
28	Caukwell and Hastenrath	1977	1974
29	Hastenrath and Caukwell	1979	1978

TABLE 4.3:5
IGY control points in the vicinity of Lewis Glacier. Marks not identified and not used in the 1973–74 survey are indicated by an asterisk. South-North (+ Y), West-East (+ X) coordinates, and elevation (h) in m

	$+Y$	$+X$	h
L 1*	1508.0	3373.9	4823.1
L 2	1450.4	3210.6	4797.2
L 3	1791.8	2884.0	4792.7
Little John*	1306.1	2577.7	4628.4
Lenana	1847.9	3622.1	4985.0
Melhuish	1630.6	2742.2	4876.5
S 3	1206.3	2745.5	4600.6
Thomson	2031.0	3159.7	4955.1
Top Hut*	1361.4	3177.5	4809.4

the Equator, and thus possess a different insolation geometry. Surface snow samples taken in February 1969 on the East side of Kibo and subjected to oxygen isotope analysis (Gonfiantini, 1970; and Table 5.5:1) yield δ-values of −3.7 to −6.8 ⁰/oo, i.e., figures quite different from those known for the polar regions (Dansgaard, 1964; Dansgaard *et al.*, 1969; Johnsen *et al.*, 1972). Samples for isotope analysis were also taken at the rim of Kibo in 1975 (Davies *et al.*, 1977a, b).

The Ruwenzori were the object of a 1952 British expedition, as part of which glaciological work was undertaken on Elena (D6) Glacier (Bergström, 1956). An extensive glaciological program was carried out by the Makerere University expeditions in the 1950s. This included measurements of temperature, humidity, wind, ablation, and runoff (Whittow *et al.*, 1963). These observations seem potentially useful for the study of heat budget forcing and water budget response, especially in terms of the daily cycle, but results have not become accessible in the professional journals.

Variations of snout positions over intervals of months to seasons have been reported in considerable detail for the Speke (C5), Elena (D6), and Savoia (D8) Glaciers (Osmaston, 1961; Whittow *et al.*, 1963; Temple, 1968). Osmaston (1961) monitored the snout recession of the Elena (D6) Glacier for the intervals between July 1952, August 1953, February 1954, and January 1958, and obtained a retreat of about 4 m over the entire 66-month period. For the Speke (C5) Glacier (Whittow *et al.*, 1963) a snout recession of several m is obtained during the three year interval from June 1958 to July 1961. Temple (1968) reports a similar retreat rate of the order of one m a^{-1} for the interval from December 1962 to December 1968. During the same period, the recession rates for the Elena (D6) and Savoia (D8) Glaciers were of the order of 0.4 and 2 m a^{-1}, respectively. Osmaston (1961) describes measurements of ice flow velocities over a six month interval on Savoia (D8) and a one month period on Elena (D6) Glacier during 1953–4. Flow rates obtained are of the order of 20 m a^{-1} for the former, and 30 m a^{-1} for the latter.

Osmaston (1961) proposes approximate water equivalents of 23, 14, and 9 cm a^{-1} for solid precipitation, ablation, and net balance, respectively, on Elena (D6) Glacier.

Valuable stratigraphic information was gathered by various expeditions on Elena (D6) Glacier. During July–August 1952 four pits were dug at different elevations in the firn area (Bergström, 1956). Layers of dirt, coarse-grained firn and ice indicating the prevalence of melting, are interpreted as forming at the times of year with decreased precipitation, namely around July–July and December–February. Bergström (1956) believes the former season to be more significant for ablation. A further pit was dug on Elena (D6) Glacier by Smith and Fletcher in December 1955 at an elevation of about 4900 m, and in June 1959 the Makerere expedition studied three pits (A, B, C) at 4650, 4900, and 4950 m (Whittow, 1960). Two distinct horizons of increased firn density and dirt content found consistently in all four pits are ascribed to June and January, but at variance with Bergström (1956) it is noted that the horizon pertaining to the latter time of year is by far more pronounced. Liquid water equivalents for pits A, B, and C, are given as 55, 58, and 44 cm for the layer June 1958 to January 1959, and as 20, 34, 47 cm for the layer January 1959 to June 1959. From the pit dug in 1955 at an elevation comparable to station B, water equivalents of 57 and 55 cm are obtained for the layers believed to correspond to the intervals January to June 1955 and June to December 1955, respectively. These results provide a useful orientation on the annual net balance and indicate a marked double variation of accumulation/ablation processes in the course of the year.

At Mount Kenya, the first measurements of ice surface velocity were performed by Troll and Wien (1949) on Lewis Glacier in April–May 1934. Terrestrial photogrammetry of a line of stakes near the equilibrium line over an interval of 104 hours yielded velocities of the order of 5 m a^{-1}. The comprehensive observation program of the Mount Kenya IGY expedition (Charnley, 1959) in 1957–8 included measurements of surface ice flow velocities at Lewis Glacier along a line similar to the one used by Troll and Wien (1949) in 1934 as well as a few other locations.

Stratigraphic descriptions of three ice pits near and above the equilibrium line of the Lewis Glacier (4) are further useful results of the IGY Mount Kenya expedition. Especially interesting with regard to the aforementioned observations in the Ruwenzori and the field program to be discussed in Chapter 5, is the occurrence of distinct ice horizons and associated dirt bands in the pit profiles. These are believed to mark the ablation seasons July–October and January–February. As a follow-up to the IGY work, Platt (1966) carried out micrometeorological measurements related to the surface heat and mass budgets. Meteorological observations over short intervals have been reported by Brinkman *et al.* (1968), Schnell and Odh (1977), and Davies *et al.* (1977a, b, 1979). The latter group also retrieved ice cores for purposes of climate interpretation.

4.5. Synthesis

The modern ice extent (Table 3:2; Map 4.1:1) at Kilimanjaro amounts to some 49×10^5 m^2. Twenty separate ice entities can be distinguished. Glaciation is more extensive on the Southern as compared to the Northern flanks, and more on the Western than the Eastern side of Kibo. The former contrast is plausible from the position of the mountain in the Southern hemisphere, while the latter asymmetry appears related to the marked local circulations associated with enhanced afternoon cloudiness.

The current ice extent in the Ruwenzori (Table 3:2, Map 4.2:1*) is estimated at 39×10^5 m^2. A total of 43 separate ice bodies are recognized at the six highest mountain massifs. Azimuth asymmetries of glaciation are not prominent, although a tendency is indicated for lower-reaching glaciers to the East. In the hyper-humid Ruwenzori, local circulations and diurnal variations of cloudiness are of subordinate importance. More vigorous glaciation to the East may reflect both moisture source and topographic factors.

At Mount Kenya (Tables 3:2 and 4.1:1, Map 4.3:1) the total ice cover is of the order of 7×10^5 m^2. Of the formerly 18 ice entities 11 are left at present. Glaciers are largest to the South and West of the peaks apparently in response to the position in the Southern hemisphere, the increased afternoon cloudiness associated with the diurnal circulations, and the orientation of the major mountain crest.

A drastic and continued glacier recession since the earliest observations at the end of the 19th century is borne out for all three high mountain regions of East Africa, although there are indications for an intermediate halt or advance in the early part of the 20th century. Historical reconstruction of long-term glacier behavior is most abundant for Mount Kenya. Lewis Glacier in particular is now the best documented glacier in all of the tropics.

CHAPTER 5

LEWIS GLACIER, MOUNT KENYA

"Die ... Aufnahmen ... hatten den Zweck, einen ausgesprochen äquatorialen Gletscher ... exakt aufzunehmen. So weit möglich, sollten dann daraus Folgerungen über seinen Eishaushalt, seine Geschwindigkeit und seine ganzen Lebensbedingungen erhalten werden. Es kam hierfür am ehesten der Lewisgletscher in Betracht ... "

(Carl Troll and Karl Wien, *Lewisgletscher*, 1949)

5.1. Design of the Field Program

The geomorphic evidence and historical documentation discussed in Chapters 3 and 4, respectively, indicate marked and largely concurrent variations of East African glaciers on both the geological and recent time scales. The recent glacier variations in particular are of interest for quantitative inference of the climatic forcing. This requires detailed information on the present glacier characteristics to be obtained by field observation. In view of the similar long-term behavior documented for many glacier ensembles, the climate-glacier coupling of even an individual specimen is of general interest.

Lewis Glacier meets various criteria of choice for a detailed field experiment: this is the largest ice body on Mount Kenya; its catchment is reasonably well defined; while the relief is pronounced, crevassing is moderate; the glacier is comparatively easily accessible; and most importantly, the continuous quantitative documentation of ice extent and thickness since the end of the 19th century is unparalleled for the entire tropics.

A first exploratory visit to the glacier occurred in July 1971. A pilot project was undertaken during my affiliation with the University of Nairobi in 1973–4. A comprehensive field project in the form of successive expeditions was carried out from December 1977 to January 1983. A limited observation program related to the glacier mass budget continued through local cooperation during the intervals between these expeditions. The field exploration of the morphology, kinematics, and mass and heat budget characteristics of the glacier forms the basis for the numerical modelling of the ice dynamics and the study of climate-glacier relationships.

The major components of the various field phases are summarized in what follows.

5.1.1. ICE SURFACE TOPOGRAPHY

Various surveying tasks took advantage of the network of control points established in the vicinity of Lewis Glacier by the IGY Mount Kenya expedition (Charnley, 1959), as listed in Table 4.3:5. During the 1973–4 pilot project, ground tacheometry of pegs laid out on the glacier provided an assessment of the surface topography and surficial ice flow velocity along selected profiles. An air photogrammetric survey flown by the Kenya Air Force on 20 February 1974 in support of the project forms the basis for a map of Lewis Glacier at scale 1 : 2500 (Caukwell and Hastenrath, 1977), which can claim greater topographic detail than any previous chart. In support of the 1977/8 expedition, the Kenya Air Force flew a survey on 13 February 1978 which resulted in a further 1:2500 map of the glacier (Hastenrath and Caukwell, 1979). Conditions in terms of flight level, aircraft, photography, and stereoplotter duplicated the 1974 mapping. As part of the 1981/2 expedition, air-photogrammetric surveys were flown by the Kenya Air Force on 11 February and by Air Survey and Developmnt GmbH on 10 March 1982. Maps at scale 1:2500 were produced from both flights but only the map of the latter survey has been published (Caukwell and Hastenrath, 1983). The maps (see Maps 4.3:9*, 4.3:10*, 4.3:11*) document in detail the variations for the two 4-year intervals 1974–8, and 1978–82.

5.1.2. BEDROCK TOPOGRAPHY AND ICE THICKNESS

Knowledge of bedrock configuration and ice thickness in conjunction with surface flow velocity permits the calculation of ice discharge through vertical cross-sections, and is also a desirable input to numerical modelling. Various independent approaches to the estimation of ice thickness were used.

Photo 5.1.2:1. Measurements with Worden gravity meter on Lewis Glacier (Photo S.H. 2 Feb. 1978).

During the 1973–4 pilot project, seismological work was undertaken along representative profiles across Lewis Glacier using a Huntec-FS3 portable seismograph, with the purpose of determining the ice thickness. To the same end geomagnetic measurements (Skinner *et al.*, 1974) were made, but this attempt was abandoned as unpromising. In the course of 1975–6, Nayan V. Bhatt continued the geomagnetic exploration of the glacier.

Seismological prospecting was resumed during our 1977/8, 1978/9, and 1979/80 expeditions. During the latter field season, an Electrotech ER-75A-12 seismograph became available. The seismological work was complemented by the gravimetric technique, using a Worden gravity meter (Photo 5.1.2:1). A further independent technique of estimating ice thickness is available from the numerical modelling of the ice dynamics. The spatial pattern of bedrock topography and ice thickness resulting from the combination of these three independent methods has been discussed in detail elsewhere (Bhatt *et al.*, 1980). Results are summarized in Section 5.2.

5.1.3. ICE SURFACE VELOCITY

During the 1973–4 pilot project lines of pegs laid out across the glacier were repeatedly surveyed by tacheometry, thus yielding the surficial ice flow velocity in the middle portion of the glacier.

Photo 5.1.3:1. Thermal drill used for placement of net balance stakes (Photo S.H. 4 March 1978).

During the 1977/8 expedition an array of 31 net balance stations (see Map 4.3:10*) was installed on the glacier using a thermal drill (Photo 5.1.3:1). At stations 1 through 15 in the upper glacier, bamboo poles of 3 m length were inserted. All other stations are made up of 2 or more wooden poles of 2 m length linked together by bands of webbing; stations 21 through 26 in the middle glacier consisting of 2, stations 31 through 35 in the middle to lower glacier of 3, and stations 44 through 45 in the lower glacier of 4 poles of 2 m each. These stations were surveyed repeatedly using an optical theodolite (Lietz T-60D) and electronic distance measuring (EDM) equipment (Beetle 500 of Precision International, USA) during the 1977/8, and again during the 1978/9, 1979/81, and 1981/2 expeditions. The novel EDM equipment allows high precision measurements of displacement rates.

A thorough new installation was accomplished in December 1981. The position of stations in January 1982 is listed in Table 5.3.1:1 and plotted in Map 4.3:11*. Stations 1, 3, 12, 4A, 6, 7, 10, 11, 81, 71, consist of a bamboo pole, stations 2 and 4B of a wooden stake, stations 51 and 13 each of two 2 m stakes linked together. Stations 22 and 25 are made up of four 2 m segments, thus totalling each a depth of 8 m. Five such segments of 2 m were inserted at stations 31, 32, 33, 42, 41, 43, which thus reach to a depth of about 10 m. The position was accurately surveyed for all stations except 1, 2, 51, and 4A.

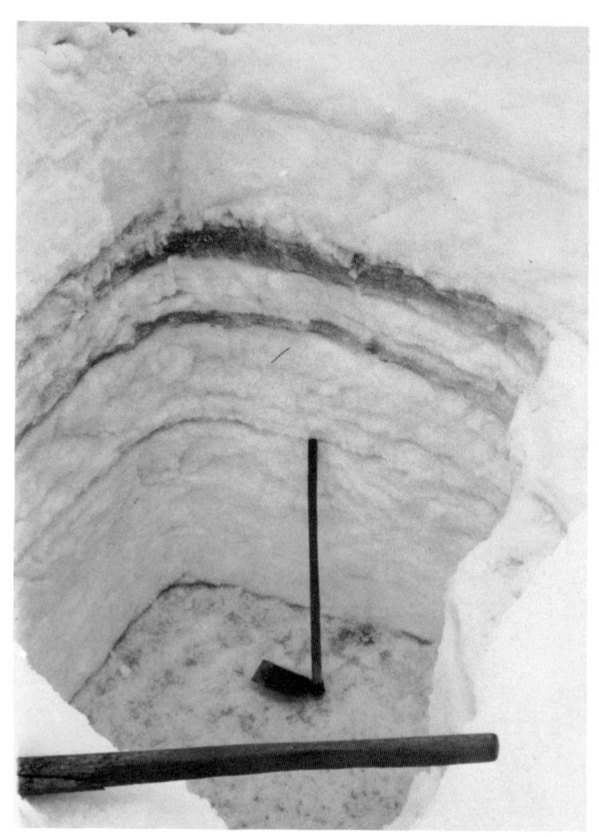

Photo 5.1.4:1. Profile in pit as site II (Photo S.H. 14 Feb. 1978).

5.1.4. NET BALANCE

The stakes described in Section 5.1.3 in connection with velocity determinations also serve the purpose of monitoring the spatial pattern of net balance. The distance from top of stake to glacier surface was measured repeatedly during our expeditions, and readings at approximately monthly intervals in the course of the intermediate years were made by local collaborators.

During various expeditions, pits were dug at strategic locations for purposes of stratigraphy (Photo 5.1.4:1). Density profiles were sampled along with a stratigraphic description. This information in conjunction with the aforementioned stake readings permits to construct the spatial pattern of net balance.

5.1.5. PRECIPITATION MEASUREMENTS

During the 1977/8 expedition accumulating precipitation gauges were installed at 'Met Station Lodge' (Roadhead) at 3050 m, at Teleki Ranger Camp at 4400 m, and at 'Radio Ridge' overlooking the glacier, at about 4800 m (Photo 5.1.5:1). The former two locations were also equipped with conventional daily gauges, the station code numbers within the National precipitation network being 9037217 and 903218, respectively. This equipment was provided by the Kenya Meteorological Department. The daily observations are reported regularly to the Kenya Meteorological Department and the Kenya Water Department. Since the beginning of 1978 precipitation measurements at the two lower locations are performed on both a daily and monthly basis, and at Radio Ridge monthly. From precipitation gaugings and net balance measurements as described in Section 5.1.4 above, the accumulation or ablation rate can be obtained as a residual. Precipitation monitoring thus provides important information for the study of the glacier mass budget.

Photo 5.1.5:1. Accumulating precipitation gauge at "Radio Ridge" (Photo S.H. 2 Feb. 1978).

Photo 5.1.6:1. Defunct V-notch weir installation of Oct. 1973 at exit of Lewis Tarn (Photo S.H. 2 Feb. 1978).

Photo 5.1.6:2. Water discharge measurements at the exit of Lewis Tarn (Photo S.H. 2 Feb. 1978).

5.1.6. WATER RUNOFF

The bedrock below Lewis Glacier consists of solid volcanic rock, such as phonolite, kenyite, and syenite. Although fractured, it is considered impermeable for practical purposes. Accordingly, the total meltwater runoff from Lewis Glacier can be monitored at the exit of the tarn below.

During the 1973–4 pilot project a V-notch weir was installed at the exit of Lewis Tarn (Photo 5.1.6:1). Discharge is computed from water level by conventional formula. Water levels were observed through several daily cycles and then at numerous, but irregular visits.

A new monitoring effort was begun during our 1977/8 expedition. Flow velocities were measured over two vertical cross-sections below the exit of Lewis Tarn (Photo 5.1.6:2). Integration over the entire cross-section yields water discharge as a function of water level. During the expedition, discharge and water level were measured through several complete daily cycles, including a wide range of runoff conditions. A water level-discharge relationship could thus be derived. Furthermore, the timing of extrema, and the relation between daily discharge totals and the discharge at the times of the extrema was established empirically. An estimate of the daily discharge total is thus possible from readings of the water level at two specific times of day. A useful safeguard is provided by readings being taken at two different cross-sections.

In the periods between our expeditions and through early 1980, a local collaborator obtained reliable runoff data on his regular visits at the beginning of each month.

5.1.7. SURFACE HEAT BUDGET

Measurements related to the surface heat budget were performed during the 1977/8 field season in particular. Experiments were designed so as to allow formulation of heat and mass budget relations in terms of simple parameters. The daily cycle of heat budget forcing and mass budget response of the glacier received particular attention.

Global radiation was monitored on a rock outcrop outside the glacier with a counting RIMCO instrument (Photo 5.1.7:1) that was read at intervals of a few hours. Short and longwave radiation fluxes were measured at similar intervals at a station established on the glacier near the equilibrium line. At this location temperature and humidity of the air were also measured with an Assmann ventilated psychrometer, and counting anemometers were read at intervals of a few hours, thus allowing crude estimates of vertical fluxes of moisture and sensible heat by the bulk-aerodynamic method. Independently, evaporation and melting during intervals of a few hours were estimated by simple lysimeters placed at various elevations on the glacier (Photo 5.1.7:2). Albedo was sampled with a simple instrument along representative traverses across the glacier at intervals of dently, sublimation and melting during intervals of a few hours were estimated by simple lysimeters (Photo 5.1.7:2). Albedo was sampled with a simple instrument along representative traverses across the glacier. Direct solar radiation was also measured by spectral bands (Hastenrath and Patnaik, 1980).

This simple observation program allows the estimate of basic heat and mass budget characteristics of Lewis Glacier.

Photo 5.1.7:1. Measurement of global radiation with RIMCO counting radiometer (Photo S.H. 1 March 1978).

5.1.8. STRATIGRAPHY

As described in Section 5.1.4., density profiles along with a stratigraphic description were sampled during various expeditions, in pits dug at strategic locations. During the 1977/8 field season, two ice cores of about 11 and 13 m, respectively, were retrieved from pits at the gentle col separating the Gregory and Lewis Glaciers. Samples were melted and shipped to the US for analysis of microparticle content, $\delta\ ^{18}O$ ratio, and total β radioactivity. Experience from the Quelccaya Ice Cap, Peru (Hastenrath, 1978; Thompson *et al.*, 1979) offers the prospect of establishing a net balance chronology. The aim is an alternate approach to climate reconstruction.

Photo 5.1.7:2. Lysimeter bowls and scale (Photo S.H. 2 Feb. 1978).

5.2. Morphology

5.2.1. ENVIRONMENT AND GEOLOGY

Lewis Glacier (Maps 4.3:1, 4.3:8*, 4.3:9*, 4.3:10*, 4.3:11*; Table 4.3:5) extends to the Southeast of the highest peaks in an approximate Northnortheast-Southsouthwest orientation from about 4975 m just below Point Lenana (4985 m) to 4600 m at the lowest point of its snout in March 1982, with a length of about 1 km. The rock threshold extending from here to Lewis Tarn was gradually vacated by the ice since the late 1950s. Lewis Tarn, which formed since 1947, is at about 4575 m. The accumulation area upward of about 4770 m is separated into a larger Western part, and a smaller Eastern branch ending in Curling Pond. A large crevasse system extends from the area of Point Thomson to the rock ridge at L2 (Maps 4.3:9*, 4.3:10*, 4.3:11*). In the main ablation area below about 4760 m, the ice stream is embedded in a wide and steep canyon between the rock buttress West of Top Hut and the faces of Points Melhuish and John.

According to Baker (1967) the peak region of Mount Kenya belongs to the central plug of the volcano. This consists of a series of partial concentric cylinders and lenses about a central core of nepheline syenite; the outer ring is made of fine-grained phonolite, locally altered and penetrated by syenite; further outward follow kenyte lavas and pyroclastics. The concentric cylinders have vertical axes, and erosion has led to a complex of precipitous peaks. Proceeding from West to East, Lewis Glacier may be underlain by the approximately North-South oriented nearly vertical contacts between nepheline syenite, phonolite, kenyte, phonolite, and again kenyte. One is led to speculate on the role of the geological structure in the origin of this glacial valley. The rocky boss in the Top Hut area is made up of porphyritic phonolite similar to that forming the buttress below and to the North of Point Thomson. This is consistent with conspicuous features of surficial morphology: the large crevasse system leading down from Point Thomson towards Top Hut finds a continuation in the large rock ridge to the West of Top Hut. This is suggestive of the crevasse system being underlain by a similarly oriented rock ledge.

As indicated in Section 5.1.2, three independent methods were applied for the determination of ice thickness and bedrock topography. The following presentation, in Section 5.2, is an extended version of the report by Bhatt *et al.* (1980). Map 5.2.1:1 shows the location of depth profiles obtained.

5.2.2. SEISMIC METHOD

5.2.2.1. Instrumentation

During February–March 1974, a Huntec FS-3 Portable Facsimile Seismograph was used (Huntec Ltd., 1958; Meidav, 1969; Hobson and Jobin, 1975). Apart from a 15-kg cable, the largest component weighs 10 kg, and transport to high altitudes is thus feasible. This is a single-channel instrument with two 14-Hz geophones and coincidence circuitry (fixed gate correlator); it records positive zero crossings of signals above an arbitrarily set threshold for both manual generation of signal by hammer or dropping of weights, and operation with explosives and shot-box. In the manual mode a sledgehammer and target

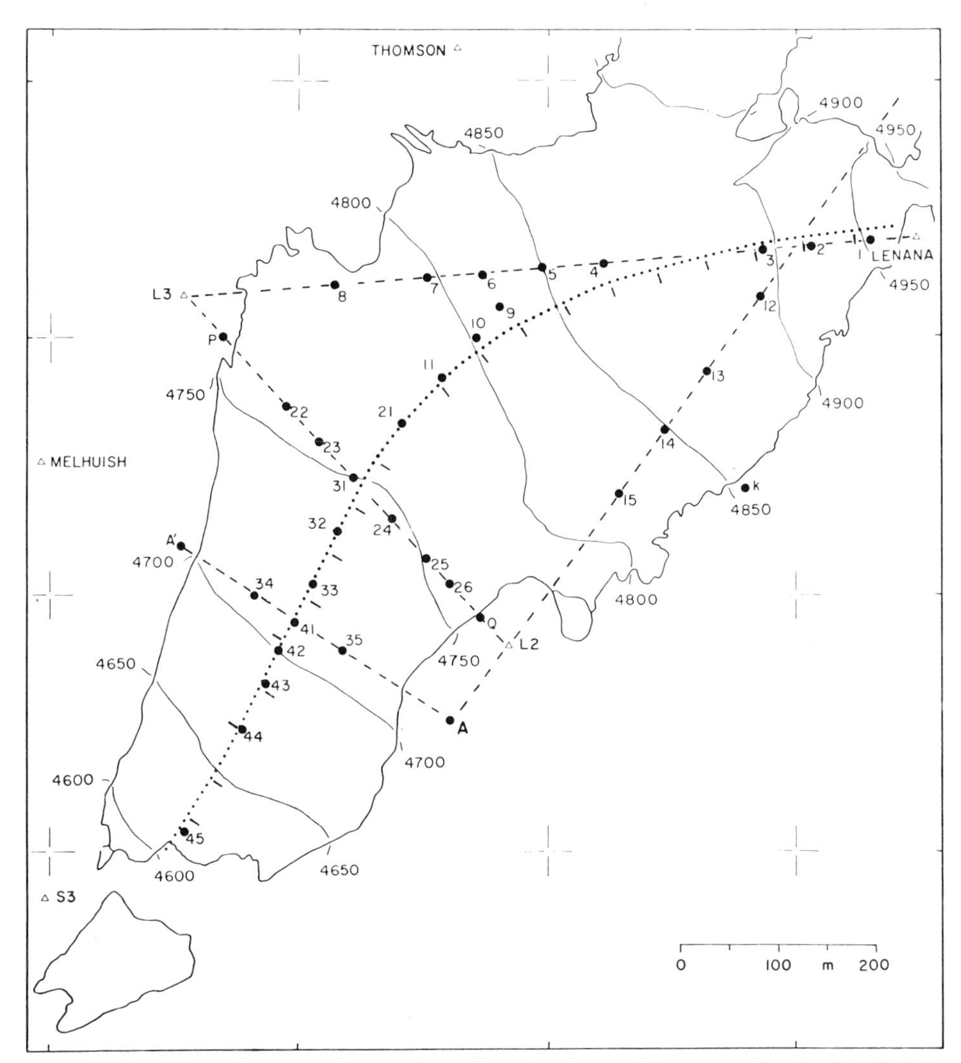

Map 5.2.1:1. Orientation map. Dots and numbers denote precision-surveyed poles installed for monitoring of surface ice flow velocity and net balance. L2, L3, A, A′, S3 are surveying points on rock outside the glacier. Thin dotted line describes modeled central flow line (Section 5.2.2.4, and profile Fig. 5.2.5:1), with tick marks at 50 m intervals indicating locations of output ice depths. Broken lines show location of profiles Figs. 5.2.5:2–5.2.5:5. Scale 1: 7500.

steel plate were used. In the explosives mode seismic electric detonators were used with charges of Cordtex wound and taped tightly around the detonator. The charges were buried in shallow hollows chipped into the ice. During this and the following season, only geophones capable of recording longitudinal waves were available.

In February 1978 and December 1978–January 1979 a later model of this system (Meidav, 1969) was used, incorporating a variable gate correlator, thus permitting velocity filtering and cancellation of random, noncoherent noise. For the latter season, shear wave geophones were available, and a vertically driven steel rod was used as the hammer target.

During December 1979–January 1980, an Electrotech ER-75A-12 interval timer was used (Electrotech, 1960s). Dimensions and weight are similar to the aforementioned equipment. 12 vertical 7-Hz geophones linked by a 15-m take-out interval cable are used, and qualitative gain control is provided by 12 transistorized amplifiers. This instrument yields 12 chart traces, shot break, and 100-Hz time breaks, all recorded on polaroid film.

Arrival times can be read to 1 ms in the Huntec and to about 2 ms in the Electrotech systems.

5.2.2.2. Theory

Reference is made to Dobrin (1960), Meidav (1969), Parasnis (1972), Mooney (1974), and Telford *et al.* (1976) for details of the theory. Of interest here are longitudinal (P) and shear (S) waves. Arrivals are plotted in a time (t) – distance (x) graph, with x denoting the distance from shotpoint to geophone. Direct waves travelling parallel to the surface are depicted as straight lines passing through the origin.

Refracted waves show up as straight line segments. The intercept of this line with the t axis, t_c, and the x coordinate at which the straight line segments of the refracted and direct waves intersect, x_c, permit calculation of the depth to the refractor. As refraction prospecting was not found to be feasible on Lewis, the pertinent theory is not considered here.

Reflected events form hyperbolas on a t–x graph, but in a t^2–x^2 plot they show up as straight lines the inverse slope of which gives the square of velocity, V^2. The depth d to the reflector can be calculated (a) from velocity V and the intercept t_0 of the hyperbola with the t-axis in a t–x graph, or (b) from velocity V and the intercept t_0 of the straight line with the t^2 axis in a t^2–x^2 plot, according to

$$d = 0.5\, V t_0 \qquad 5.2.2.2{:}(1)$$

The velocity corrected for dip of bedrock is obtained from split or reversed spread by (Meidav, 1969)

$$V = \frac{\Delta x\, 2^{1/2}}{(\Delta t^2_{\mathrm{up}} + \Delta t^2_{\mathrm{down}})^{1/2}} \qquad 5.2.2.2{:}(2)$$

where Δt^2_{up} and $\Delta t^2_{\mathrm{down}}$ are obtained from x^2–t^2 plots of the up-dip and down-dip spreads.

The normal move-out (Telford *et al.*, 1976, p. 263)

$$\Delta t_n = \frac{x^2}{2V^2 t_0}\left[1 - \left(\frac{x}{4d}\right)^2\right] \qquad 5.2.2.2{:}(3)$$

Errors in the determination of ice thickness are related to (a) instrument accuracy and taping of sensing ensemble; (b) slope of surface along spread. The combined effect of these error sources is estimated to stay with 10% of the ice depth. More serious, and not amenable to formal error theory, are possible incorrect interpretations of arrivals.

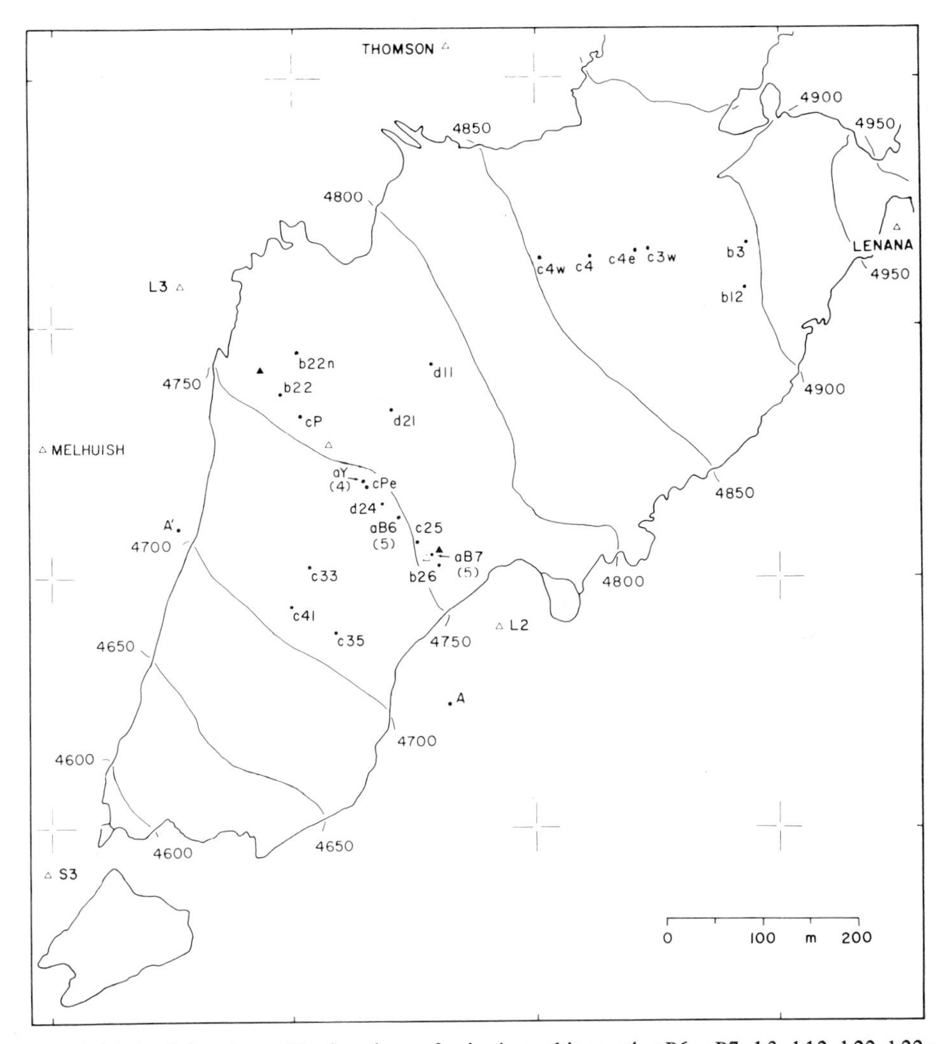

Map 5.2.2.3:1. Seismology. The locations of seismic probings, a4, aB6, aB7, b3, b12, b22, b22n, b26, c3w, c4, c4e, c4w, c25, c33, c35, c41, cP, cPe, d11, d21, d24, are indicated by dots, with a, b, c, d, denoting the 1974, 1978, 1978/79, and 1979/80 field seasons, respectively. The decrease of ice thickness in m from February 1974 to February 1978 (source: Hastenrath and Caukwell, 1980) is given in brackets for the 1974 locations. End points of explosives and hammer refraction profiles, Fig. 5.2.2.4:1, shot in February–March 1974 are shown by filled and open triangles, respectively. Scale 1:7500.

5.2.2.3. Observation Program

The February-March 1974 work consisted mostly of refraction surveys with the hammer source, yielding profiles of 70–80 m. A 225-m profile was shot using explosives along the line L2–L3. The refraction profiles and the reflection stations of interest here are shown in Map 5.2.2.3:1.

In February 1978 and December 1978–January 1979, work concentrated on reflection profiles of 100-m extent or less. Relevant reflection stations are plotted in Map 5.2.2.3:1.

During the latter season, horizontal geophones were used and hammer blows were directed horizontally on a vertically driven pipe. Direction of impacts was reversed at every impact point in order to record phase reversals of the horizontal shear wave (Mooney, 1974). After some experimentation, a geophone separation of 3.6 m, a correlator gate width of 4 m and hammer impacts at 5 m intervals were used. Continuous profiling was carried out along two lines, Map 5.2.2.3:1, with hammer impacts at 15 and 20 m from the geophone center, and the entire array being moved at steps of 10 m.

In December 1979–January 1980, the twelve-channel Electrotech ER-75A-12 was used, with 10-m geophone intervals and hammer impacts at the end geophones. Only three reverse profiles were recorded, as shown in Map 5.2.2.3:1.

Difficulties encountered in field work on Lewis Glacier are concerned with electrical contacts, geophone plants, battery output, failure of polaroid film, energy transfer from hammer impacts, and noise arising from wind, rockfall, and possibly movement in crevasses.

5.2.2.4. Evaluation

Time–distance plots of the attempted refraction work during February–March 1974 are shown in Fig. 5.2.2.4:1. First arrivals of about 3700 m s^{-1} were obtained, but there were no indications of *P* wave refraction from bedrock. Profiles on moraine and phonolite rock to the East of the glacier yielded velocities of 2000 and 2500 m s^{-1}, respectively. These values suggest a velocity inversion at the glacier-bedrock interface, which would explain the absence of arrivals interpretable as refraction from the glacier bed. *P* wave refraction was obtained, however, at station 3 in the upper glacier, presumably due to the density transition from firn to ice; the velocity in the former being about 2500 m s^{-1}. As the refraction technique was not found promising, our subsequent work concentrated on reflection.

Table 5.2.2.4:1 lists acoustic properties of the media encountered at Lewis Glacier. While derived from *P* waves, Table 5.2.2.4:1 is also of interest for work with shear waves. For all three possible interfaces between media, impedance contrasts *C* are conducive to good reflections. Reflection coefficients *R* are indicative of the fraction of incident energy that may be reflected. Impedance contrasts at ice/air and ice/water interfaces would be larger than unity, giving rise to phase reversals in the seismic signals (Telford *et al.*, 1976, p. 252). This can be envisaged for shear waves encountering a crevasse, whereas shear waves do not transmit through water. Phase reversals of the reflected *S* event are occasionally encountered in both the Electrotech analog, and the Huntec facsimile records.

An example illustration of the records obtained at the numerous reflection stations, time–distance plots at stations 11 and 21 obtained in December 1978 with Huntec-FS3, and in December 1979 with the Electrotech ER-75A-12 instruments are shown in Fig. 5.2.2.4:2. The straight-line plots depict direct arrivals with distinctly different velocities for *P* and *S* waves. Consistent with the instrumentation as described in Sections 5.2.2.1 and 5.2.2.3, Fig. 5.2.2.4:2 shows *P* arrivals for the Electrotech but not the Huntec system. Representative velocities of about 3700 and 1670 m s^{-1}, respectively, were adopted from the numerous measurements in various portions of Lewis Glacier. The signals arranged along hyperbolic lines and conforming to the normal moveout, Eq.

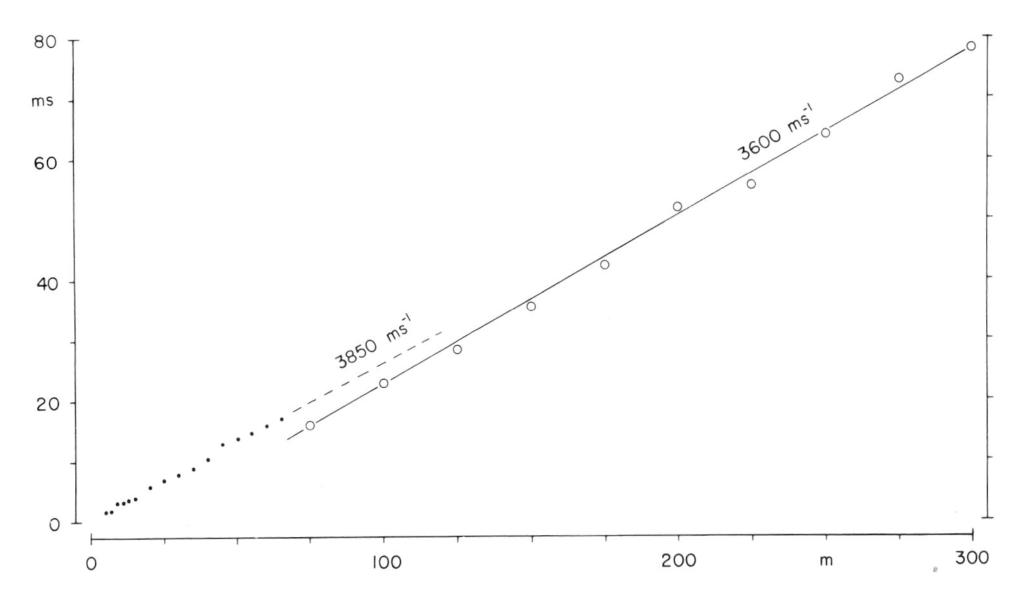

Fig. 5.2.2.4:1. Time-distance plots for easternmost hammer segment (dots, broken line) and explosives profile (circles, solid line) directed E to W along line L2–L3.

TABLE 5.2.2.4:1

Acoustic properties derived from P wave velocities. Density ρ in 10^3 kg m^{-3}; P velocity V in m s^{-1}; acoustic impedance $Z = \rho V$ in 10^6 kg m^{-2} s^{-1}; impedance contrast $C = Z_1/Z_2$, and reflection coefficient $R = (Z_2 - Z_1)/(Z_2 + Z_1)$, dimensionless

(a)	ρ	V	Z
snow	0.5	2450	1.2
ice	0.9	3740	3.4
water	1.0	1500	1.5
phonolite (rock)	2.6	2300	6.0
moraine, kenyite	2.5	2000	5.0

(b)	C	R
snow/ice	0.36	0.22
ice/phonolite	0.57	0.27
ice/moraine	0.68	0.19

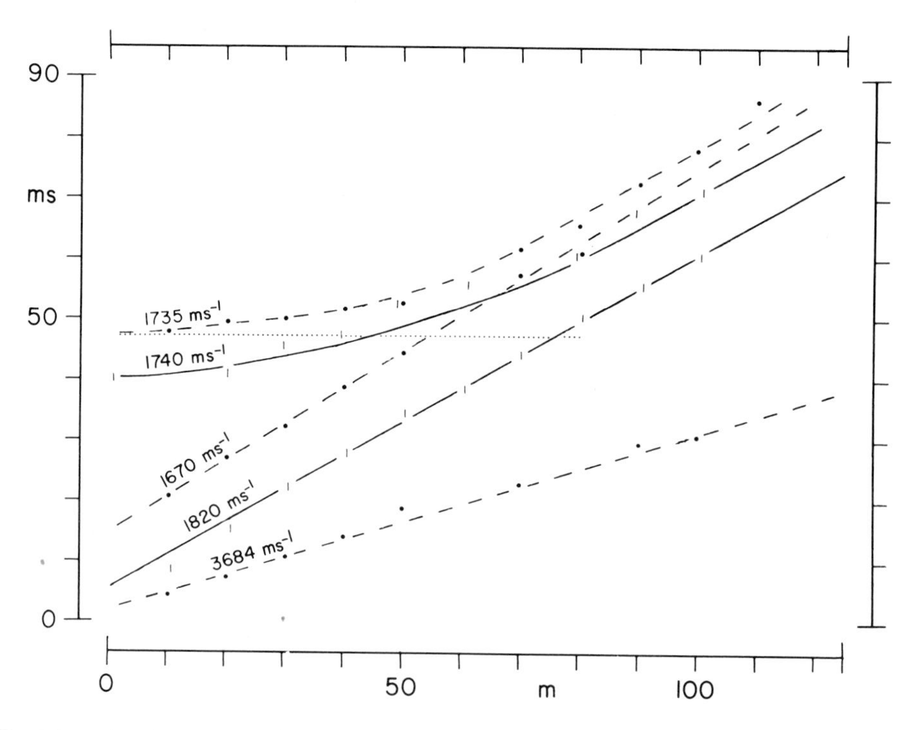

Fig. 5.2.2.4:2. Time-distance plots near stations 11 and 21 obtained in December 1978 (Huntec-FS3), solid lines and vertical dashes, and in December 1979 (Electrotech ER-75A-12), broken lines and dots. Normal moveout, Eq. 5.2.2.2:(3), referring to December 1979 hyperbola is entered as dotted line.

5.2.2.2:(3), up to a distance of about 90 m are interpreted as reflected shear waves. These are of particular interest for ice thickness determinations.

Results from the two instruments as plotted in Fig. 5.2.2.4:2 are similar, except that both the lines of direct S and reflected S arrivals are offset against each other. Among the possible causes are movement of the stake network with reference to bedrock and changes in crevassing between the two field seasons.

From t–x graphs such as exemplified by Fig. 5.2.2.4:2 and the corresponding plots in a t^2–x^2 coordinate system, depth to bedrock was computed as described in Section 5.2.2.2, using the representative shear wave velocity of 1670 m s^{-1}. The intersect times t_0 in the two plots differ little, and their arithmetic mean was used in Eq. 5.2.2.2:(1). The depths thus obtained are plotted in Map 5.2.5:1 and Figs. 5.2.5:1–5.2.5:5.

5.2.3. GRAVIMETRY

5.2.3.1. Instrumentation

A Worden gravity meter from Texas Instruments Inc. (1960) with a scale constant of 0.10051 mgal/scale division was used. Dependence of the dial constant on temperature is accounted for by repeated temperature measurements. The weight, only a few kg, was found to be an asset for work at high altitude.

5.2.3.2. Theory

For the basic theory reference is made to Parasnis (1972, pp. 38–83), Hammer (1939), and Hammer and Heck (1941). The corrected gravity difference between a station and a reference base is

$$\Delta g_{corr} = \Delta g_{obs} + 0.3086\, h - 0.04191\, h\rho + T\rho \qquad \text{[mgal]} \qquad 5.2.3.2{:}(1)$$

where the vertical distance h, in meters, is positive if the station is above the base and negative if it is below. The density of the bedrock, ρ, is in g cm^{-3}. Corrections for tidal variations and instrument drift are included in the first right-hand term, which denotes the observed difference from a reference base. The second right-hand term is the 'free-air' and the third the Bouguer correction. The last term on the right-hand side (which is always positive) is the terrain correction. The left-hand term is the Bouguer anomaly.

For purposes of ice thickness determination, observations are reduced to a reference point outside the glacier, assuming a representative rock density for ρ. The Bouguer anomaly thus obtained is then regarded as resulting from part of the height difference h in the third right-hand term of Eq. 5.2.3.2:(1) corresponding to ice rather than rock. In terms of the density difference between rock and ice, $\Delta\rho$, the ice thickness becomes

$$\Delta h = \Delta g_{corr}/(0.04191\, \Delta\rho) \qquad \text{[m]} \qquad 5.2.3.2{:}(2)$$

Based on numerous samples from various parts of Mount Kenya, a representative rock density of 2.65×10^3 kg m^{-3} is used here, and an ice density of 0.9×10^3 kg m^{-3}. Thus, according to Eq. 5.2.3.2:(2), a residual anomaly of one mgal corresponds to an ice thickness, Δh, of about 14 m.

5.2.3.3. Observation Program

During the 1977–8 expedition, a network of stakes was installed on the glacier for monitoring of ice surface velocity and net balance (Map 5.3:4). These poles were precision surveyed by optical theodolite and electronic distance measuring (EDM) equipment. Some of these locations also served as stations in the gravimetric survey (Map 5.2.3.3:1), since horizontal coordinates and elevation were accurately enough established to determine the last three right-hand terms of Eq. 5.2.3.2:(1).

Four reference stations P, Q, K, A′, were established on rock near the ice rim, and accurately surveyed (Map 5.2.3.3:1). These were linked with each other by gravimeter measurements alternating in close temporal succession between two neighboring stations. Ultimately, measurements at all stations could thus be related formally to one reference

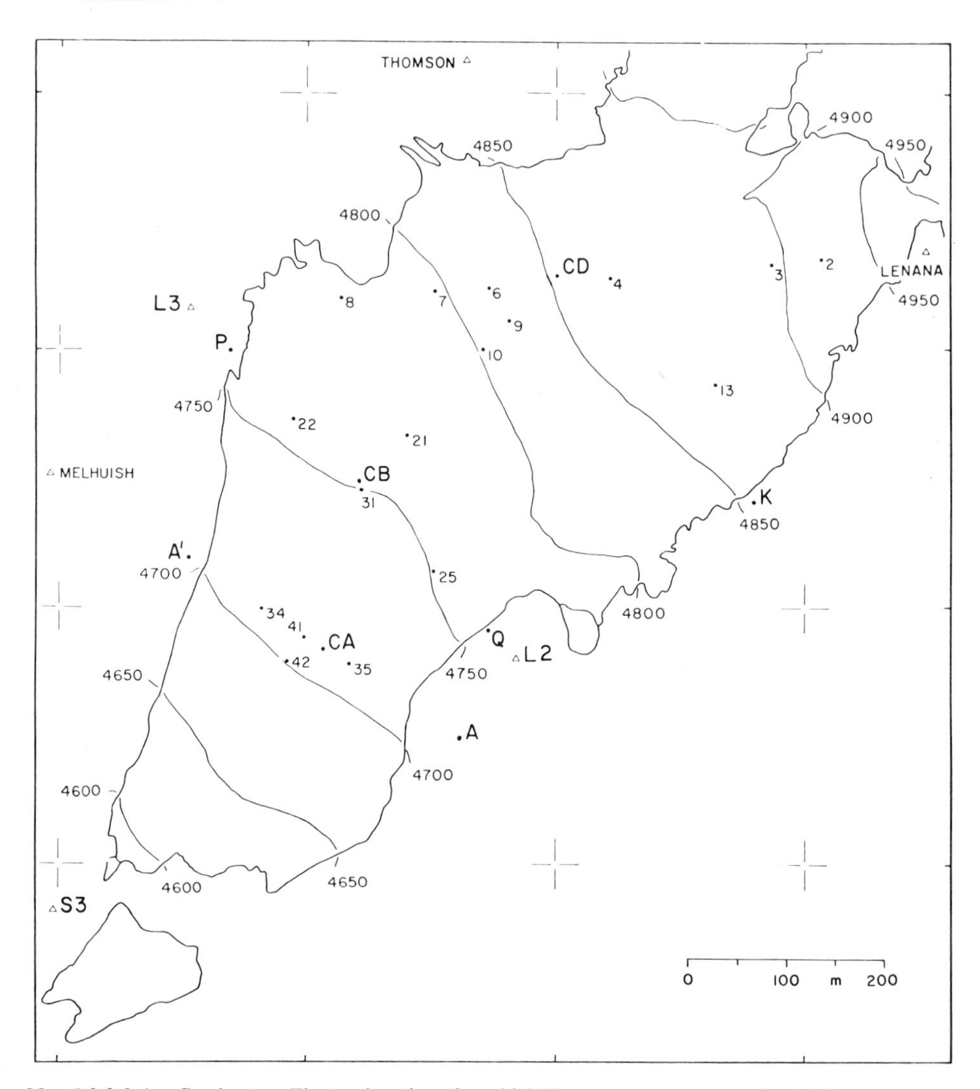

Map 5.2.3.3:1. Gravimetry. The ten locations for which Hammer zones were evaluated from maps are Lenana, K, CD, L3, CB, L2, A′, CA, A, S3. The four gravimeter reference stations on rock near the ice rim are P, Q, K, A′. The seventeen gravimeter stations on the ice are 2, 3, 4, 6, 7, 8, 9, 10, 13, 21, 22, 25, 31, 34, 35, 41, 42. Scale 1 : 7500.

station, namely P. Gravimeter readings at stations on the ice were similarly intertwined with measurements at one or more reference stations, with readings at each location being reported at intervals of about one hour or less. Thus the ice stations 2, 3, 13 were referred to reference station K; ice stations 6, 7, 9, 10, 21 to reference station P; ice stations 4, 6, 8 to reference stations P and K; ice stations 22, 25, 31 to reference stations P and Q; and ice stations 34, 35, 41, 42 to reference station A′. The rapid repetition of readings at each station allowed elimination of tidal variations and instrument drift, thus yielding a value for the first right-hand term in Eq. 5.2.3.2:(1).

5.2.3.4. *Evaluation*

An estimate of the first right-hand term in Eq. 5.2.3.2:(1) is obtained as described in Section 5.2.3.3. The second and third right-hand terms can be evaluated together, inasmuch as a constant value is assumed for rock density ρ (Section 5.2.3.2); the elevation of each station being available from the topographic survey.

The most crucial term in Eq. 5.2.3.2:(1) is the terrain correction $T\rho$. This was evaluated by the procedure advanced by Hammer (1939). The excellent 1:5000 map of the peak area (Forschungsunternehmen Nepal-Himalaya, 1967) covers the immediate vicinity of the glacier, out to between 500 and 1500 m, where local relief is most pronounced. A 1:25 000 topographic map (Survey of Kenya, 1971) is available to a radius of about 8 km, and 1:50 000 sheets (Survey of Kenya, 1955–72) to 40 km and beyond. A 1:1000 000 map of Kenya (Shell, 1968) offers adequate topographic information for more distant areas. These maps served as a basis for estimation of terrain corrections contributed by Hammer zones B to Q, which cover an area of radius 100 km. However, only zones out to J, with an outer radius of 6550 m, were found relevant. Map evaluations were performed and terrain corrections computed for ten locations on and near the glacier, including the rock reference stations K and A′ (Map 5.2.3.3:1).

From this information, terrain corrections were derived for the other two rock reference stations and the 17 gravimeter stations on the ice (Map 5.2.3.3:1). To this end, linear interpolations were performed along the straight line joining the gravimeter station and the two closest of the ten terrain correction base locations. In the few cases where these three points deviate from an exact straight line, the perpendicular projection of the gravimeter station onto the straight line between the nearest pair of terrain correction base locations was used. Thus the terrain correction base locations LE and CD served for the gravimeter stations 2, 3, 4; base locations CD and L3 for stations 6, 7, 8; base locations L3 and CB for stations 22, P; base locations CB and L2 for stations 25, 31, Q; base locations CD and CB for stations 9, 10, 21; base locations A′ and CA for stations 34, 41, 42; base locations CA and A for station 35; base locations CD and K for station 13.

With each of the four right-hand terms thus determined, the left-hand term of Eq. 5.2.3.2:(1) was obtained for all gravimeter stations on both ice and rock, with regard to reference station P. The spatial pattern of rock-related gravity anomalies is reflected in the resulting Δg_{corr} at the rock reference stations, while both this effect and the presence of an ice layer contribute to Δg_{corr} at the ice gravimeter stations. For the ice gravimeter stations, the former component was eliminated by linear interpolation between a pair of rock reference stations. Thus the ice gravimeter stations 6, 7, 8, 9, 10, were reduced with respect to reference stations P and K; the ice gravimeter stations 21, 22, 25, 31, with regard to the reference stations P and Q; the ice gravimeter stations 34, 35, 41, 42, with respect to reference station A′, according to the cross-glacier variation of Δg_{corr} indicated by the pair of reference stations P and Q. Such a reduction procedure was not found possible for the gravimeter stations 2, 3, 4, 13, since they are located rather far from appropriate rock reference stations. However, inference was attempted from the available observations. To that end, a best-guess depth of 17 m at ice gravimeter station 2 was used, as this is located close to the ice rim. Stations 3, 4, and 13 were then reduced with regard to stations P and 2. The ice thickness value resulting for station 3 appears unrealistic, while good spatial consistency is indicated by the other two stations. The ice thickness estimates thus obtained for 16 of the 17 gravimeter station locations is plotted in Map 5.2.5:1 and Figs. 5.2.5:1–5.2.5:5.

Errors in the estimation of ice thickness by the gravimetric method shall be considered with reference to Eq. 5.2.3.2:(1). An overall instrumental tolerance of 0.2 scale divisions would correspond to an error of 0.02 mgal in the first right-hand term. An accuracy of 0.1 m in the surface topography leads to uncertainties of 0.03 and 0.01 mgal in the second and third right-hand terms. The most important term, namely the terrain correction $T\rho$, however, is fraught with the largest errors by far. Differences in terrain correction between ice and reference stations are of the order of 2–4 mgal. An uncertainty in the assumed rock density ρ of 0.05×10^3 kg m^{-3} corresponds to 0.03 mgal. An uncertainty in the mean elevation of individual segments of Hammer zones of the order of 10 m would contribute an error of about 0.5 mgal. Accordingly the error in the computation of the terrain correction is estimated to be of the order of 0.6 mgal. The overall error of about 0.7 mgal corresponds to an uncertainty of ice thickness estimates by the gravimetric method Eq. 5.2.3.2:(2) of the order of 10 m.

5.2.4. ICE DYNAMICS

5.2.4.1. Theory

The Lewis Glacier bedrock topography was also estimated by numerical modeling and analytical calculation. The former method involves the short time step, dynamic model of Allison and Kruss (1977) which has been extended to include the possibility of computing the bedrock elevation along a central line. In this model the vertically averaged deformational velocity V_i is calculated using a straightforward power flow law

$$V_i = k\,\tau_b^n\,Z \qquad 5.2.4.1{:}(1)$$

where τ_b is the centerline basal shear stress and Z is the ice depth. Values ranging from 1 to 4 have been reported for the empirically determined constant n, with values in the range 3–4 for stresses greater than 1 bar, and close to 1 for stresses up to about 0.5 bar. Basal stresses at Lewis are of the order of but, if anything, somewhat less than 1 bar, and hence an intermediate n of 2 is used (Paterson, 1969, p. 91; Budd, 1969, pp. 18–20). This is as found in Budd (1975) and is close to the value employed by Budd and Jenssen (1975). However, as the value of n is not strictly defined for a valley glacier, the effect on computed ice depth of varying n from 2 to 3 is considered below. The constant k is set to 0.16 bar^{-2} a^{-1} for $n = 1$ and to 0.14 bar^{-3} a^{-1} for $n = 3$, which are representative values for temperate ice (Budd and Jenssen, 1975).

Budd and Jenssen (1975) suggest that for ordinary glaciers the large scale average velocity distribution may be computed assuming deformation only with no basal slip. The surface deformational velocity V_s is computed employing

$$V_s = \frac{n+2}{n+1}\,V_i \qquad 5.2.4.1{:}(2)$$

which results from approximate integration, for a power flow law, of the expression

$$V_i = \frac{1}{Z} \int_0^Z \mathrm{v}\, dz \qquad 5.2.4.1{:}(3)$$

where v is the deformational velocity at depth z.

The basal stress formulation established by Budd (1970) for other than small scale features is

$$\tau_b = \tau_c - 2 \frac{\partial}{\partial x} (Z \overline{\sigma'_x}) \qquad 5.2.4.1{:}(4)$$

where τ_c is the centerline downslope stress, $\overline{\sigma'_x}$ the vertically averaged longitudinal stress deviator and the x-axis is parallel to the modeled line at any point. However, the second right-hand term in Eq. 5.2.4.1:(4) was found to be unimportant for this modeling, and hence the approximation $\tau_b \approx \tau_c$ was made. The frictional effect of the valley walls is embodied in the stress shape factor s (Nye, 1965) of the equation

$$\tau_c = s \rho g Z \sin \alpha \qquad 5.2.4.1{:}(5)$$

where ρ is the ice density (0.9×10^3 kg m^{-3}), g the gravitational acceleration (9.806 m s^{-2}), and the ice surface slope $\tan \alpha$ is a function of the surface elevation E, i.e.

$$\tan \alpha = - \frac{\partial E}{\partial x} \qquad 5.2.4.1{:}(6)$$

The surface deformational velocity V_s is calculated by an iteration cycle using the above equations of 5.2.4.1 in the sequence (6), (5), (4), (1) and (2), with the second right-hand term in Eq. (4) set to zero.

Ice depth variation over a time step, which is 0.1 year in this case, is established using the continuity formulation

$$\frac{\partial Z}{\partial t} = A - \frac{1}{W} \frac{\partial}{\partial x} \left(C_v V_s \cdot \frac{m}{m+1} W Z \right) \qquad 5.2.4.1{:}(7)$$

where A is the net balance input as a function of elevation and W is the surface width. The term $C_v V_s$ is the mean cross-section velocity $\overline{V}$; the cross-section velocity ratio C_v relates $\overline{V}$ to the centerline surface velocity V_s (Nye, 1965). The valley power m defines a valley shape of form $W \alpha Z^{1/m}$ such that $(m/m + 1) WZ$ is the cross-section area.

The ice depth and surface velocity are determined at grid points spaced evenly, 50 m apart for the Lewis, down the modeled central line (Map 5.2.1:1). In the Allison and Kruss (1977) study the ice depth was added to an input bedrock profile to produce surface elevation values. Here the bedrock topography is itself computed by subtraction of the modeled ice depth from input observed surface elevation. Ice depths corresponding to the known glacier retreat rate and estimated magnitudes of valley power m are computed for a range of stress shape factor s and cross-section velocity ratio C_v values, with the depths accepted being those giving the best fit between modeled and measured surface velocities.

The boundaries of the upper glacier, above 4800 m, are near to vertical as they are predominantly flowline or ice divide. Hence a rectangular cross-section shape, i.e. $m \to \infty$, was used. At lower elevations study of the exposed valley walls suggests that m be set to 2. The best velocity correspondence was given by s and C_v values of about 0.9 and 0.7 in the upper and 0.8 and 0.7 in the lower glacier, respectively. The ice thickness and bedrock elevation obtained at each grid point along the modeled line are included in Map 5.2.5:1 and Fig. 5.2.5:1.

The analytical approach is based on the same theoretical foundation as the numerical model. Combining Eqs. 5.2.4.1:(1), (2), and (5), with the simplifying assumption that $\tau_b \approx \tau_c$, leads to the ice depth expression

$$Z = \left[\frac{V_s\,(n+1)/(n+2)}{k(10^{-5}\ s\,\rho\,g \sin\alpha)^n} \right]^{1/(n+1)} \qquad 5.2.4.1{:}(8)$$

with constant values as stated above. Measured surface velocities V_s and slopes tan α (Eq. 10.2.4.1:(6)) plus stress shape factor s obtained during the modeling complete the required input.

Ice depths at all motion stakes (Map 5.2.5:1) were determined using measured V_s, map-derived α and the s of the nearest grid point. For stations near the centerline, analytical and numerical model depths agree closely. However, the analytic technique also provides results in other regions of the glacier, allowing construction of the further bedrock profiles included in Figs. 5.2.5:2. to 5.2.5:5.

The errors in the calculated ice depth resulting from measured/derived input variables are similar for both the modeling and analytical techniques. As can be seen from Eq. 5.2.4.1:(8), a large surface slope makes the ice depth Z comparatively insensitive to n. Thus, the technique is relatively favorable for a steep valley glacier such as Lewis.

Calculations with Eq. 5.2.4.1:(8) show that changing n from 2 to 3 or k from 0.16 to 0.14 or 1.8 a^{-1} bar $^{-n}$ alters the depth Z by a few percent. The stress shape factor s cannot be determined better than within ± 0.1. The error in the measurement of ice surface flow velocity V_s is estimated at 0.05 m a^{-1}. The available 1:2500 map with a 10-m contour spacing (Caukwell and Hastenrath, 1977) allows assessment of surface slope with a tolerance of 5%.

The effect of sliding on the glacier bed has not been included in this study. The various criteria leading to the adoption of a deformational model for Lewis are given in Kruss (1981) and are not detailed here. However, a discussion of the possible importance of basal sliding is warranted. The effect of a sliding velocity greater than zero can be appreciated from Eq. 5.2.4.1:(8) in which, all other things being equal, V_s would be reduced by an amount equal to the sliding velocity. Thus, a sliding velocity of 25% of the total surface velocity would lead to ice depths 10% below those calculated for no slip, while a 50% sliding component would reduce depths by about 20%. However, the sliding velocity at Lewis is not believed to be large.

Errors in the input such as those considered above lead to an uncertainty in ice depth of about 20%, which is equivalent to a mean over the glacier of the order of 5 m.

5.2.5. SYNTHESIS OF TECHNIQUES

Estimates obtained from the three independent methods are plotted together in Map 5.2.5:1 and Figs. 5.2.5:1–5.2.5:5. At most stations the agreement is close, and the smooth line bedrock profiles in Figs. 5.2.5:1–5.2.5:5 could be drawn comfortably within the error tolerances suggested in Sections 5.2.2–5.2.4. Exceptions are the gravimeter estimates at stations 6 and 9 in profiles Fig. 5.2.5:2 and 5.2.5:1, respectively. These were discarded as too large in the spatial context with neighboring stations. Otherwise, the three methods furnish plausible spatial patterns, in that they indicate a decrease of ice thickness from the central portions towards the rim of the glacier, and results are compatible within error tolerances.

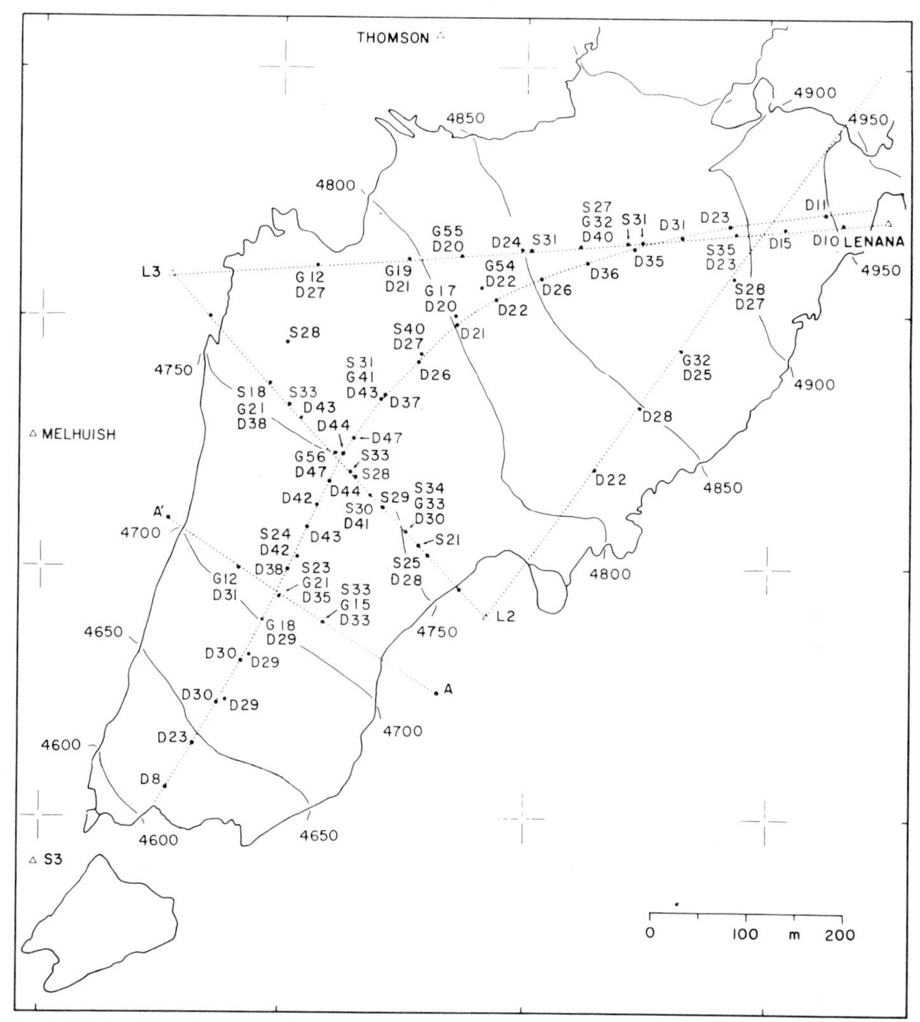

Map 5.2.5:1. Map of ice thickness estimates in m. Symbols S, G, D, refer to seismology, gravimetry, and dynamic estimation, respectively. 1974 seismic measurements are reduced to 1978 datum, see Map 5.2.2.3:1. Dotted lines show location of profiles Figs. 5.2.5:1–5.2.5:5. Scale 1:7500.

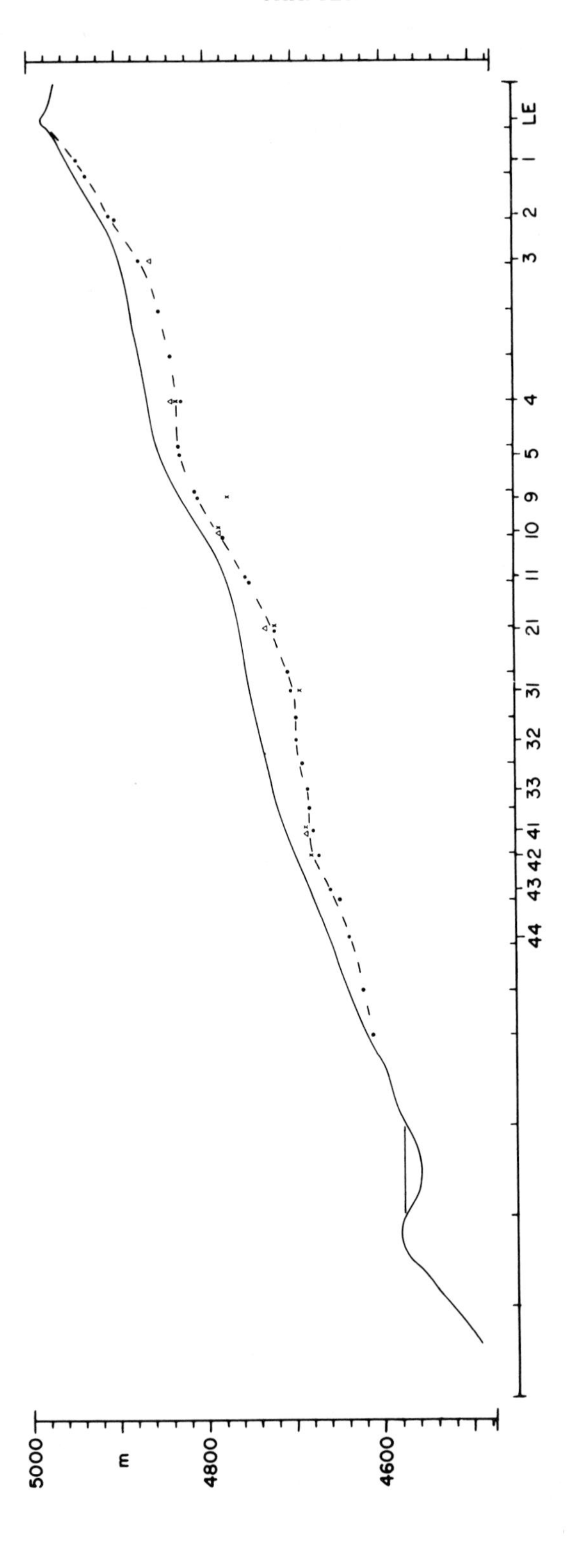

Fig. 5.2.5:1. Longitudinal profile of observed surface (solid) and inferred bedrock (broken) topography from Point Lenana along modeled central flowline (Maps 5.2.1:1 and 5.2.5:1) to Lewis Tarn. Triangles, crosses, and dots denote depth estimates by the esismic and gravimetric techniques, and dynamic estimation, respectively. Scale 1: 750.

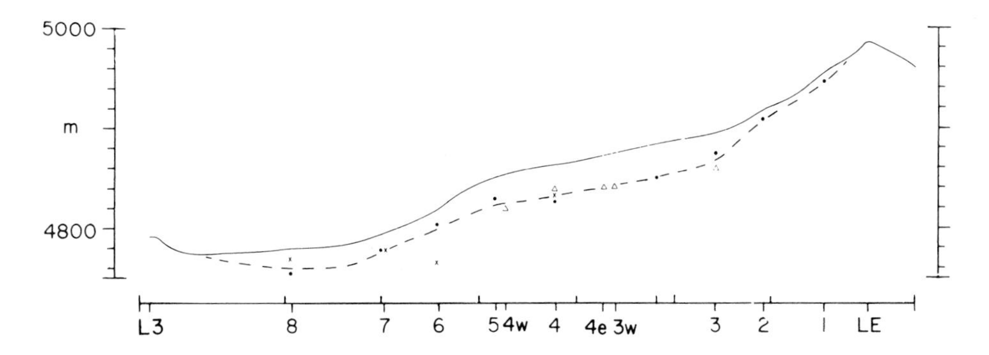

Fig. 5.2.5:2. Surface and bedrock profile along line Lenana – L3 (Maps 5.2.1:1 and 5.2.5:1). Symbols and scale as in Figs. 5.2.5:1–5.2.5:5.

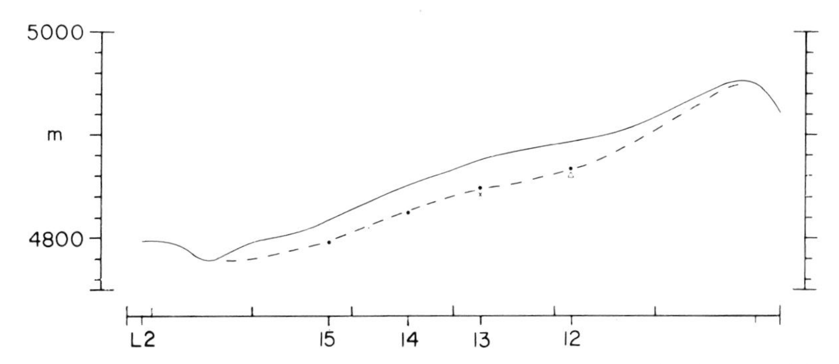

Fig. 5.2.5:3. Surface and bedrock profile along line Lenana – L2 (Maps 5.2.1:1 and 5.2.5:1). Symbols and scale as in Figs. 5.2.5:1–5.2.5:5.

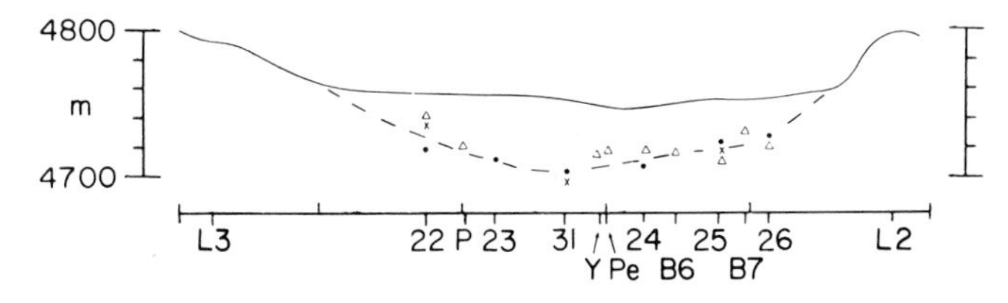

Fig. 5.2.5:4. Surface and bedrock profile along line L2–L3 (Maps 5.2.1:1 and 5.2.5:1) Symbols and scale as in Figs. 5.2.5:1–5.2.5:5.

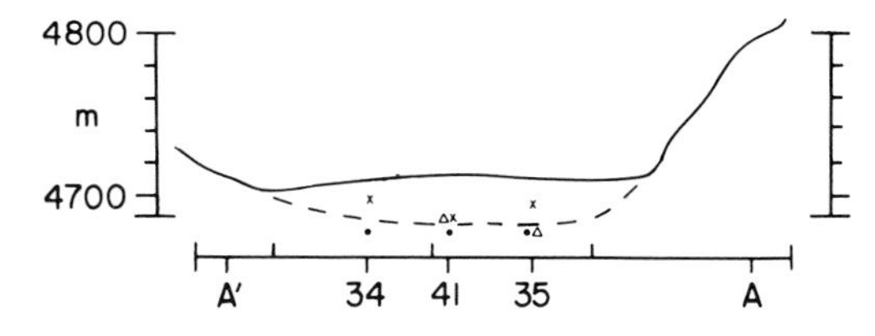

Fig. 5.2.5:5. Surface and bedrock profile along line A–A′ (Maps 5.2.1:1 and 5.2.5:1). Symbols and scale as in Figs. 5.2.5:1–5.2.5:5.

5.2.6. BOTTOM TOPOGRAPHY AND ICE VOLUME

From the profiles Figs. 5.2.5:1–5.2.5:5 values of bedrock topography were read at all the station locations and at points spaced in 50-m intervals. These data were plotted in map form and contour analyzed. Map 5.2.6:1 is the resulting map of subglacial bedrock topography.

The bedrock profiles in Figs. 5.2.5:1–5.2.5:3 and the Map 5.2.6:1 show a ledge in the area between Point Thomson and the ridge on which L2 is situated. This is consistent with the geologic structure and rock morphology outside the glacier, as described in Section 5.2.1. Fig. 5.2.5:4 suggests greatest depth in the Western part of the upper transverse profile. Small depths in the lower transverse profile, Fig. 5.2.5:5, are commensurate with the proximity to both the lateral boundaries and the snout of the glacier. Steep valley walls are indicated for the portion of the glacier below about line L2–L3.

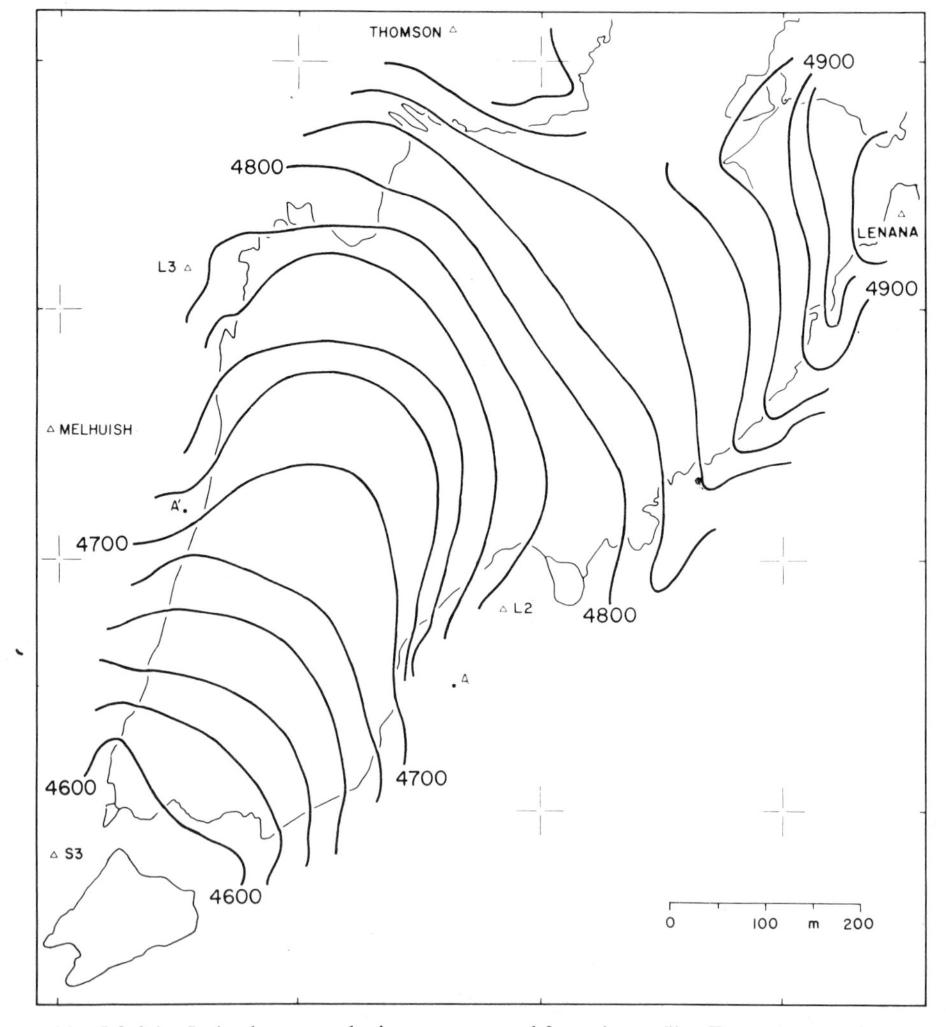

Map 5.2.6:1. Bedrock topography in m, constructed from the profiles, Figs. 5.2.5:1–5.2.5:5.

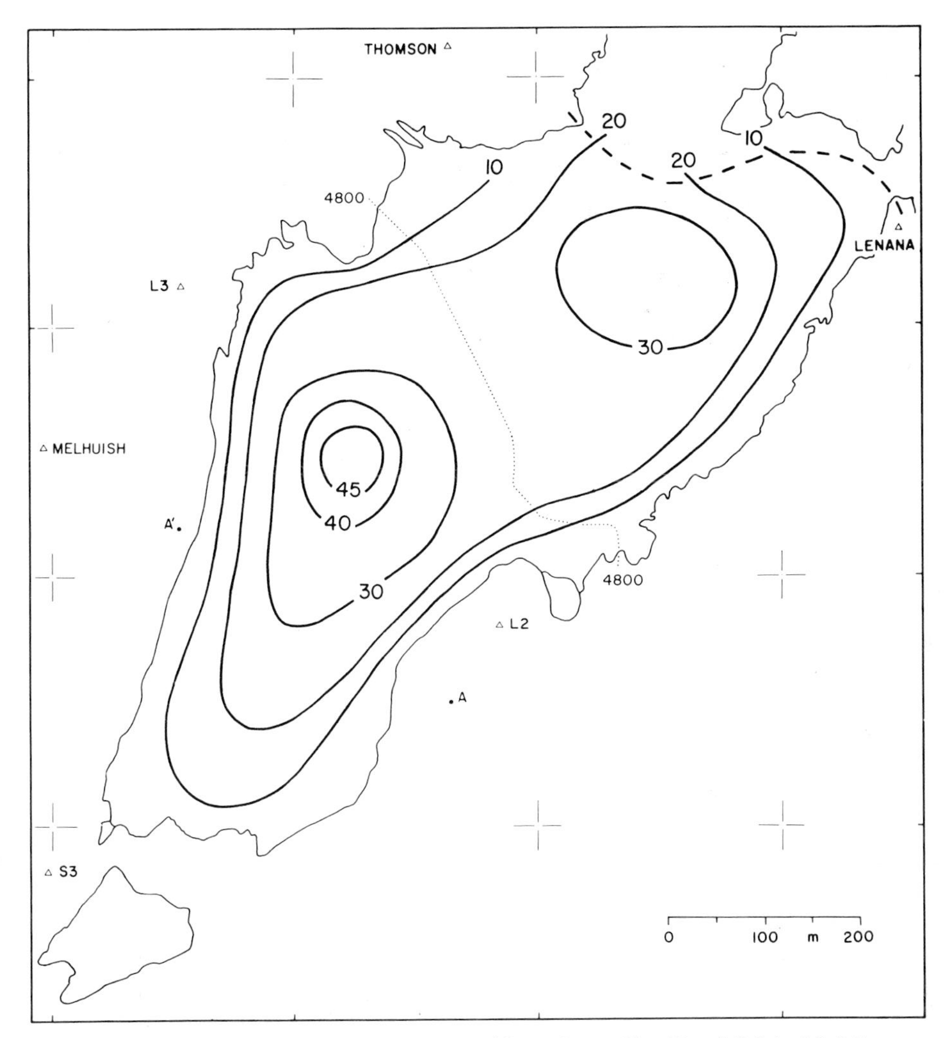

Map 5.2.6:2. Ice thickness in m, constructed from the profiles, Figs. 5.2.5:1–5.2.5:5.

The lower portion of the longitudinal profile, Fig. 5.2.5:1, beyond Lewis Tarn is drawn from Forschungsunternehmen Nepal-Himalaya (1967). H. Loeffler (personal communication 1973) measured the depth of the tarn at 16.7 m in 1960.

In a procedure analogous to the construction of the bedrock topography, ice thickness values were read from the profiles Figs. 5.2.5:1–5.2.5:5. Point values obtained were contour analyzed, with the glacier rim providing the line of zero thickness. The resulting map of ice thickness, Map 5.2.6:2, is consistent with the surface and bedrock topographies (Hastenrath and Caukwell, 1979; and Maps 4.3:10* and 5.2.6:1), within the tolerance of graphical procedures.

Ice volume was obtained from Map 5.2.6:2 by planimetering. As the 4800 m surface contour is close to the equilibrium line, this is now chosen as separation between the 'lower' and 'upper' glacier. For the datum February 1978, estimates of ice volume amount to 36, 26, and 62 $\times$ 10^5 m^3, for the lower, upper, and the entire glacier, respectively.

5.2.7. LONG-TERM VARIATIONS OF ICE EXTENT AND VOLUME

Fig. 5.2.7:1, Map 5.2.7:1, and Table 5.2.7:1 illustrate the variations in the extent of Lewis Glacier since the end of the last century. Information for the turn of the century is limited to the account of Gregory (1894), and the report and photographs of Mackinder's (1900) expedition. Maps are available for 1934 at scale 1:13 333 (Troll and Wien, 1949), for 1958 at scale 1:2500 (Charnley, 1959), for 1963 at scale 1:5000 (Forschungsunternehmen Nepal-Himalaya, 1967), for 1974 at scale 1:2500 (Caukwell and Hastenrath, 1977), for 1978 at scale 1:2500 (Hastenrath and Caukwell, 1979), and for 1982 at scale 1:2500 (Caukwell and Hastenrath, 1983).

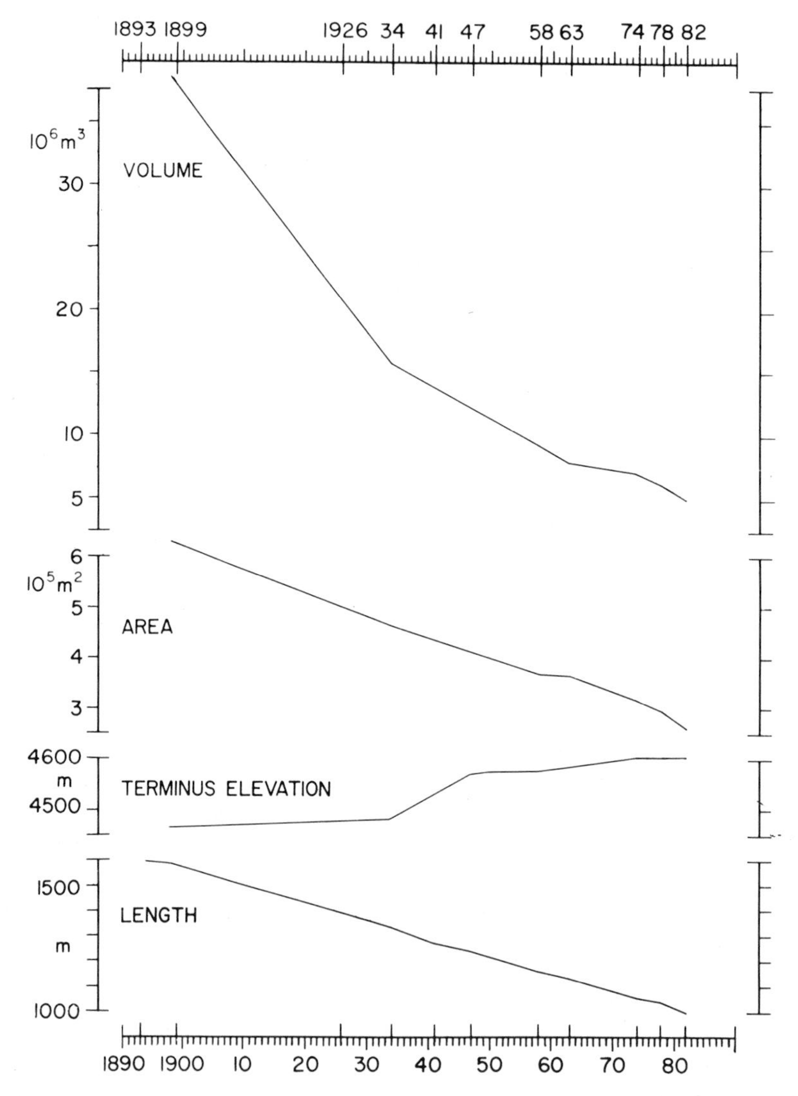

Fig. 5.2.7:1. Secular variation of Lewis Glacier 1899–1980. Volume in 10^6 m^3; area in 10^5 m^2; length of glacier and elevation of terminus in m. Sources as for Table 5.2.7:1.

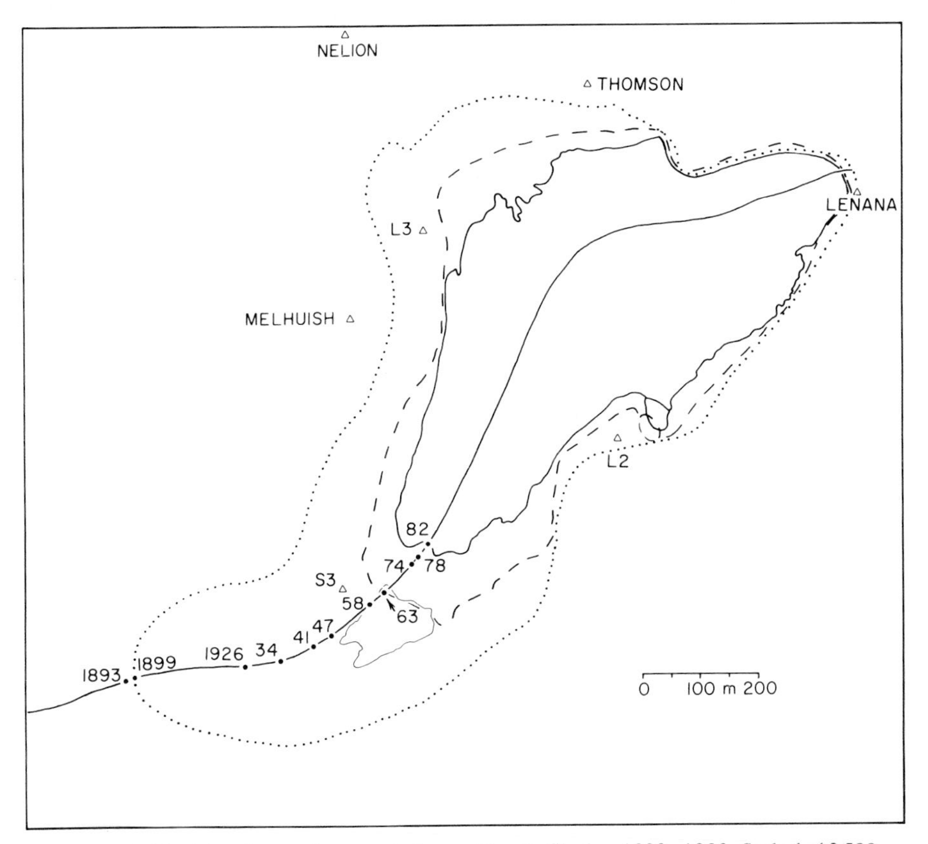

Map 5.2.7:1. Terminus and area variations of Lewis Glacier, 1899–1982. Scale 1:12 500.

The length of the glacier decreased approximately linearly from about 1590 m in 1899 to only 995 m in 1982. In this interval the terminus elevation shifted upward by about 135 m in an irregular fashion determined by the bedrock topography. The area decreased nearly linearly from about 630 in 1899 to only 261 $\times$ 10^3 m^2 in 1982, the bulk of the change belonging to the lower glacier. Most interesting in terms of the mass and heat budget are the changes in ice volume. The volume shrinkage has followed a strongly nonlinear pattern, with rate of volume decrease being largest in the early part of the century, and gradually becoming smaller for the more recent intervals. For the interval 1963–74 there are indications of an increase in thickness in limited areas of the upper glacier (Map 5.2.7:2). During the time spans 1974–8 and 1978–82, for which more reliable evaluation was possible, such an increase of thickness is not apparent (Maps 5.2.7:3 and 5.2.7:4).

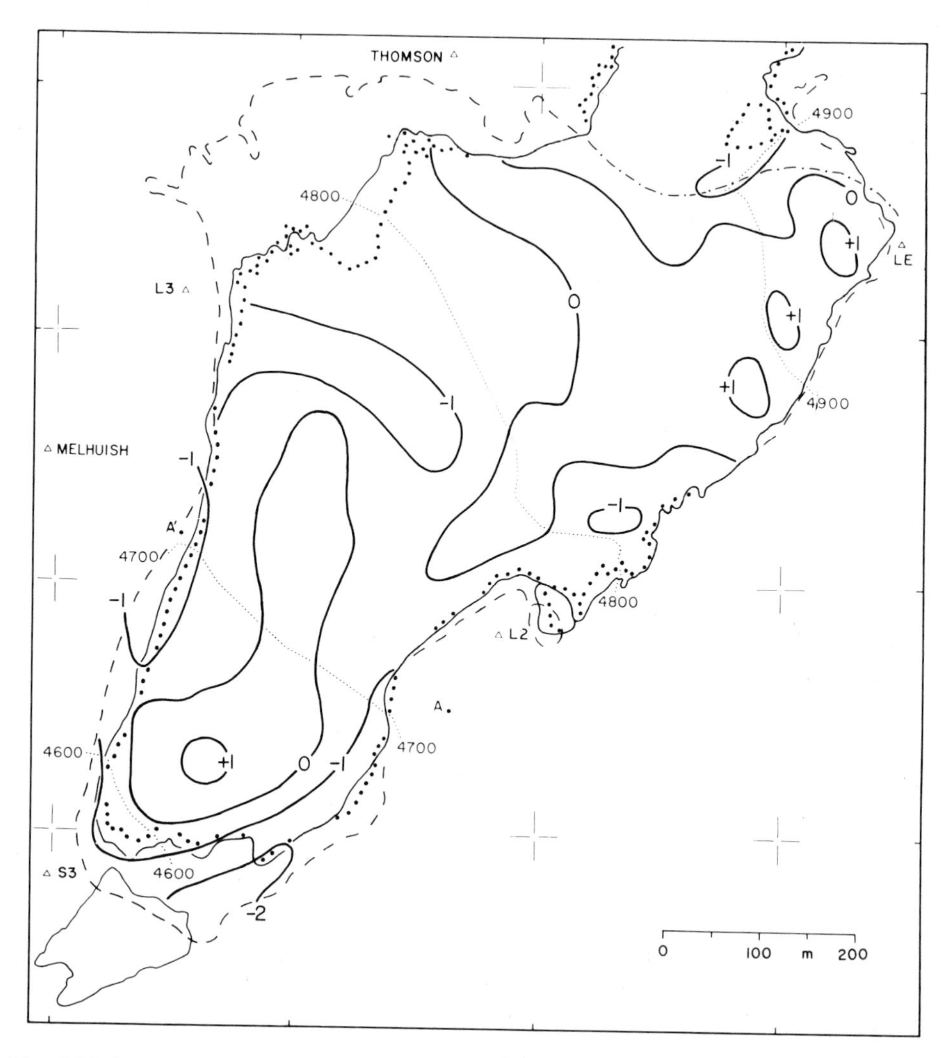

Map 5.2.7:2. Changes in ice thickness, 1974 minus 1963, in m. Ice rim in 1963, 1974, and 1978, is shown by broken, solid, and large dotted lines, respectively. 1978 height contours are entered as thin dotted lines. Scale 1 : 7500.

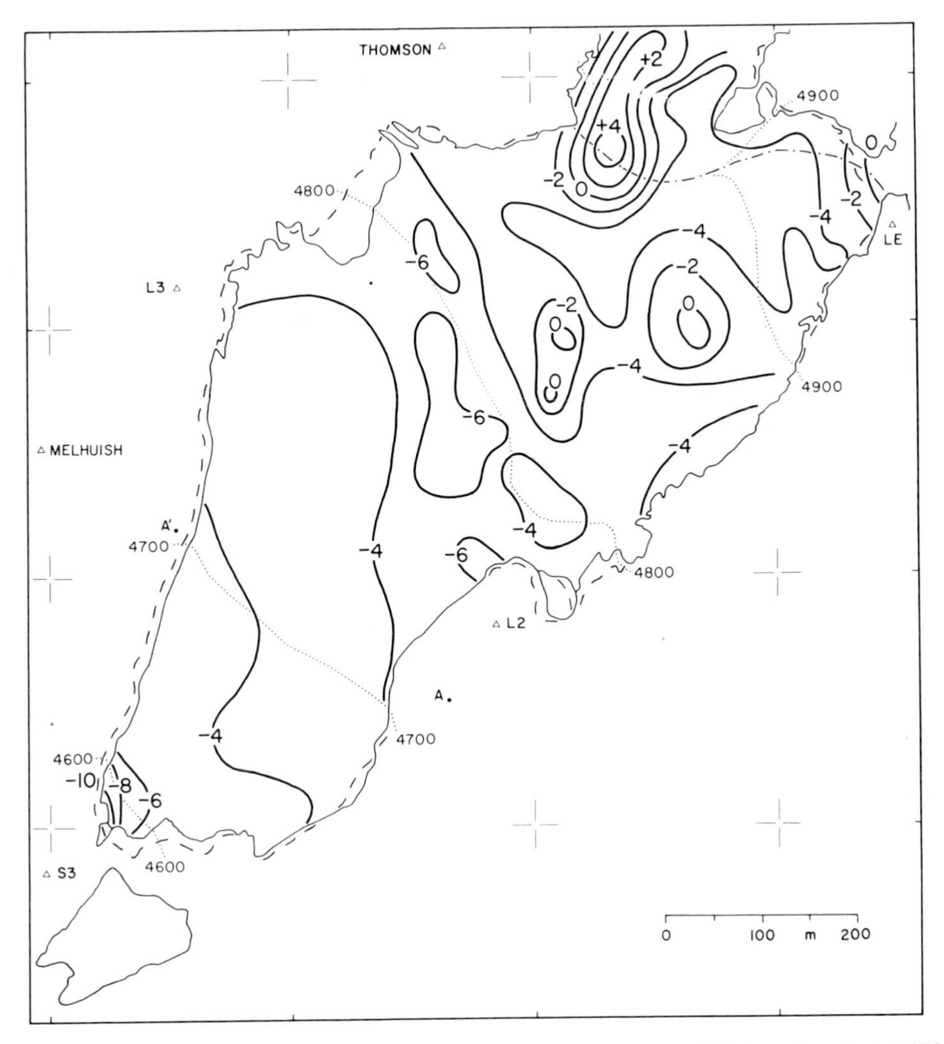

Map 5.2.7:3. Changes in ice thickness, February 1978 minus February 1974, in m. Ice rim in 1978 is shown as solid, and in 1974 as broken line. 1978 height contours are entered as thin dotted lines. Scale 1 : 7500.

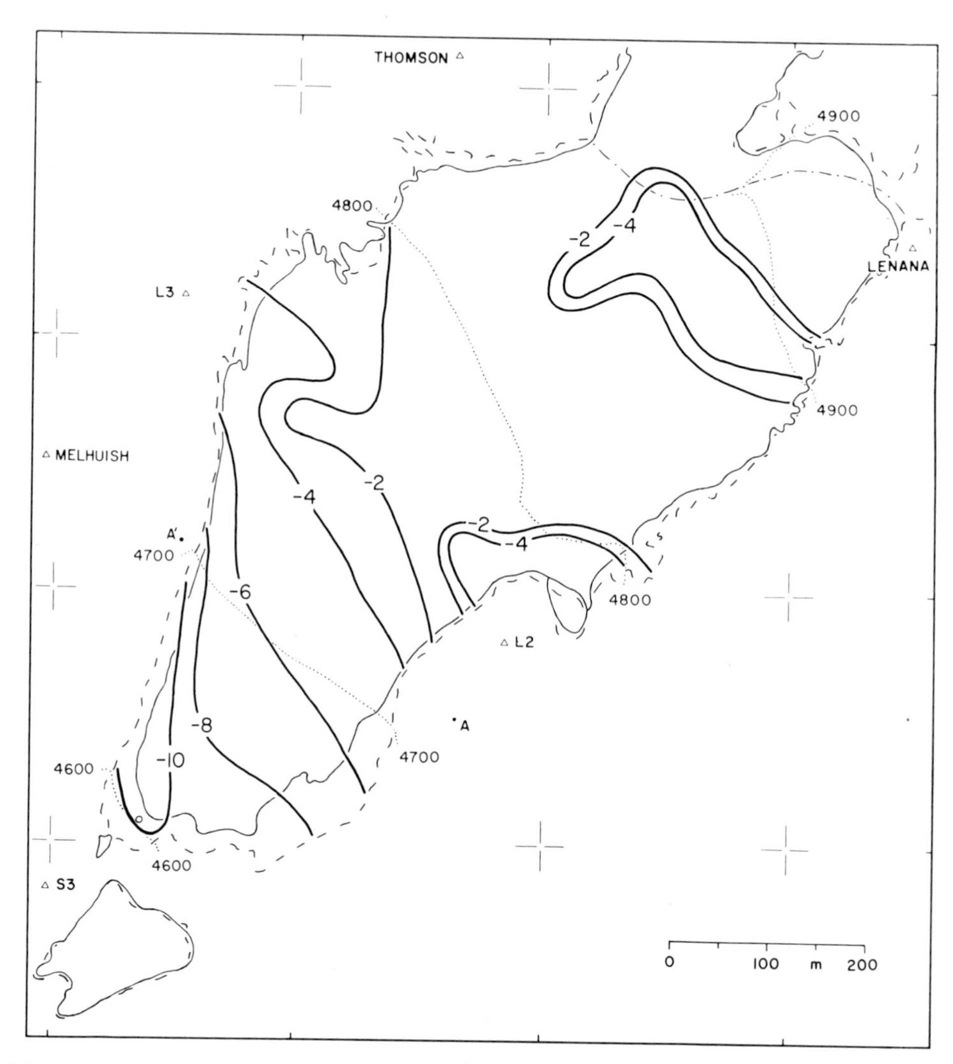

Map 5.2.7:4. Changes in ice thickness, March 1982 minus February 1978, in m. Ice rim in 1982 is shown as solid, and in 1978 as broken line. 1978 height contours are entered as thin dotted lines. Scale 1 :7500.

TABLE 5.2.7:1

Area of Lewis Glacier by 50 m elevation bands, in 10^2 m^2. Sources are: photographs of Mackinder expedition (archive of Mountain Club of Kenya) and Forschungsunternehmen Nepal-Himalaya (1967) for 1899; Troll and Wien (1949) for 1934; Charnley (1959) for 1958; Forschungsunternehmen Nepal-Himalaya (1967) for 1963; Caukwell and Hastenrath (1977) for 1974; Hastenrath and Caukwell (1979) for 1978; Caukwell and Hastenrath (1983) for 1982

	1899			1934			1958			1963			1974			1978			1982		
m	Total	W	E	Total	W	E	Total	W	E	Total	W	E	Total	W	E	Total	W	E	Total	W	E
5000																					
				40			40			40			40			37			28		
4950																					
				200	162	38	195	159	36	191	157	34	189	155	34	156	126	30	140	115	25
4900																					
				710	550	160	705	548	157	700	537	163	594	438	156	592	442	150	586	422	146
4850																					
				900	670	230	780	578	202	775	583	192	539	406	133	500	368	132	438	325	113
4800																					
				930	870	60	792	762	30	779	703	76	743	708	35	705	663	42	628	593	35
4750																					
				490			435			415			415			442			430		
4700																					
				540			365			329			328			320			247		
4650																					
				470			278			322			196			184			129		
4600																					
				220			30			29			16			14			1		
4550																					
				75																	
4500																					
				25																	
4450																					
4400																					
Σ	6270			4600			3620			3580			3060			2950			2609		

5.3. Kinematics

The present section is based in part on Hastenrath and Kruss (1979, 1982) and Kruss and Hastenrath (1983).

5.3.1. OBSERVATIONS

The first surface velocity measurements on Lewis Glacier are due to Troll and Wien (1949) in May 1934. They laid out rocks in a straight line across the glacier, in the vicinity of the equilibrium line, and used terrestrial photogrammetry to determine the displacement over a 104-hour interval. Their results are depicted in Map 5.3.2:1.

The IGY Mount Kenya Expedition performed velocity measurements in various portions of the upper glacier by tacheometric surveys of stakes repeated over a period of about 5 months in 1957–8. Their results are also plotted in Map 5.3.2:1.

During the 1973–4 pilot project two lines of stakes were installed across the middle

TABLE 5.3.1:1

Location of net balance stakes, surveyed on 25–26 January 1982. From coordinates, line L2–L3 has azimuth of 316° 16′09″. Asterisks denote stations newly installed in December 1981

From L2 to stake	Orientation	Horizontal distance (m)	Elevation (m)
03	76°40.2′	456.63	4892.9
4B	66°43.1′	412.74	4875.5
12	79°40.6′	419.04	4890.8
13*	79°39.6′	327.07	4868.6
From L3 to stake			
06*	314°59.6′	318.53	4828.9
07	310°54.7′	239.48	4792.3
10	323°12.7′	287.66	4795.8
11	332°45.1′	264.41	4775.4
22	02°46.7′	150.02	4751.0
25	02°04.0′	353.51	4746.1
71 (*)	346°59.1′	221.89	4759.0
81 (*)	354°41.8′	277.89	4752.0
31	03°20.4′	249.58	4744.3
32*	13°06.6′	278.51	4731.8
33	21.00.3′	319.35	4716.8
41*	30°10.6′	373.51	4688.4
42*	26°51.6′	344.91	4702.1
43	32°54.9′	397.44	4676.2

TABLE 5.3.1:2

Surface ice flow velocities measured over the intervals (a) 25–29 January to 24–26 February 1978; (b) 25–29 January 1978 to 29 December 1978–5 January 1979; (c) 28 December 1978–5 January 1979 to 26–27 December 1979; (d) 25–29 January 1978 to 26–27 December 1979; (e) 26–27 December 1979 to 29–30 December 1981; (f) 26–27 December 1979 to 25–16 January 1982; (g) 29–30 December 1981 to 25–26 January 1982. Speed S in cm a^{-1}, and direction D in degrees counted clockwise from North

Station	a		b		c		d		e		f		g	
No.	Jan 78 – Feb 78		Jan 78 – Jan 79		Jan 79 – Dec 79		Jan 78 – Dec 79		Dec 79 – Dec 81		Jan 79 – Dec 79		Dec 81 – Jan 82	
	S	D	S	D	S	D	S	D	S	D	S	D	S	D
01	98	332	39	336	88	236	46	260	–	–	–	–	–	–
02	61	162	–	–	–	–	–	–	–	–	–	–	–	–
03	143	247	158	247	179	245	168	246	104	246	105	267	176	263
04	269	242	–	–	–	–	–	–	–	–	–	–	–	–
05	223	236	–	–	–	–	–	–	–	–	–	–	–	–
06	343	228	–	–	–	–	–	–	–	–	–	–	351	237
07	223	232	–	–	–	–	237	225	221	221	216	221	093	201
08	87	184	106	204	44	238	71	215	–	–	–	–	–	–
09	440	231	430	242	447	235	438	238	–	–	–	–	–	–
10	359	225	–	–	–	–	334	231	289	229	290	229	335	242
11	304	223	327	224	328	220	327	222	316	223	310	293	179	210
12	153	272	–	–	–	–	–	–	146	249	143	249	061	273
13	216	249	234	246	227	241	229	244	–	–	–	–	186	234
14	267	205	247	220	190	234	216	226	–	–	–	–	–	–
15	224	272	199	201	183	213	186	210	–	–	–	–	–	–
21	485	220	362	220	349	216	355	218	–	–	–	–	–	–
22	236	204	–	–	–	–	197	213	171	219	171	219	165	219
23	431	215	–	–	–	–	350	218	–	–	–	–	–	–
24	462	225	399	217	401	214	399	216	–	–	–	–	–	–
25	408	224	372	222	332	219	338	221	303	220	304	220	330	219
26	286	231	–	–	–	–	303	223	–	–	–	–	–	–
31	542	233	472	233	–	–	391	212	344	214	352	213	563	207
32	511	204	423	213	398	216	410	215	–	–	–	–	292	215
33	543	213	450	212	461	206	453	208	400	209	402	208	478	200
34	210	212	435	215	409	210	419	212	–	–	–	–	–	–
35	421	206	431	208	445	203	438	206	–	–	–	–	–	–
41	317	210	498	217	453	215	463	215	–	–	–	–	270	218
42	401	210	457	221	483	215	466	217	–	–	–	–	461	210
43	450	223	434	220	413	208	421	214	339	216	338	216	306	217
44	431	215	393	218	340	203	362	210	–	–	–	–	–	–
04B	–	–	–	–	–	–	–	–	–	–	–	–	072	237
71	–	–	–	–	–	–	–	–	–	–	–	–	264	245
81	–	–	–	–	–	–	–	–	–	–	–	–	(703)	(217)

portion of the glacier (Map 4.3:9*), and ice flow velocities were determined by tacheometric surveys repeated over periods of 3 and 6 months. Results are likewise entered in Map 5.3.2:1.

During the 1973–4 pilot project, the remnants of a meteorological shelter of the IGY Expedition were located. With reference to the IGY map (Charnley, 1961), a displacement over a 16-year period could thus be determined. The corresponding motion vector is also plotted in Map 5.3.2:1.

During the December 1977–March 1978 expedition, an array of 31 stakes was installed on the glacier (Map 4.3:10*), as described in Section 5.1.3. These were surveyed at various subsequent epochs, thus yielding the velocity estimates summarized in Table 5.3.1:2. This includes the values resulting from the thorough new installation of the stake network in December 1981, described in Section 5.1.3. (Table 5.3.1:1.) Ice velocity estimates are furthermore plotted in the Maps 5.3.2:1, 5.3.2:3, and 5.3.4:4 and Figs. 5.3.2:1–5.3.2:7.

The aforementioned velocity determinations are fraught with errors of differing magnitude, dependent on technique and time interval. Uncertainties in the various measurements are estimated to be as follows: 3 m a^{-1} for the 1934 photogrammetry; about 1 m a^{-1} for the 1957 and 1973–4 tacheometric surveys; about 1.5 m a^{-1} for the 1958–74 displacement estimate; and a few cm a^{-1} for the EDM measurements during 1978–82 (Fig. 5.3.2:1).

5.3.2. VELOCITY AND MASS FLUX PATTERN

Map 5.3.2:1, depicting ice surface flow velocities obtained from measurements in 1934, 1957–8, 1973–4, and with reference to the 1958–74 period, shows a reasonable spatial consistency. Secular variations in velocity will be discussed in Section 5.3.4.

The monitoring program initiated during the 1977–8 expedition provides a detailed spatial pattern of ice flow characteristics. Surface velocity, measured for the array of motion stakes during the interval January 1978 to January 1979 is plotted in Map 5.3.2:1 by direction and amount, while Map 5.3.2:2 illustrates the flowline pattern as constructed from the 1974 surface topography. Included in Fig. 5.3.2:1 are observed velocity, calculated mass flux and velocity, and observed surface and computed bedrock topography as profiles along the modeled line (Section 5.2.4).

The model of ice dynamics (ref. Kruss, 1981, 1983a; Hastenrath and Kruss, 1979) permits calculation of ice mass response to changes in net balance. An initial input net balance variation was modified by trial and error until a good fit between computed and observed glacier retreat was obtained. This basic approach was also employed by Allison and Kruss (1977). As well as glacier extent, longitudinal profiles of ice depth, surface velocity, and total mass flux are calculated. This allows an estimation of changes in these parameters with time.

In order significantly to reduce small velocity perturbations, a filter involving an averaging over approximately 100 m is applied to the central line velocities (see Maps 5.3.2:1, 5.3.2:3, 5.3.2:4) and Figs. 5.3.2:1 to 5.3.2:4; the filtered observed and modeled profiles agree closely (Fig. 5.3.2:2). The calculated mass flux is commensurate with the computed velocity, bedrock, and valley shape (ref. Section 5.2.4). Due to the down-glacier decrease in cross-sectional area, the maximum velocity is at an elevation well below the largest ice discharge.

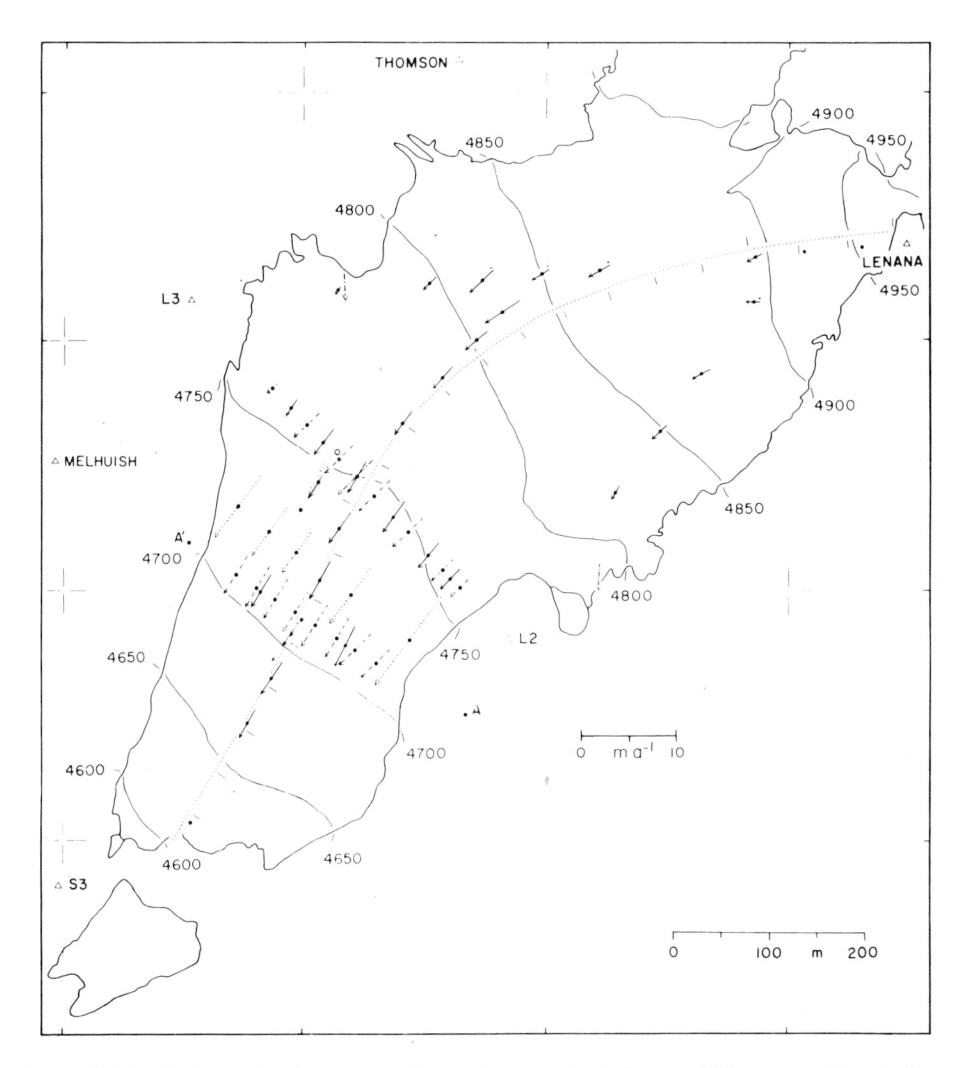

Map 5.3.2:1. Surface velocities measured at various epochs during the 20th century. Dots indicate the location of targets, and arrows the direction and magnitude of motion. Velocity scale in m a^{-1} is ten times the horizontal scale in m (i.e. 10 m a^{-1} would be represented as a 100 m arrow). Dotted, dash-dotted, broken, and solid arrows refer to measurements in 1934 (Troll and Wien, 1949), in 1957 (Charnley, 1961), in 1973–74, and during 1978–80, respectively. Solid arrows with small dot at right-hand side of tail denote measurements limited to January–February 1978 Open circle and dot indicate the 1958 and 1974 locations of a meteorological shelter and its remnants. The corresponding motion vector, shown by solid line shaft with perpendicular short barbs to the right, is plotted at the midpoint of this distance. The dotted line represents the modeled central line, with tick marks at 50 m intervals indicating longitudinal distance. 1978 height contours in m are entered as thin solid lines. Scale 1:7500.

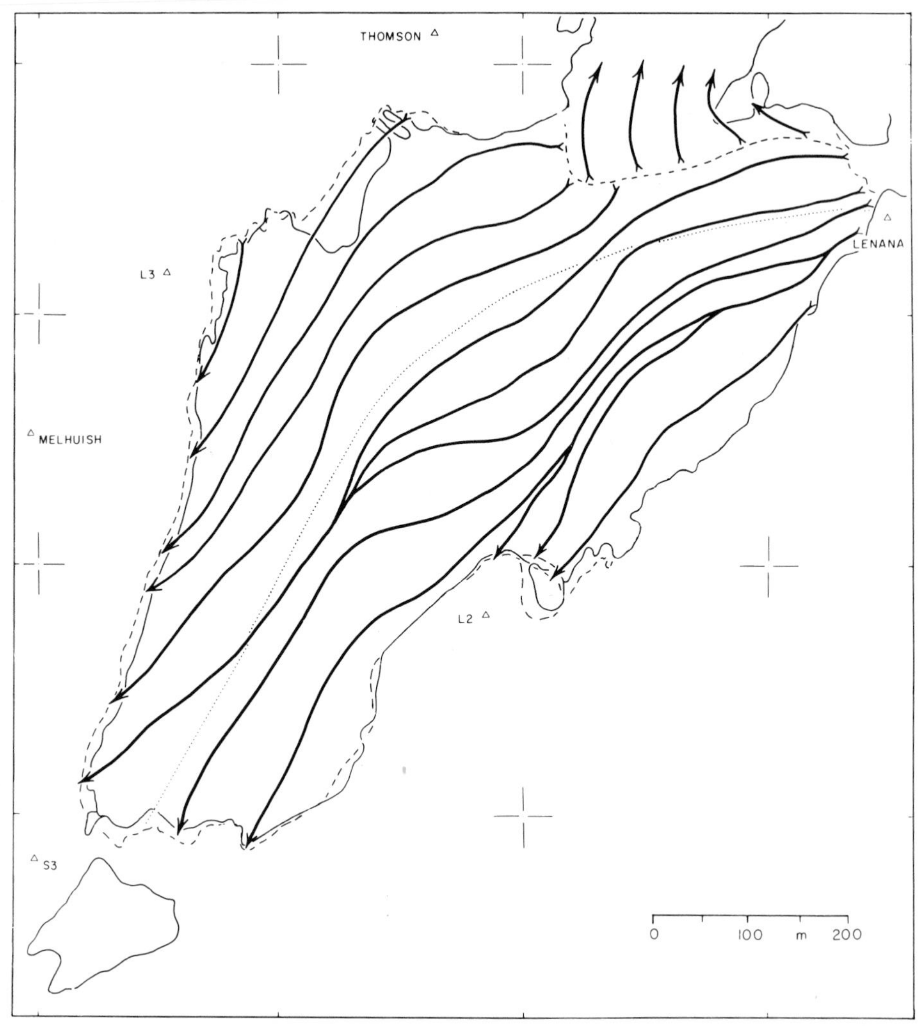

Map 5.3.2:2. Approximate flowline pattern constructed from (1974) surface topography. Dotted line represents the modeled central line. Ice rim in 1974 and 1978 is delineated by solid and broken lines, respectively. Scale 1:7500.

Velocity estimates for the period since the 1979/80 expedition (Table 5.3.1:2) are plotted in Map 5.3.2:4. The pattern is overall very similar to the earlier measurement intervals (Table 5.3.1:2, Maps 5.3.2:1, 5.3.2:3), although the values are in general less. This tendency for a slowdown is further illustrated in Figs. 5.3.2:3 and 5.3.2:4, and will be considered in Section 5.3.4. Fig. 5.3.2:5 also shows intra-annual variations of velocity.

This discussion has so far concentrated on the horizontal velocity pattern. The vertical component of surface velocity in particular is of interest in relation to the vertical net balance profile (Hastenrath, 1983a). This can be determined through the procedure described by Paterson (1969, pp. 65–8) for the central flowline. The various EDM surveys provide the horizontal and vertical coordinates of the foot point of net balance stakes, as

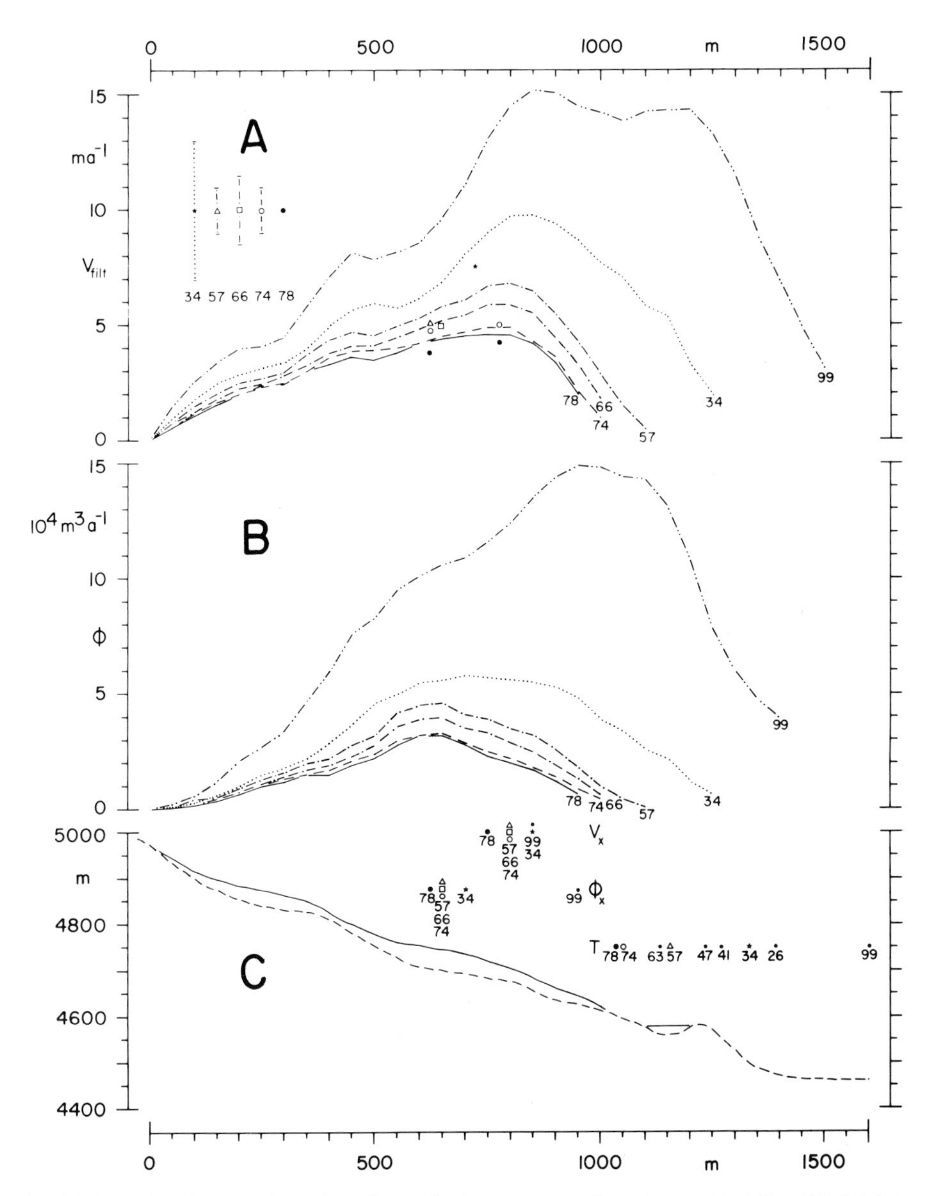

Fig. 5.3.2:1. Secular variation of surface velocity and mass flux along central line. Modeled longitudinal profiles are shown of filtered velocity V_{filt}, and filtered mass liquid water equivalent flux ϕ, for various epochs as follows: 1899 double dot-dashed, 1934 dotted, 1957 dash-dotted, 1966 double dash-dotted, 1974 broken, 1978 solid line. In the graph of V_{filt}, observed velocities are entered as asterisk for 1934, triangle for 1957, square for 1958 to 1974 mean displacement rate (labelled 1966), open circle for 1974, and large dot for 1978–80 (labelled 1978). Vertical lines to the left indicate the estimated uncertainties of measurements. In the bottom part, bedrock topography is depicted as broken, and tarn and 1978 ice surface topography by solid line. Modeled maxima of surface velocity V_x and mass flux ϕ_x are denoted by the aforementioned symbols for the 1899, 1934, 1957, 1966, 1974, and 1978 epochs. The observed terminus positions T are furthermore indicated by small dots for the additional epoch 1926, 1941, 1947, and 1963.

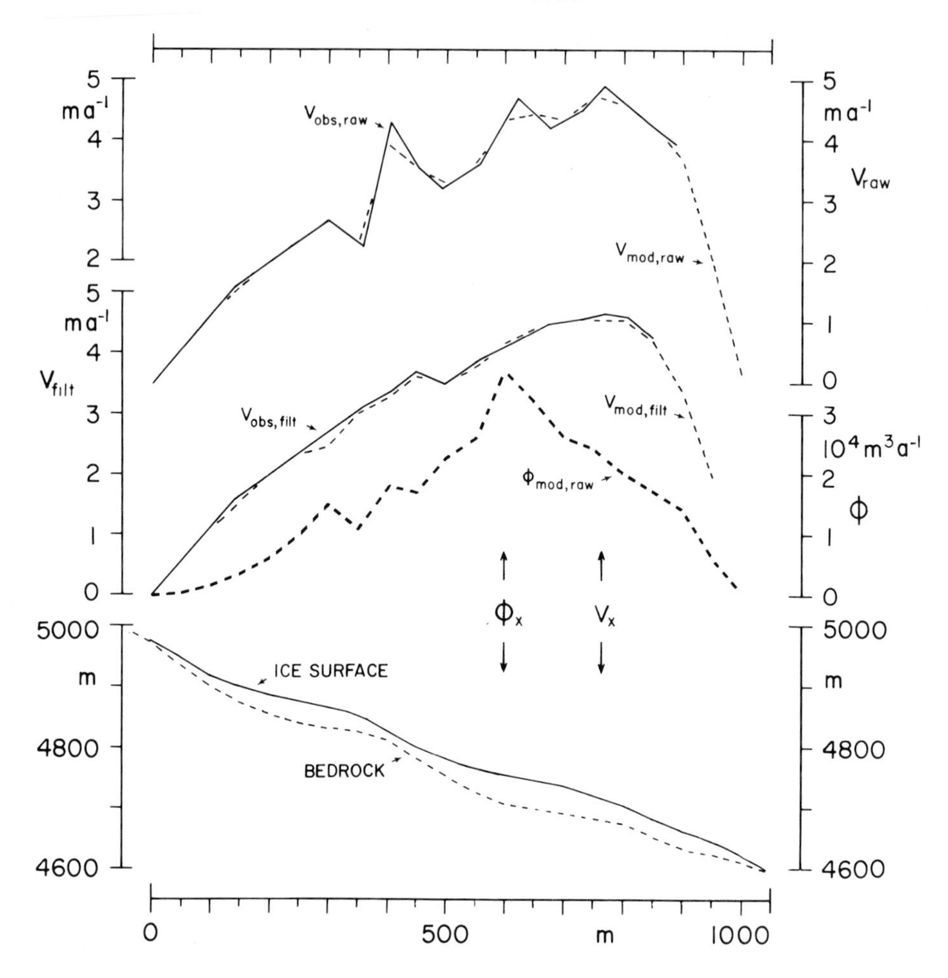

Fig. 5.3.2:2. Profile along central line in 1978. Top part of graph shows 'raw' surface velocities, while middle part includes 'filtered' surface velocities and liquid water equivalent 'raw' mass flux. Solid lines denote observed, and broken lines modeled values, with heavy broken line referring to mass flux. In the bottom part, solid and broken lines signify measured surface and calculated bedrock topography. Maximum mass flux ϕ_X and surface velocity V_X are identified for comparison with Maps 5.3.2:3 and 5.3.3:1.

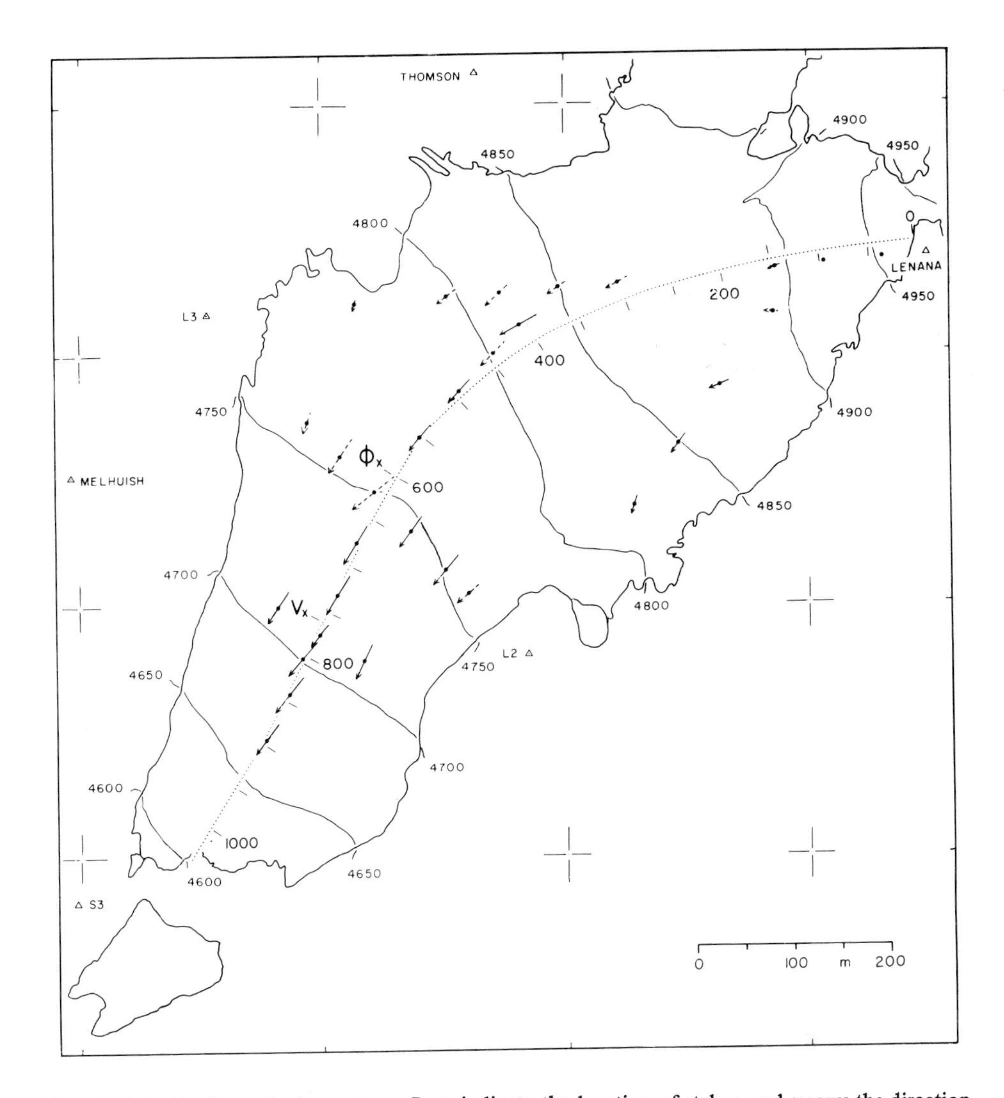

Map 5.3.2:3. Surface velocity pattern. Dots indicate the location of stakes, and arrows the direction and magnitude of motion. Velocity scale in m a^{-1} is ten times the horizontal scale in m (i.e. 10 m a^{-1} would be represented as a 100 m arrow). Solid arrows refer to the interval from February 1978 to December 1978/January 1979 and broken arrows from January to February 1978. The dotted line represents the modeled central line, with tick marks at 50 m intervals indicating longitudinal distance, and Φ_X and V_X maximum mass flux and surface velocity, respectively, as in Fig. 5.3.2:2. 1978 height contours (in m) are entered as thin solid lines. Scale: 1:7500.

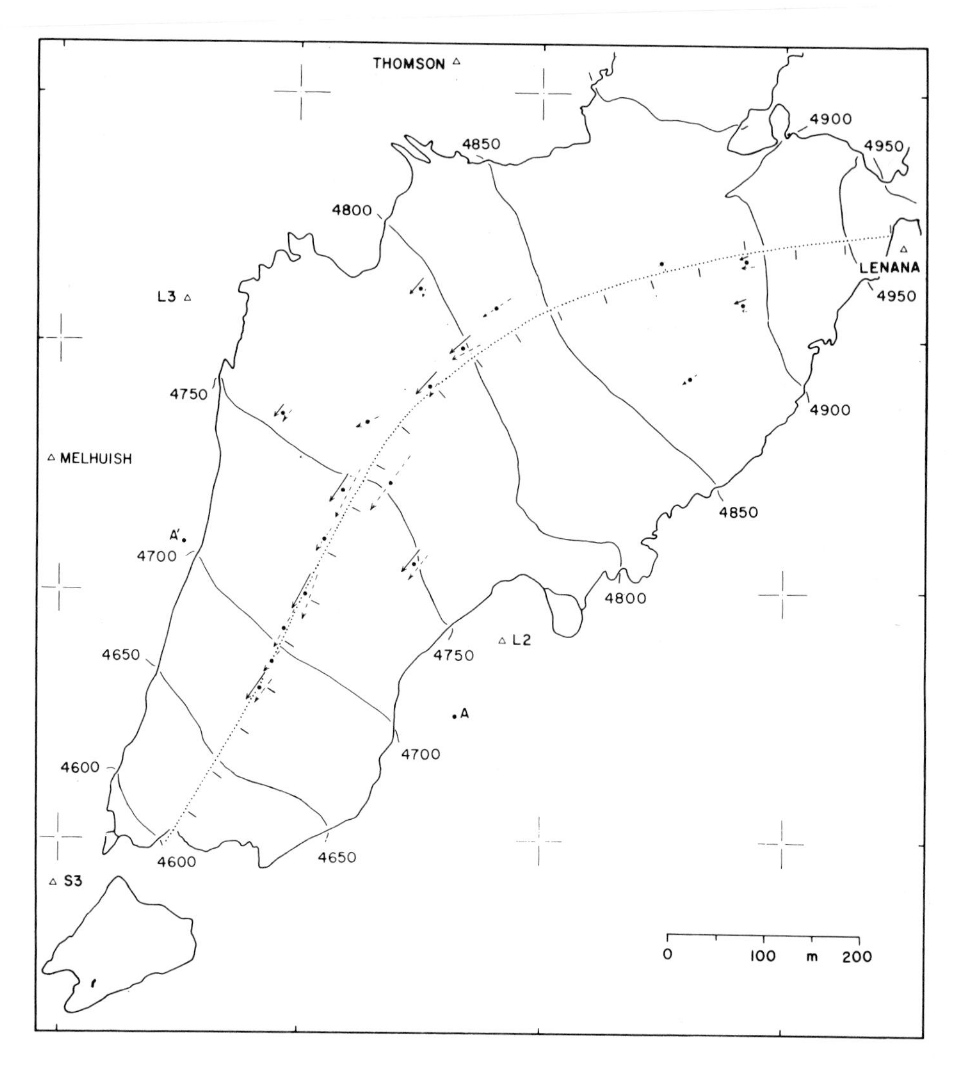

Map 5.3.2:4. Surface velocity map of Lewis Glacier. Arrows indicate the magnitude and direction of the horizontal ice velocity resulting from surveys over two intervals. Dashed and solid arrows give two year mean velocities for 80/81 (26–27 December 1979 to 25–26 January 1982) and one month values over Jan 82 (29–30 December 1981 to 25–26 January 1982), respectively. Velocity scale in m a^{-1} is ten times the horizontal scale in m (i.e. 10 m a^{-1} would be represented as a 100 m arrow). Positions (dots) are for 25–26 January 1982. The base map at scale 1:7500 is the 1978 glacier and the dotted line defines a central longitudinal line.

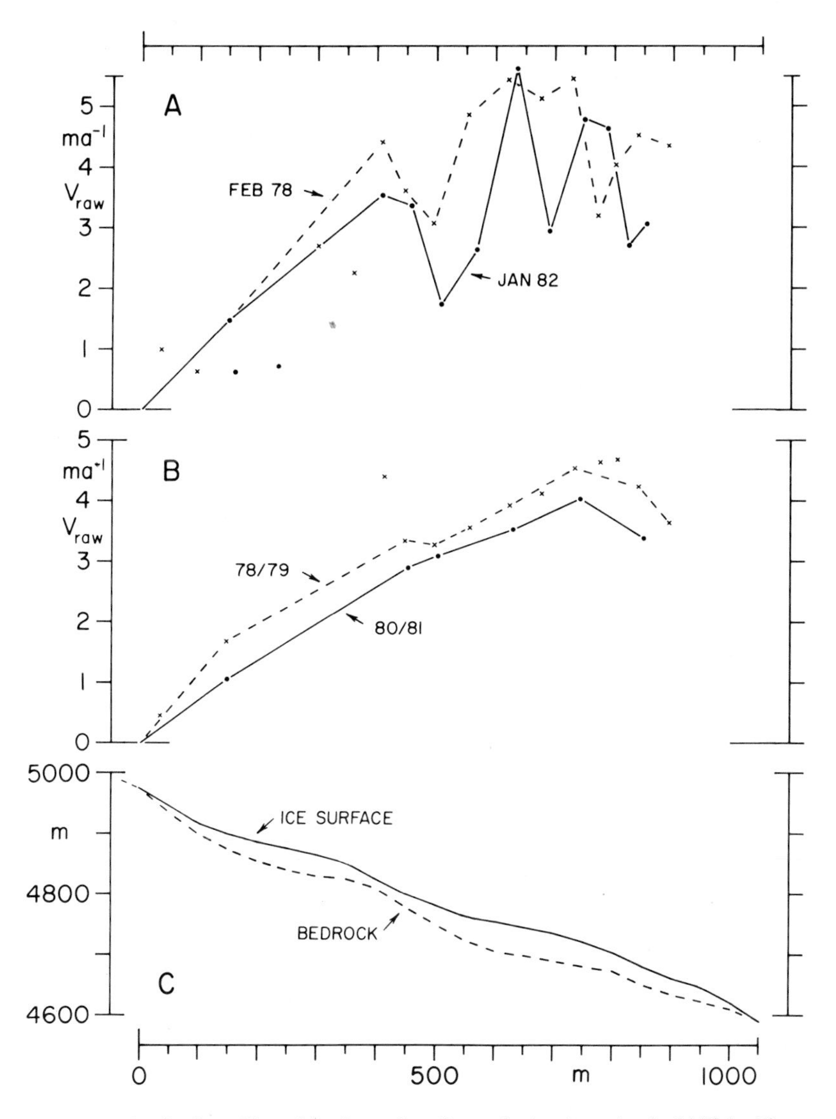

Fig. 5.3.2:3. Longitudinal profiles of horizontal surface velocity (raw data). (a) Velocities measured over one month during Feb 1978 (25–29 January to 24–26 February 1978; crosses, dashed line) and Jan 82 (29–30 December 1981 to 25–26 January 1982; dots, solid). (b) Two year mean velocities for 78/79 (25–29 January 1978 to 26–27 December 1979; crosses, dashed) and 80/81 (26–27 December 1979 to 25–26 January 1982; dots solid). (c) Longitudinal profiles of 1978 surface elevation (solid) and bedrock (dashed). All profiles are with respect to down glacier distance along the central line defined in Maps 5.3.1:1, 5.3.2:2, 5.3.2:3. To aid in the comparison of differing epochs, only locations with velocity data at both times are connected.

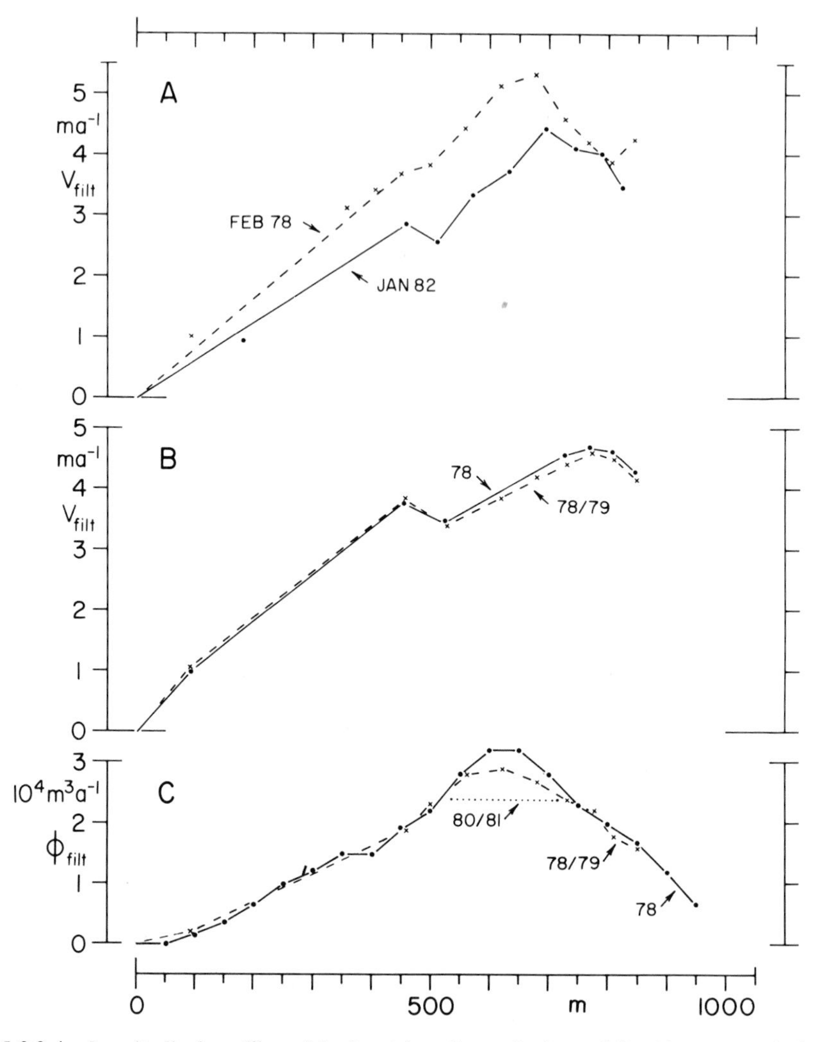

Fig. 5.3.2:4. Longitudinal profiles of horizontal surface velocity and liquid water equivalent mass flux (filtered); (a) filtered velocities over one month for Feb 78 (25–29 January to 24–26 February 1978; crosses, dashed line) and Jan 82 (29–30 December 1981 to 25–26 January 1982; dots, solid); (b) filtered 78 (25–29 January 1978 to 28 December 1978–5 January 1979; dots, solid); and two year mean velocities for 78/79 (25–29 January 1978 to 26–27 December 1979; crosses, dashed); (c) filtered mass fluxes for 1978 (modelled annual, dots, solid), 78/79 (crosses, dashed), and 80/81 (26–27 December 1979 to 25–26 January 1982; dotted) in liquid water equivalent units. The 80/81 line gives the best-estimate magnitude of the 80/81 maximum mass flux and also defines the boundaries of its location. All profiles are with respect to down glacier distance along the central line defined in Maps 5.3.1:1 to 5.3.2:4. Only locations with velocity data at both times are connected; all mass flux points connected.

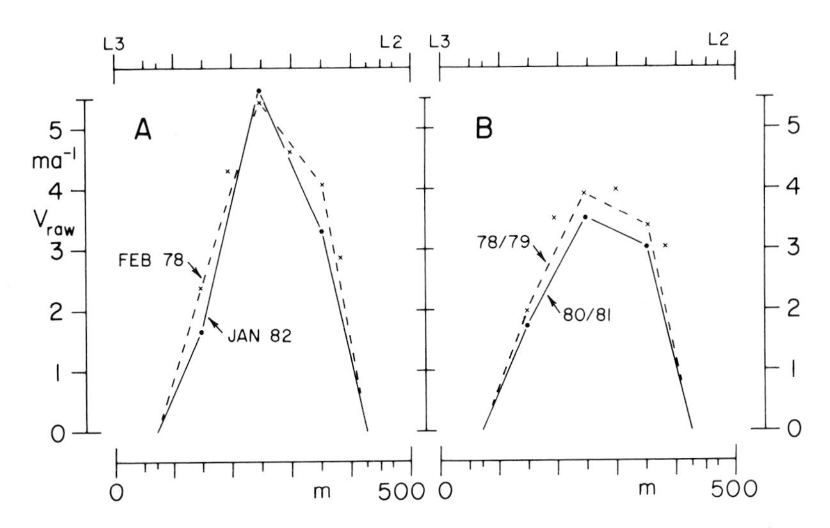

Fig. 5.3.2:5. Transverse profiles of horizontal surface velocity (raw). (a) Velocities over one month for Feb 78 (25–29 January 1978 to 24–26 February 1978; crosses, dashed line) and Jan 82 (29–30 December 1981 to 25–26 January 1982; dots, solid). (b) Two year mean velocities for 78/79 (25–29 January 1978 to 26–27 December 1979; crosses, dashed) and 80/81 (26–27 December 1979 to 25–26 January 1982; dots, solid). All profiles are along a line between control points L3 and L2 in Fig. 1; distance is from L3. Only locations with velocity data at both times are connected.

shown in Table 5.3.1:1 for the datum 25–6 January 1982. Moreover, the distance from a reference mark on the stake to the glacier surface is available from the net balance measurements (Tables 5.4.3:1 and 5.4.3:2). Thus the vertical and horizontal displacement rates of the reference mark in an absolute coordinate system can be calculated. Let Δz and Δs denote the displacement of the stake mark in the vertical and in the horizontal along the central line (Maps 5.2.1:1, and 5.3.2:1–5.3.2:4), respectively, during the time interval Δt, and $\tan \alpha$ the surface slope. Then the vertical component of surface flow counted positive downward,

$$v = \frac{1}{\Delta t}(\Delta z - \Delta s \tan \alpha) \qquad 5.3.2{:}(1)$$

and the horizontal component of surface flow in the plane described by the central line

$$u = \frac{\Delta s}{\Delta t} \qquad 5.3.2{:}(2)$$

The vertical and horizontal surface flow components were calculated for the net balance stakes located near the central line. For stakes 3, 10, 11, 31, 43, evaluation was possible for the four-year period 25–9 January 1978 to 25–6 January 1982. For stakes 21, 32, 41, 42, calculations were limited to the two-year interval 26–7 December 1979 to 29–30 December 1981, and for stake 1 to the one year period 25–9 January 1978 to December 1978–5 January 1979.

The resultant surface flow velocity in the longitudinal-vertical plane in depicted in part A of Fig. 5.3.2:6. The flow vector is directed downward in the upper and upward in the lower glacier. This feature appears enhanced in part B of Fig. 5.3.2:6, which is

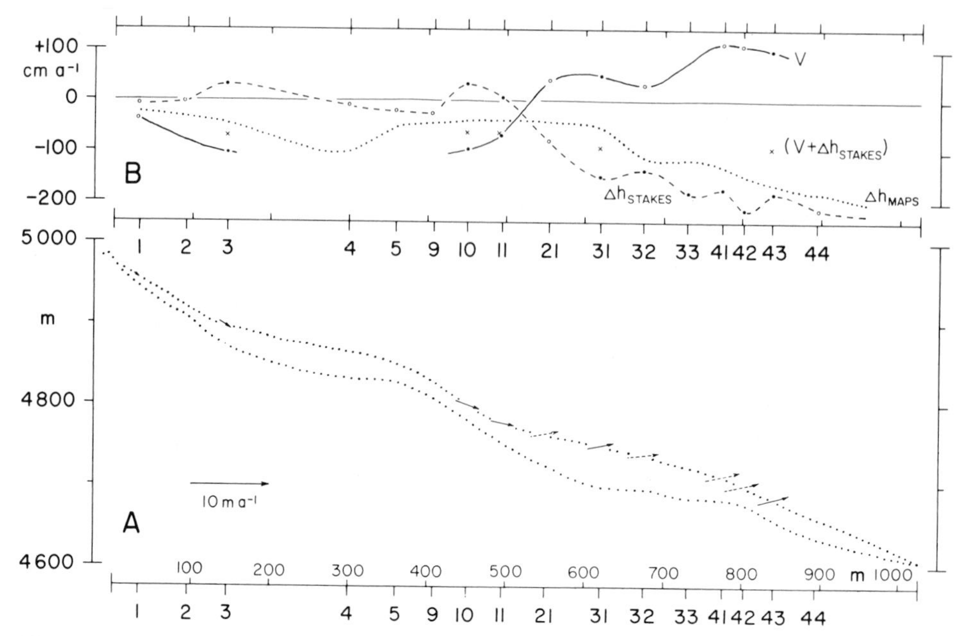

Fig. 5.3.2:6. Relation of vertical flow component, net balance, and surface lowering along the longitudinal axis of the glacier (dotted line in Maps 5.2.1:1, 5.3.2:1–5.3.2:4) Horizontal scale in m is counted from the highest part of the glacier down the longitudinal axis and station numbers are as in Maps 4.3:10* and 4.3:11*. (A) Surface flow velocity in longitudinal–vertical plane. Surface and bedrock topography are without vertical exaggeration. Velocity vectors in m a^{-1} are plotted at ten times the space scale in m (i.e. 1 m a^{-1} would be represented as a 10 m arrow). Solid arrows at stations 3, 10, 11, 31, 43, refer to the four year interval 25–29 January 1978 to 25–26 January 1982. Broken line arrows at stations 21, 32, 41, 42, belong to the two year period 26–29 December 1979 to 29–30 December 1981, and the arrow at station 1 is based on the one year span 25–29 January 1978 to 28 December 1978–5 January 1979 (sources: Tables 5.3.1:2, 5.4.3:1, 5.4.3:2 and analogs of Table 5.3.1:1 for various earlier surveys). **(B)** Variation of vertical flow component, height changes at stake network, and lowering of surface topography along the longitudinal axis, in cm a^{-1}. For the vertical flow component V, here plotted as positive upward (solid), dots and open circles indicate stations with information for the four-year and two (one) year periods described in **(A)**, respectively (sources as described in (A)). Height changes at the stake network Δh_{STAKES} (Broken line) refer to the budget years 1 March 1978 to 1 March 1982, with dots denoting the stations for which complete observations are available from Table 5.4.3:2; otherwise the curve is plotted from Map 5.4.3:6. Observed time change of surface topography Δh_{MAPS} (dotted line) is plotted for the interval February 1978 to March 1982, but dots denote stations for which values are for the two year interval described in **(A)** above (sources: Map 5.2.7:4 and analogs of Table 5.3.1:1 for earlier surveys). Crosses indicate the surface lowering obtained as difference $(V + \Delta h_{STAKES})$ obtained from the vertical velocity (solid line) and the height change at the stake network (broken line).

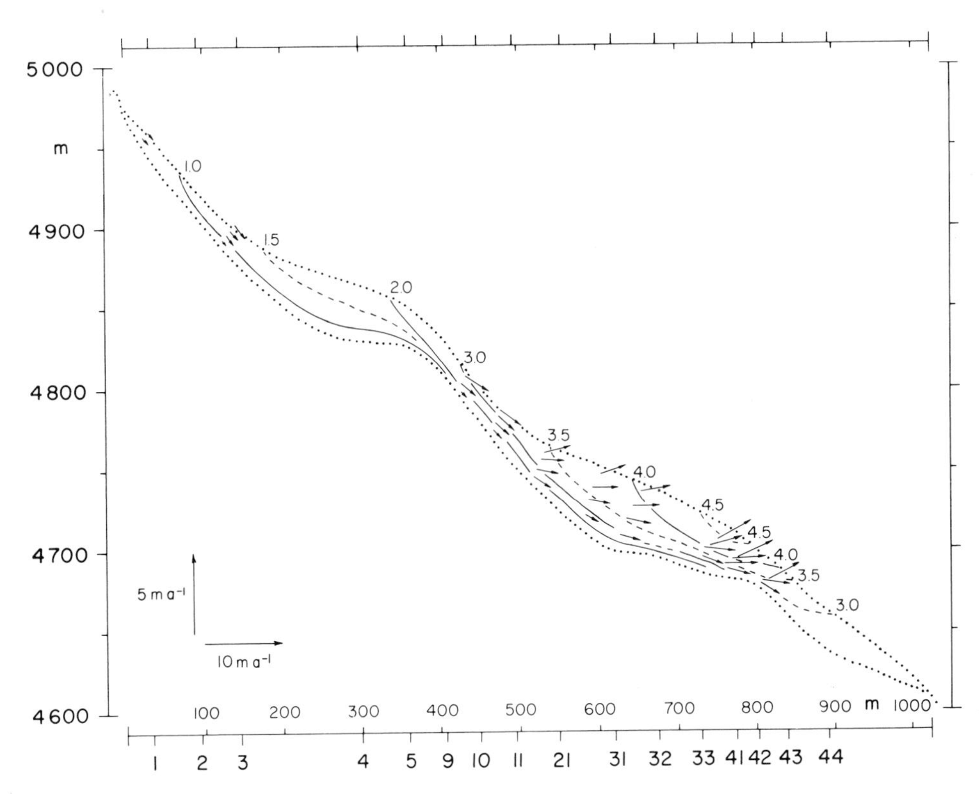

Fig. 5.3.2:7. Ice flow pattern in longitudinal–vertical plane. Horizontal scale in m is counted from the highest point of the glacier down the longitudinal axis and station numbers are as in Maps 4.3:10* and 4.3:11*. Dotted lines denote surface and bedrock topography, and solid and broken lines isotachs in m a^{-1}. Velocity vectors are plotted at ten times the space scale in m (i.e. 10 m a^{-1} would be represented as a 100 m arrow). Vertical exaggeration is double. Surface velocity vectors are the same as in part A of Fig. 5.3.2:6, but note the difference in orientation and length of arrows due to the vertical exaggeration (source: Fig. 5.3.2:6 and Eq. 5.3.2:(3)).

further relevant to the relation between net balance, surface lowering, and flow dynamics discussed in Section 5.4.3.

Although velocity measurements are as a rule limited to the surface, it is desirable to obtain a notion of the ice flow pattern in the interior of the glacier. This is illustrated in Fig. 5.3.2:7, constructed from the information contained in part A of Fig. 5.3.2:6 as described in the following. The surface velocity vectors (Fig. 5.3.2:6) are decomposed into the components parallel and perpendicular to the surface. The velocity component perpendicular to the surface is assumed to decrease in a linear fashion perpendicularly away from the surface to zero at the bedrock. Concerning the velocity component V parallel to the surface, the decrease in the direction perpendicular to the surface is, drawing on Eqs. 5.3.4.1:(1), 5.2.4.1:(3), and 5.2.4.1:(5) (see also Budd, 1969, p. 193), described by

$$V_s - V_z = \frac{n+2}{n+1} k(s \rho g \sin \alpha)^n z^{n+1} \qquad 5.3.2{:}(3)$$

where the subscripts s and z refer to the surface and a perpendicular distance z from the surface towards the bedrock; the other symbols are as introduced in Section 5.2.4.1. For k and n the values of 0.16 $bar^{-2}\,a^{-1}$ and 2, respectively, are used as in Section 5.2.4.1. For the stress shape factor s, values of 0.9 and 0.8 are adopted for the upper (>4750 m) and lower glacier, respectively. The values for ice density ρ and gravitational acceleration g are as in Section 5.2.4.1. The surface slope $\tan \alpha$ is read from the Maps 4.3:10* and 4.3:11*. At each of the ten stations shown in Figs. 5.3.2:6 and 5.3.2:7 the velocity components perpendicular and parallel to the surface were thus calculated for points at 0.25, 0.50, and 0.75 for the distance measured perpendicularly from the surface to bedrock. The resultant vectors are plotted in Fig. 5.3.2:7 by direction and amount, along with the isotach pattern in the longitudinal–vertical plane.

Complementing Fig. 5.3.2:6, the longitudinal–vertical cross-section in Fig. 5.3.2:7 shows the submergent flow in the upper and the emergence in the interior of the lower glacier, as well as the onion-skin arrangement of isotach sheets, with lowest values near bedrock and in the upper glacier, and a maximum near the surface in a portion of the lower glacier well above the terminus. The Lewis Glacier is a rare example where this conceptually plausible flow pattern in the glacier interior (Paterson, 1965, p. 65) has been quantitatively substantiated from observation.

Fig. 5.3.2:7 is also pertinent to the age of ice within Lewis Glacier. Estimates of representative residence times are desired in relation to the secular glacier recession (Section 5.2.7) and in support of climatic ice core studies (Section 5.5). With this motivation a 'bulk residence time' of ice in the upper glacier was first calculated by dividing the ice volume upward of the line L2–L3 near 4750 m (Maps 4.3:9*, 4.3:10*, 4.3:11*) by the estimated mass flux across a vertical plane at that line, which yielded a value of about 150 years (Hastenrath, 1975). The comprehensive field program on Lewis Glacier since the beginning of 1978 permits an update of this exercise. For the datum February 1978, the ice volume upward of the 4750 m contour is estimated at about 3×10^6 m^3 (Section 5.2.6, Map 5.2.6:2). The maximum mass flux of 3×10^4 $m^3\,a^{-1}$ is found at about this elevation (Figs. 5.3.2:1, 5.3.2:2, 5.3.2:4). The ratio yields a 'bulk residence time' in the upper glacier (>4750 m) of about 100 years.

Complementing these bulk considerations, Fig. 5.3.2:7 offers a better appreciation of trajectories and age of ice in the interior of Lewis Glacier. For the following evaluation assume near steady-state, so that trajectories approximately coincide with streamlines. As a first example, consider an ice element at an initial surface location near 300 m on the longitudinal axis, or somewhat above station 4. The trajectory for this ice element is characterized by submergence, a closest proximity to the bedrock between 400 and 500 m (around station 10), then emergence and a surfacing near 600 m or somewhat above station 31. Integration by finite steps between isotachs yields a total travel time from the initial location to the final surfacing location of the order of a century. As a second example, consider an ice element in an initial surface location at 125 m on the longitudinal axis, or somewhat above station 3. Over its trajectory in the upper glacier the element submerges, coming closest to the bedrock around 400 m on the longitudinal axis. The longitudinal coordinate of 500 m, which corresponds approximately to the 4750 m surface contour, is reached after nearly 4 centuries. The trajectory surfaces below the longitudinal coordinate of 700 m or near station 33, with a total travel time of 4–5 centuries.

The travel time from a surface location in the uppermost extremity of the glacier to the 4750 m surface contour of nearly 4 centuries should be compared with the 'bulk residence time' for ice in the upper glacier of the order of 1 century as estimated above. Note that for the ice elements located at the surface along the longitudinal axis, the travel time would average around 2 centuries (being 4 centuries for the upper extremity of the glacier and zero at the 4750 m surface contour). Moreover, the ice flows faster along the centerline than to either side of it. With these qualifications, the travel time derived from Fig. 5.3.2:7 and the estimate of 'bulk residence time' do not seem inconsistent.

The above considerations also point to the inevitable shortcomings of residence time calculations. Lewis Glacier is not in steady-state, so that flowlines differ from trajectories. Furthermore, at earlier epochs in this century, the glacier was longer and thicker, so that the travel distance from the upper extremity of the glacier to the snout region was longer, but the faster flow had the reverse effect on the travel time. Finally, the time scale of climate and glacier variations is comparable to or smaller than the range of typical residence times. All these factors hamper the estimation of residence time from flow considerations.

5.3.3. CREVASSE PATTERN

Map 5.3.3:1 shows the location of crevasses in 1974 and 1978. In view of the photography at slightly different hours of day and differing surface snow cover conditions, a discussion of the small crevasses is not warranted.

Most spectacular is the cross-glacier arc of crevasses in the upper portion of the glacier, which seemingly align with a steep rock threshold manifested by adjacent rock outcrops outside the ice rim. These huge crevasses are apparently quite active. During the 1973–4 project, stakes placed at the rims of one crevasse were observed to move apart a few meters in the course of several weeks. Map 5.3.3:1 suggests that during 1974–8 crevasses in the upper part of the glacier have formed anew at locations somewhat higher than mapped in 1974. In other parts of the glacier there are apparent displacements towards both lower and higher elevations.

An ice ridge in the upper glacier has in its Eastern portion a cliff facing northwestward, and in its Western portion a cliff facing in approximately the opposite direction.

Transverse crevasses prevail in all of the upper and middle glacier. However, orientation changes distinctly over a narrow band further downslope, and longitudinal crevasses dominate the aspect at lower elevations. In the middle portion of the glacier, a large longitudinal crevasse developed in the course of 1974–8.

As is apparent from Maps 5.3.2:1, 5.3.2:3, and 5.3.3:1, the transition from predominantly transverse crevasses in the upper and middle, to a longitudinal orientation in the lower portions of Lewis Glacier takes place in the region of highest flow velocity. Nye (1952) and Paterson (1969, pp. 113–17) have discussed the orientation of crevasse systems in relation to flow characteristics; transverse and longitudinal crevasses are indicative of extending and compressive flow, respectively. The theory explains major features in the crevasse pattern of Lewis Glacier, in that transverse/longitudinal crevasses are found where velocity increases/decreases in the direction of flow.

The location of crevasses in 1978 and 1982 is compared in Map 5.3.3:2. Regions of prominent crevasse formation in the upper glacier, which are seemingly favored by the bedrock topography, stand out although the exact location of crevasses differs between

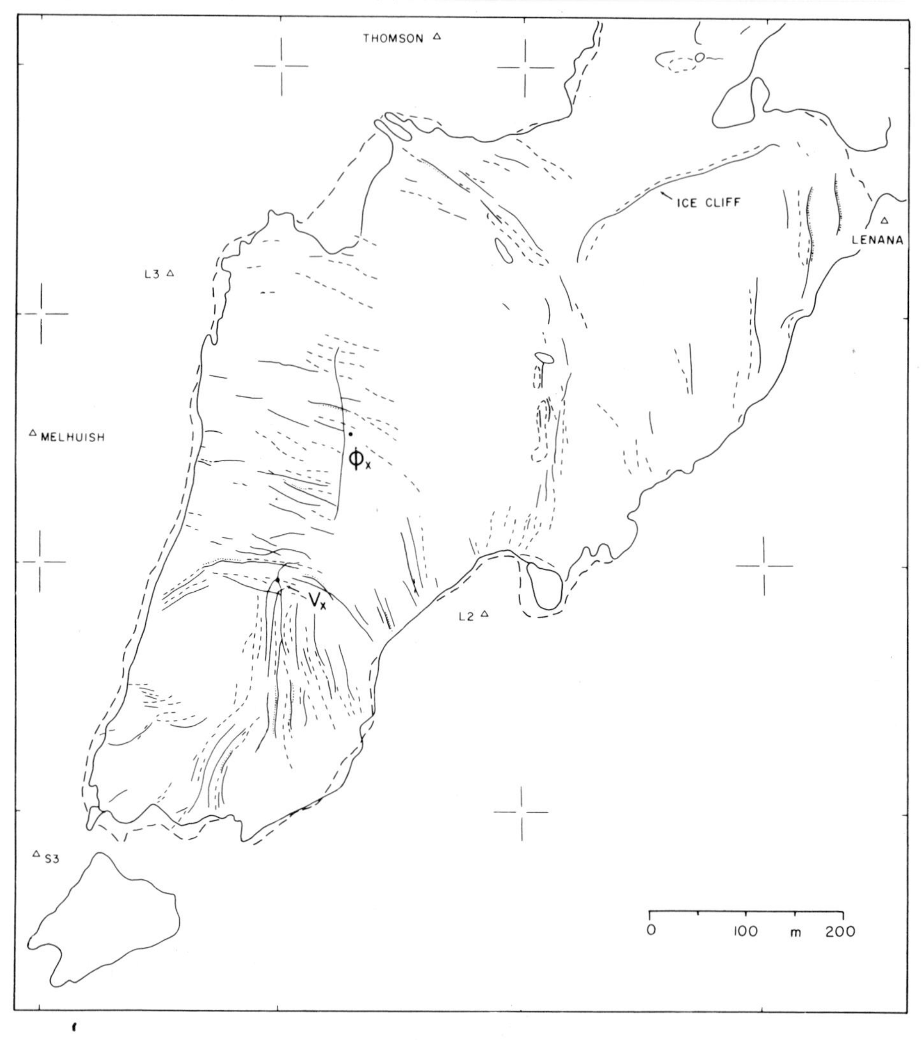

Map 5.3.3:1. Crevasse pattern of Lewis Glacier: 1978 solid and 1974 broken lines. Location identifications Φ_X and V_X indicate maximum mass flux and velocity, respectively (ref. Fig. 5.3.2:2). Point Thomson, Lenana, Melhuish, L3, L2, S3 are control points outside the glacier. Ice rim in 1974 and 1978 is delineated by solid and broken lines, respectively. Scale 1 : 7500.

the 1974, 1978, and 1982 maps (Maps 4.3:9*, 4.3:10*, 4.3:11*, 5.3.3:1, 5.3.3:2). Large ice holes are more prominent in 1982 than at the earlier epochs. From 1978 to 1982, the large Southwest to Northeast oriented ice cliff in the upper glacier moved northwestward (Map 5.3.3:2).

The 1974 and 1978 aerial surveys document the development of a large longitudinal crevasse in the middle glacier (Map 5.3.3:1). However, this had disappeared by 1982 (Map 5.3.3:2). Changes in the crevasse pattern merit attention in that they may reflect changes in internal flow characteristics. Remote sensing by air photogrpahy offers the potential for inference on the velocity pattern of other tropical glaciers.

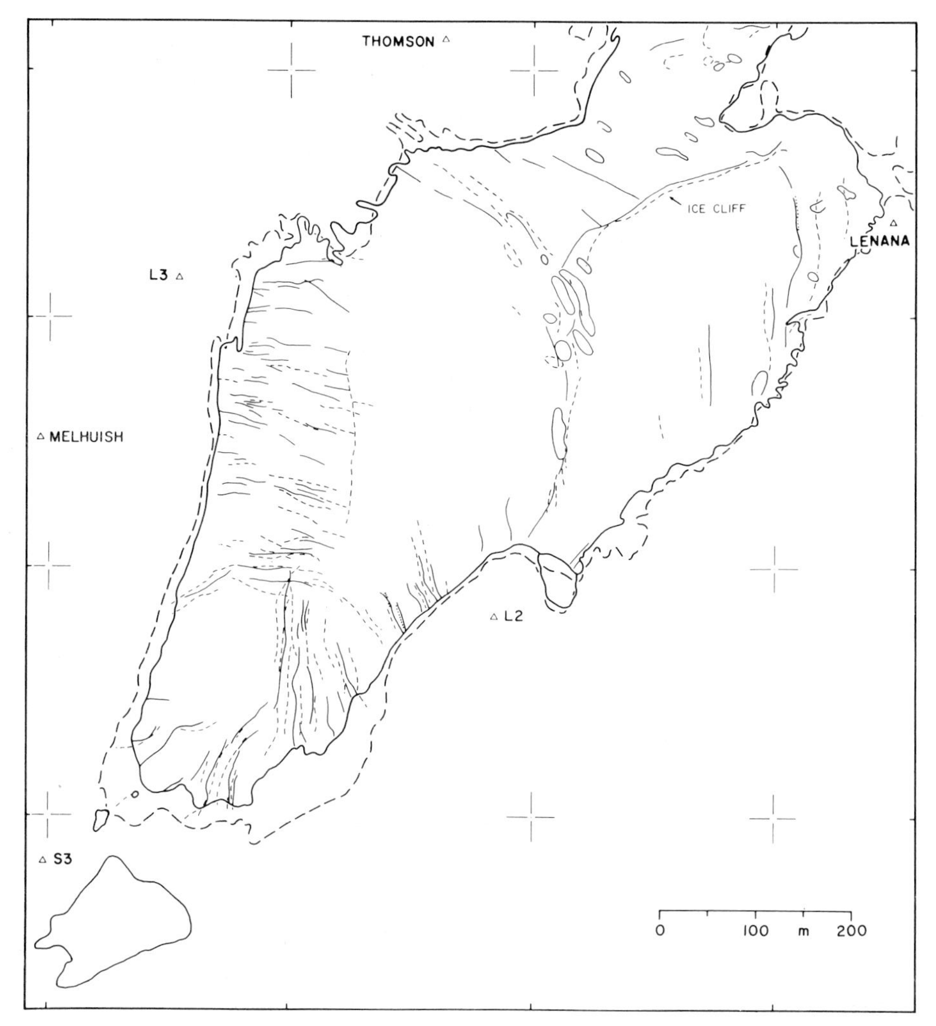

Map 5.3.3:2. Changes in crevasse pattern, 1982 solid and 1978 broken lines. Symbols as for Map 5.3.3:1. Scale 1:7500.

5.3.4. LONG TERM VARIATIONS OF VELOCITY

Map 5.3.2:1 shows surface velocities obtained from our measurements in the course of 1977–80 along with observations at various earlier epochs in the course of this century. A good overall agreement of spatial pattern from the various surveys is seen. There is an indication of a gradual decrease of velocities with time.

Fig. 5.3.2:1 depicts the mass flux and surface velocity along the longitudinal line identified in Map 5.3.2:1 as obtained from numerical modeling for the epochs 1899, 1934, 1957, 1966, 1974, 1978. At various points along this central line, actual observations are entered for comparison. Uncertainties of measurements, as estimated in Section 5.3.1 are also shown. Where the location of surveying targets was not close to the central line, values given were derived in line with the systematic increase of velocities from the boundaries towards the middle of the glacier. Fig. 5.3.2:1 indicates agreement between modeled and observed velocities within error tolerances. It furthermore illustrates the systematic tendencies for overall decrease of velocity and mass flux and upward displacement of maxima associated with the secular terminus retreat.

Given the drastic recession since the end of the last century, the future changes of Lewis Glacier are of considerable interest. These will depend on the climatic and more specifically the net balance conditions of the present and the near future. A computer simulation for the epoch 1895 was performed using as input the observed vertical net balance profile during the 1979/80 budget year. This was a dry year with strongly negative net balance for all portions of the glacier, in contrast to the 1978/9 balance year which approached equilibrium conditions. Note that the same general trends albeit less pronounced, result from modelling using the 1978/9 net balance.

A very drastic decrease of velocity by 1985 was calculated. In fact, the computed glacier deteriorated to such an extent that the model began to break down, so that output is informative but not reliable. With reference to the 1978 datum, by the mid-1980s the maximum velocity decreased to less than half; the velocity maximum became flat and wandered some 100 m up the glacier; and the maximum mass flux decreased to much less than half, with its position moving less than 50 m up the glacier. During this period, the terminus also receded about 50 m.

Concomitant with such a shift of the velocity maximum, one may expect some further upward displacement of the transition zone between longitudinal crevasses in the lower, and transverse crevasses in the upper glacier, but less well defined contrasts in crevasse orientation than that detailed for the 1978 glacier (Map 5.3.3:1). Continued strongly negative net balance along with vanishing ice movement would set the stage for rapid overall thinning of ice, with terminus retreat becoming increasingly determined by the *in situ* net balance rather than the mass economy and flow dynamics of the glacier as a whole. The condition of the glacier in the mid 1980s in expected to lie between these model results and the observed 1978–80 conditions. Based on numerical modeling of the future behavior of the Lewis, it is likely that the glacier will continue to retreat at the late 1970s rate, or even more rapidly, into the mid 1980s. A major overall decrease in the surface velocity is also indicated, with the maximum velocity by 1985 being reduced by as much as half the 1978 value. Further, the computer simulation suggests the velocity maximum will become increasingly less sharply defined and will move quite significantly up glacier. Further changes of Lewis Glacier depend on the past history and the net balance in the course of 1980–5. With this qualification, the model calculations indicate a possible extreme scenario of velocity, mass flux, crevasse, and terminus conditions for the mid 1980s.

Subsequent to this numerical experiment using as input information through 1980 (Hastenrath and Kruss, 1982), further field observations of velocity were obtained (Table 5.3.1:2, Map 5.3.2:4, Figs. 5.3.2:3–5.3.2:5), that allow a verification midway into the forecast (Kruss and Hastenrath, 1983). It is noted that the glacier average net balance of the budget years 1980/1 and 1981/2 was much less negative than for 1979/80 (Tables 5.4.3:5, 5.4.3:7, 5.4.3:9, 5.4.3:11). This would correspond to more moderate velocity changes to 1985 than treated in the numerical model.

The longitudinal and transverse profiles Figs. 5.3.2:3–5.3.2:5 all illustrate a general velocity decrease into the early 1980s. Table 5.3.1:3 summarizes the changes in mean velocity and mass flux between the intervals depicted in Figs. 5.3.2:3 and 5.3.2:4.

It is relevant to consider here the intra-annual variation apparent from comparing the one-month curves in part (a) to the two-year mean profiles in part (b) of Fig. 5.3.2:3. Filtered versions of three of the profiles are contained in parts (a) and (b) of Fig. 5.3.2:4. The transverse profiles of Fig. 5.3.2:5 show much more peaked curves for the two one month intervals than for the two year periods. The variation of velocity in the course of the year may be due in part to change in the contribution of basal sliding. It is noteworthy that, despite the surface lowering, mean and maximum velocities were faster in Jan. '82 than for '80/81. Also, the maxima for the single months are shifted away from a region of steeper slope into the area of deepest ice. The months January and February are at the height of the dry season when meltwater production at the Lewis is greatest. The vicinity to Curling Pond may also be a factor. This apparent intra-annual variability in velocity is relevant in ascertaining long-term tendencies of velocity decrease.

The maximum time period separating comparable observations since 1978 is the almost four years between the one month values (part (a) of Figs. 5.3.2:3 and 5.3.2:4) for Feb. '78 (25–9 January to 24–6 February 1978) and Jan. '78 (29–30 December 1981 to 25–6 January 1982), which is somewhat more than half the seven-year forecast interval. Table 5.3.1:3 shows a velocity reduction of about 40% over this four-year interval, well on the way towards the decrease approaching 50% forecast by 1985. The time elapsed between the midpoint of the 1978 and 1979/80 periods (part (b) of Fig. 5.3.2:4) is less than six months, and the velocity reduction is accordingly small. The midpoints of the '78/9 and '80/1 periods (part (b) of Fig. 5.3.2:3) are two years apart. The velocity reduction of 16% over this time span corresponds to a value of 55% through 1985.

The velocity maximum in 1982 appears rather less well defined than in 1978, as expected. However, a shift of the maximum down glacier is also indicated, contrary to prediction. The maximum mass flux (Table 5.3.1:3 and part (c) of Fig. 5.3.2:4) diminished by about 25% from 1978 to 1980/1. This also compares well with the predicted reduction to less than half for 1985. The prediction of changes in mean velocity, velocity maximum, and maximum mass flux, based on numerical simulation is thus largely verified mid-way to the 1985 forecast endpoint.

TABLE 5.3.1:3

Mean velocity defined by the lines in Figs. 5.3.2:3 and 5.3.2:4 and maximum mass flux. Refer to Figure captions for details of time intervals. (a) longitudinal mean velocities and the percentage decreases between epochs. (b) maximum mass fluxes estimated, in liquid water equivalent, and corresponding net reduction since 1978

	(a) mean velocity cm a^{-1}	%	(b) maximum mass flux 10^3 m^3 a^{-1}	%
Feb 78	334			
Jan 82	272	19		
1978	299		32	
78/79	297	1	29	9
78/79	290			
80/81	244	16	24	25

5.4. Mass and Heat Budget

5.4.1. BASIC THEORY

The heat budget equation for a control volume of the glacier with walls extending from the bedrock vertically up into the atmosphere (Fig. 5.4.1:1) can with reference to horizontal unit area be written

$$SW\downarrow - SW\uparrow - LW\uparrow\downarrow = Q_e + Q_m + Q_s + Q_t + (B - G - T - F - H - R) \qquad 5.4.1{:}(1)$$

Here the left-hand terms denote downward and upward directed shortwave and net longwave radiation. The first four right-hand terms signify latent heat flux at the glacier–air interface (positive upward), heat used for melting throughout the ice column except at its base, sensible heat flux at the glacier–air interface (positive upward), and transient heat storage in the ice. The terms within the bracket on the right-hand side are, respectively, heat used in basal melting, geothermal heat flux, and basal frictional heating at the bedrock–glacier interface, net heat contribution by the freezing of meltwater inflow

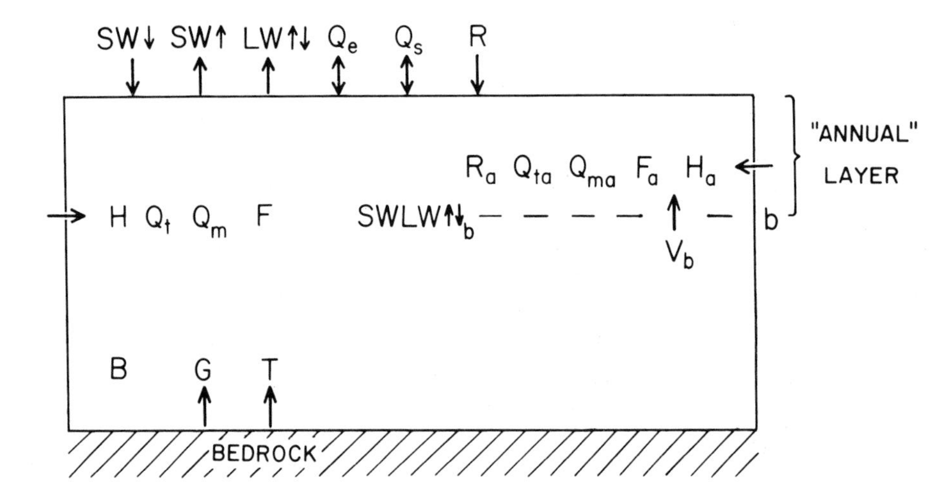

Fig. 5.4.1:1. Heat budget scheme for control volume of glacier with vertical walls extending from bedrock to atmosphere. $SW\uparrow\downarrow$ downward directed shortwave radiation at the surface; $SW\uparrow$ upward directed shortwave radiation; $LW\uparrow\downarrow$ net longwave radiation; Q_e latent heat flux at glacier-air interface; Q_s sensible heat flux at glacier-air interface; Q_m heat used in melting throughout ice column, except at base; Q_t transient heat storage within the ice; B heat used in basal melting; G geothermal heat flux; T basal frictional heating; F net heat contribution by freezing of meltwater inflow and refreezing of melt produced within the ice column; H net energy gain associated with horizontal heat transfers; R heat contributed by rain as result of water-ice temperature difference and/or the freezing of rain water. The subscript 'a' refers to the 'annual layer' bounded by the maximum and minimum surface heights in the course of the balance year, and the subscript 'b' to the base of this layer. Accordingly: $SWLW\uparrow\downarrow_b$ net allwave radiation at base of the 'annual layer'; Q_{ta} transient heat storage within the 'annual' layer; F_a net heat contribution by freezing of meltwater inflow and refreezing of melt produced within the 'annual' layer; H_a net energy gain associated with horizontal heat transfers within the 'annual' layer; V_b upward heat transfer across the base of the 'annual' layer; R_a portion of R contributed to the 'annual' layer.

and the refreezing of melt produced within the ice column, net energy gain associated with horizontal heat transfers, and heat contributed by rain as a result of the water–ice temperature difference and/or the freezing of rain water. Terms that would appear with a positive/negative sign on the left-hand side of Eq. 5.4.1:(1) amount to a heat gain/loss to the glacier.

Terms within the bracket on the right-hand side cannot be readily measured but are small. Only estimates of general magnitude are offered here. The heat used in basal melting B broadly corresponds in magnitude to the sum of the terms G and T. The geothermal heat flux and the basal frictional heating T are of the order of 5×10^{-2} and 1×10^{-2} W m^{-2}, respectively (Budd, 1969, p. 75, 61; Berg, 1949, p. 236). Regarding the heat contribution by freezing of liquid water in the course of the daily cycle, a phase change of 10 kg m^{-2} of water over 12 hours would correspond to 76 W m^{-2}. This term thus plays some role in the heat budget variation between day and night, as discussed in Section 5.4.12. In estimating the magnitude of heat gain associated with horizontal heat transfers, consider a change of the horizontal temperature gradient of 0.1°C/100 m per 100 m of horizontal distance, an ice depth of about 30 m, and a thermal conductivity of 2 W °K^{-1} m^{-1} (Sellers, 1965, p. 127; Berg, 1949, p. 231). This would correspond to a value of G of the order of 10^{-3} W m^{-2}. Regarding the heat contribution by rain, consider liquid precipitation of 50 mm a^{-1}, and a temperature difference between rain water and ice of the order of 1°C. The cooling of the water to the temperature of the ice, and the subsequent freezing would correspond to heat additions of the order of 10^{-2} and 5×10^{-1} W m^{-2}, respectively. The terms within the bracket on the right-hand side are thus small compared to the other terms of Eq. 5.4.1:(1).

The terms on the left-hand side can be regarded as the radiational forcing, and the first four right-hand terms as the major heat budget response of the glacier.

A layer near the glacier surface bounded by the maximum and minimum surface heights in the course of the net balance year is of particular interest, because it is related to the net balance of the glacier as determined from a stake network installed on the glacier (Section 5.4.3). The heat budget equation of this 'annual' layer (Fig. 5.4.1:1) can with reference to horizontal unit area be written

$$\begin{aligned} &SW\downarrow - SW\uparrow - LW\uparrow\downarrow - SWLW\uparrow\downarrow_b \\ &\quad = Q_e + Q_{ma} + Q_s + Q_{ta} - (F_a + H_a - V_b + R_a) \end{aligned} \qquad 5.4.1{:}(2)$$

Here the subscript a refers to the 'annual' layer, and the subscript b to the base of his layer, V_b is the upward heat transfer across the base, and R_a the portion of R contributed to the 'annual' layer. Otherwise, the notation is as defined for Eq. 5.4.1:(1). Reference is also made to Fig. 5.4.1:1. Although temperature remains almost always at 0°C at depths as shallow as 0.5 m, vertical gradients conducive to upward heat transfer occur with surface cooling during the night. Consider a surface temperature of −5°C during 6 hours of the night, constant temperature of 0°C at 0.5 m depth, and a thermal conductivity of 2 W °K^{-1} m^{-1}. This would correspond to a daily average vertical heat flux in the surface layer V of about 5 W m^{-2}. Although the vertical heat flux in the surface layer and at the base of the 'annual' layer are not generally identical, these considerations illustrate the smallness of V_b. R_a may account for the larger portion of R, but even this was shown to be small.

The mass continuity for a control volume of the glacier with reference to horizontal unit area (Fig. 5.4.1:2) is expressed in three budget equations, for ice, liquid, and both phases combined.

For the continuity of mass in solid form

$$P_s + P_{lr} + F_{ilr} - \frac{Q_{ei}}{L_s} - \frac{Q_{mm}}{L_m} + F_{is} - F_{os} = \Delta_s \qquad 5.4.1{:}(3)$$

For the continuity of mass in liquid form

$$P_l - P_{lr} - F_{ilr} - \frac{Q_{ew}}{L_e} - \frac{Q_{em}}{L_e} + \frac{Q_{mm}}{L_m} + F_{il} - F_{ol} = \Delta_l \qquad 5.4.1{:}(4)$$

Addition of Eqs. 5.4.1:(3) and 5.4.1:(4) yields the budget equation of mass in both phases combined

$$P_s + P_l - \left(\frac{Q_{ei}}{L_s} + \frac{Q_{ew}}{L_e} + \frac{Q_{em}}{L_e}\right) + F_{is} - F_{os} + F_{il} - F_{ol} = \Delta_s + \Delta_l \qquad 5.4.1{:}(5)$$

The meaning of terms is as follows: P_s, P_l, and P_{lr}, denote, respectively, precipitation in solid (subscript s) and liquid (subscript l) forms, and the portion of liquid precipitation which is subsequently frozen; L_e, L_m, L_s, are the latent heats of evaporation, melting, and sublimation; Q_{ei}, Q_{ew}, Q_{em}, and Q_{mm}, signify the amounts of heat used in the sublimation of ice (subscript *ei*), in the evaporation of water not having been in solid phase within the control volume (subscript *ew*), in the evaporation of meltwater (subscript *em*), and in the melting of ice (subscript *mm*), respectively; F_{is}, F_{os}, F_{il}, F_{ol}, denote the lateral inflow (subscript i) and outflow (subscript o) of ice (subscript s) and water (subscript l), respectively; F_{ilr} signifies the portion of the lateral inflow of water subsequently frozen; Δ_s and Δ_l represent the time rate of change of ice (subscript s) and water (subscript l) mass. Δ_s corresponds to the net balance as obtained from successive mappings of the glacier surface topography.

For the continuity of mass in solid form in the 'annual' layer

$$P_s + P_{lra} + F_{ilra} - \frac{Q_{ei}}{L_s} - \frac{Q_{mma}}{L_m} + F_{isa} - F_{osa} = \Delta_{sa} \qquad 5.4.1{:}(6)$$

Here the subscript a refers to the 'annual' layer. Otherwise, the notation is as defined above. Reference is also made to Fig. 5.4.1:2. The right-hand term, Δ_{sa}, corresponds to the net balance as measured from a stake network.

For the glacier as a whole $\Delta_s \leqslant \Delta_{sa}$, because of possible melting at depths greater than the annual layer, i.e. $Q_{mm} > Q_{ma}$. For portions of the glacier, $(\Delta_{sa} - \Delta_s)$ depends on (a) the relative importance of melting and freezing processes as expressed by the second, third, and fifth left-hand terms of Eqs. 5.4.1:(3) and 5.4.1:(6); and (b) mass redistribution related to the ice flow characteristics of the glacier, as reflected in the sixth and seventh left-hand terms. Effects referred to under (a)/(b) are expected to provide the largest positive/negative contribution to $(\Delta_{sa} - \Delta_s)$ in the lower glacier.

Of foremost interest for the ice budget of the glacier are Eqs. 5.4.1:(3) and 5.4.1:(6). P_{lr}, F_{ilr}, P_{lra}, are considered as small compared to the solid precipitation P_s. For portions of the glacier, the quantities $(F_{is} - F_{os})$ and $(F_{isa} - F_{osa})$ can be substantial, but for the glacier as a whole only the first, fourth, and fifth left-hand terms of Eqs. 5.4.1:(3) and 5.4.1:(6) are important. These are considered in Sections 5.4.2 and 5.4.10. In the following, 'sublimation' will be used to denote a change from the solid to the vapor phase, regardless of a possible intermediate liquid phase.

The first two right-hand terms of Eqs. 5.4.1:(1) and 5.4.1:(2) and the fourth and fifth terms of Eqs. 5.4.1:(3) and 5.4.1:(6) link the heat and mass budgets of the glacier.

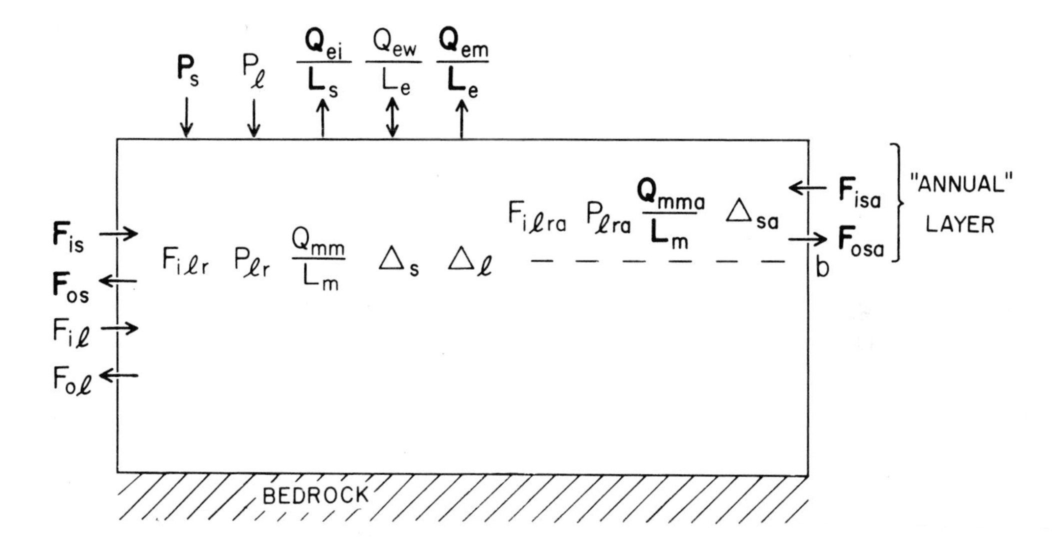

Fig. 5.4.1:2. Mass budget scheme for control volume of glacier, with vertical walls extending from bedrock to atmosphere. P_s precipitation in solid form; P_l precipitation in liquid form; P_{lr} portion of liquid precipitation subsequently frozen; L_e latent heat of evaporation; L_m latent heat of melting; L_s latent heat of sublimation; Q_{ei} net heat used in the sublimation of ice; Q_{ew} heat used in the evaporation of water not having been in solid phase within the control volume; Q_{em} net heat used in the evaporation of meltwater; Q_{mm} net heat used in the melting of ice; F_{is} lateral inflow of ice; F_{os} lateral outflow of ice; F_{il} lateral inflow of water; F_{ilr} portion of lateral inflow of water subsequently frozen; F_{ol} lateral outflow of water; Δ_s time rate of change of ice mass; Δ_l time rate of change of water mass. The subscript 'a' refers to the 'annual layer' bounded by the maximum and minimum surface heights in the course of the balance year. Accordingly: P_{lra} portion of liquid precipitation subsequently frozen in the 'annual' layer; Q_{mma} heat used in melting of ice in the 'annual' layer; F_{isa} lateral inflow of ice in the 'annual' layer; F_{osa} lateral outflow of ice in the 'annual' layer; F_{ilra} portion of lateral inflow of water in the 'annual' layer subsequently frozen; Δ_{sa} time rate of change of ice mass in the 'annual' layer, i.e. net balance. Terms pertaining to the continuity of ice mass are shown in bold symbols.

5.4.2. PRECIPITATION

Accumulating precipitation gauges have been installed by the Kenya Water Department (KWD) in various locations on the mountain. In terms of their vicinity to Lewis Glacier (Map 5.4.2:1), the following gauges are of interest here: gauge no. 3 at 4200 m near Klarwill Hut; gauge no. 9 at 4800 m near Top Hut; and gauge no. 4 at 4520 m near Simba Col. However, these gauges have evidently not been read according to a firm schedule. In addition to the rare gaugings by KWD personnel, measurements have been made occasionally by interested visitors. Since 1974 readings of KWD gauges have become particularly sporadic. Table 5.4.2:1 lists precipitation totals over large intervals defined by coincident dates of measurement at the three gauges. Observations cover about 20 years at Klarwill's Hut, and about 15 years at the other two sites. Precipitation maybe somewhat smaller at Austrian Hut than at Klarwill's Hut, while the largest amounts are found at Simba Col on the North side of the mountain. From the data in Table 5.4.2:1, the long-term average annual precipitation in the area of Lewis Glacier is estimated at somewhat more than 800 mm.

As part of a feasibility study for a baseline monitoring station under the sponsorship of the World Meteorological Organization and the United Nations Environmental Programme (Schnell and Odh, 1977; Davies *et al.*, 1979), meteorological stations have been installed at four locations around 4200 m. However, these are too far away from the peak region to be of immediate use in glacier studies.

Regular and detailed precipitation measurements start with the beginning of 1978, when we installed gauges provided by the Kenya Meteorological Department (KMD), as described in Section 5.1.5 and Map 5.4.2:1. Observations during 1978–83 are summarized in Table 5.4.2:2.

The concurrent measurements with both daily and accumulating gauges at Roadhead and Teleki Ranger Camp are of interest in relation to Austrian Hut, where only the accumulating gauge is in use. The upper, receptor portion of the accumulating gauges is painted black, so that precipitation in solid form melts quickly and flows into the inner container where it is protected from evaporation. Inspection of the gauge shortly after fresh snowfall has shown this design to be satisfactorily effective. The inner container of the KMD design has a narrow snout and no oil film is used against evaporation, in contrast to the KWD gauges. One might expect slightly smaller monthly totals for the accumulating as compared to the daily gauges as a result of evaporation. On the other hand, the daily gauges may yield an under-estimate, because a certain amount of water will stick to the walls of funnel and container at each daily measurement, resulting in a deficit of water in the measuring glass. Observations at Roadhead Met Station Lodge and Teleki Ranger Camp presented in Table 5.4.2:2 bear out non-systematic and small differences of monthly totals obtained by the two gauge types. This suggests that the monthly precipitation totals reported for the accumulating gauge at Austrian Hut also provide reliable estimates of precipitation in the area of Lewis Glacier. Observations during numerous visits indicate that precipitation over most of Lewis Glacier occurs almost exclusively in solid form.

In accordance with the seasonality of large-scale circulation patterns and rainfall regimes as discussed in Chapter 3, Table 5.4.2:2 shows two precipitation seasons extending mainly from March to June and from September to December as in most of Central

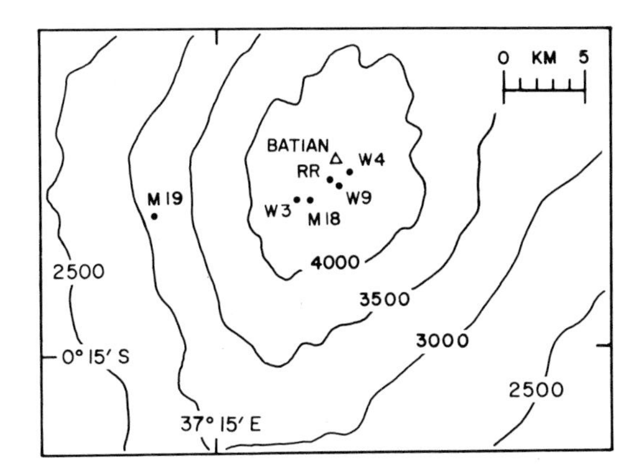

Map 5.4.2:1. Precipitation gauges at Mount Kenya. W3, W9 and W4 denote accumulating precipitation gauges of Kenya Water Department at Klarwill's Hut at 4200 m, near Austrian Hut at 4800 m, and at Simba Col at 4520 m respectively. M19 and M18 signify location of daily and accumulating gauges of Kenya Meteorological Department/Mount Kenya National Parks at Road Head Met Station Lodge at 3050 m (KMD no. 9037217) and at Teleki Ranger (Camp at 4200 m (KMD no. 9037218), respectively, RR refers to accumulating gauge on 'Radio Ridge' near Austrian Hut at 4800 m.

TABLE 5.4.2:1

Precipitation totals in mm obtained at accumulating gauges of Kenya Water Department. See Map 5.4.2:1 for station locations

Gauge	3 Klarwill's Hut 4200 m	9 Austrian Hut 4800 m	4 Simba Col 4520 m
22 Mar 1959			
	3880	–	–
5/6 Feb 1964			
	7036	7528	9725
24/25 Oct 1972			
	6680	4420	7200
23/24 Sept 1979			
	685	955	1030
17/19 Feb 1981			

Kenya. Table 5.4.2:3 lists rainfall index series for Central Kenya, (a) for the semesters March–August and September–February representative of the two rainy seasons, (b) for July–August, and (c) for January–February, the latter two being periods of minimum precipitation activity. Reference is also made to Fig. 2.5:5. However, the series of detailed precipitation measurements at Mount Kenya, Table 5.4.2:2, are still short and permit only limited comparison with the interannual variation of rainfall in Central Kenya at large.

TABLE 5.4.2:2

Precipitation totals in mm obtained at daily and at accumulating gauges of Kenya Meteorological Department during 1978–83. See Map 5.4.2:1 for station locations

Road Head/Met Station Lodge, 3050 m, KMD No. 9037217

	J	F	M	A	M	J	J	A	S	O	N	D	Year
Daily													
1978	22	217	248	284	160	80	34	84	141	266	132	180	1848
79	216	134	69	193	102	50	29	12	84	131	163	61	1244
80	0	20	38	179	266	51	2	55	60	53	161	46	931
81	34	44	204	461	290	119	142	114	211	176	139	104	2038
82	50	49	39	257	343	92	124	140	46	202	237	105	1684
83	25	82	51	152	114	75	119	82	146	249	199	133	1427
84	8	23											
Accumulating													
1978	–	--	–	253	–	75	43	79	164	278	84	182	(1158)
79		358		399	75		119		87	75		214	1327
80	36	20			458			32	?	107	158	39	850
81	27	32	241	474	160	104	150	150	60?	190	118	87	(1793)
82	85	48	48	208	355	111	135	148	80	123	237	120	1698
83	11	66	61	275	121	96	115	104	143	248	229	147	1633
84	28	24											

Teleki Ranger Camp, 4200 m, KMD No. 9037218

Daily

1978	–	–	–	164	64	24	40	29	111	139	76	82	(729)
79	110	48	74	96	144	37	22	14	47	66	157	33	848
80	21	27	49	64	126	22	1	24	8	82	118	44	586
81	23	40	264	151	125	25	19	95	56	71	71	46	986
82	25	37	33	165	134	66	7	43	26	156	162	65	919
83	22	49	20	167	84	42	34	81	57	108	120	140	924
84	15	43											

Accumulating

1978 – – ←– 149 →← 150 →←— 51 →←— 20 →←— 32 →← 21 →← 105 →← 119 →←— 70 →← 83 → (800)

79 ←———— 209 →←——— 229 ———→← 63 →←————— 36 —————→← 39 →←—— 55 →←——— 143 ———→ 774

80 ←-0→← 20 →←28 →←———— 253 ————————→←— 56 →←—— 16 →←—— 79 →←— 127 →←— 36 —→ 615

81 ←—— 19 →←—— 28 →←– 237 →←- 158 →←— 91 →←— 36 →←—— 20 →←— 77 →←—— 40 →←——178 →←— 70 →← 56 → 1010

82 ←—— 35 →←—— 44 →←—— 56 →← 140 →←— 177 →←— 44 →←—— 16 →←— 48 →←— 66 →←—— 79 →←— 128 →← 115 → 948

83 ←—– 24 →←—— 40 →←–– 45 →← 186 →←— 48 →←— 48 →←—— 20 →←— 77 →←—— 88 →←– 139 →←— 115 →← 138 → 968

84 ←—— 20 →←—— 43 →←——

Radio Ridge/Austrian Hut, 4800 m

Accumulating

1978 – ←—62——→←265 →← 150 →←— 55 →←—— 28 →←— 20→←—— 25 →←—— 96 →←— 123 →←—— 70 →←— 117 ——→ (1011)

79 ←———— 120 ———→←—— 241 ———→← 63 →←——— 54 ———→← 86 →←——— 127 ———→ 691

80 ←—— 24 →←——— 40→←——— 285 ————→←—19→←—— 14 →←— 131 →←— 171 →←—— 55 ——→ 739

81 ←—— 36 →←—— 15 ——→←200 →←—190 →←– 135 →←—— 32 →←— 85 →←—130 →←— 120 →←— 150 →←— 150 →←—— 52 ——→ 1295

82 ←—— 27 →←—— 40 ——→←51 →←— 79 →← 240 →←—— 36 →←— 29 →←— 56 →←—— 54 →←— 104 →←— 140 →←—134 ——→ 990

83 ←—— 4→←—— 28 ——→←15 →←—144 →← 49 →←—— 28 →←— 6 →←— 77→←—— 38 →←— 30 →←— 120 →←—120 —→ 659

84 ←— 13→←— 36 —→←

TABLE 5.4.2:3

Precipitation index series for Central Kenya. All-station averages of normalized (i.e. expressed as percentage of standard deviation) departure were computed from unpublished data of Kenya Meterological Department for station group Kiambu, Machakos, Muranga, Nakuru, Ngong, Nyeri. (a) the semesters March–August and September–February are representative of the two rainy seasons, while (b) July–August and (c) January–February are the periods of minimum precipitation activity

Nominal date	(a) Rainy seasons	(b) July–August	(c) January–February
1 March 80			– 66
	– 80		
1 Sept 79		– 43	
	+ 50		
M 79			+208
	+ 73		
S 78		– 41	
	+ 77		
M 78			+110
	+ 38		
S 77		+ 32	
	+100		
M 77			– 18
	– 56		
S 76		– 12	
	– 47		
M 76			– 74
	– 66		
S 75		+ 46	
	+ 6		
M 75			–125
	– 85		
S 74		+229	
	+ 86		
M 74			– 92
	– 54		
S 73		– 37	
	– 80		
M 73			+ 61
	+ 28		
S 72		– 64	
	– 44		
M 72			+ 25
	– 51		
S 71		+ 11	
	+ 26		
M 71			– 46
	–102		
S 70		– 58	
	+ 57		
M 70			+ 29

TABLE 5.4.3:1b
Readings of net balance stakes during 1979, in cm

Station No.	Jan 1–5	Feb 4	March 1	May 1	May 26–27	June 1	Sept 1	Sept 21–22	Oct 1	Nov 1	Dec 26–31
1	+ 73	+ 42	–	–	+ 9	+ 5	–	– 44	– 61	– 81	– 96
2	–	–	–	–	–	–	–	–	–	–	new 0
3	+247	–	–	(+252)	–	(+249)	–	(+189)	+190	+176	+186
4	–	–	–	–	–	–	–	–	–	–	new 0
5	–	–	–	–	–	–	–	–	–	–	new 0
6	–	–	–	–	–	–	–	–	–	–	new 0
7	–	–	+103	+ 95	+125	+ 98	–	+ 87	+ 82	+117	+104
8	+150	–	(+125)	+ 80	+ 79	+ 84	–	+ 55	+ 56	+ 35	+ 58
9	+ 71	+ 60	–	+ 40	+ 75	+ 46	–	+ 31	+ 36	+ 17	– 20
10	–	–	+ 97	(+106)	+125	(+111)	–	+118	+124	+135	+118
11	+176	+151	+127	+137	+142	+147	–	+132	+133	+ 99	+ 61
12	–	–	–	–	–	–	–	–	–	–	new 0
13	+145	–	–	+ 6	+ 5	+ 6	–	+ 20	+117	+ 91	+ 90
14	+105	–	–	+ 57	+ 57	+ 58	–	+ 74	+ 47	+ 28	– 2
15	+102	–	–	–	–	–	–	–	–	–	– 90
21	+ 94	+ 80	– 11	+ 5	+ 7	+ 7	–	– 13	– 14	– 40	– 60
22	–	–	– 45	– 85	– 84	– 85	–	– 75	– 86	–100	–141
23	–	–	– 42	– 43	– 41	– 43	–	–100	–101	–123	–168
24	–	+ 9	– 54	– 46	– 43	– 48	–	– 80	– 80	–120	–162
25	+ 24	– 2	– 3	– 62	– 55	– 63	–	– 92	– 93	–136	–175
26	–	– 3	– 77	– 58	– 58	– 58	–	–103	–103	–141	–198
31	–	+ 16	– 69	– 64	– 64	– 64	– 94	–121	–120	–148	–195
32	+ 20	– 20	– 98	–	– 92	– 92	–167	–151	–156	–181	–228
33	+ 19	–	–135	–141	–145	–142	–176	–	–213	–239	–284
34	– 41	–	–169	–	–	–156	–	–	–219	–245	–336
35	– 18	– 67	–156	–	–	–	–	–	–212	–220	–285
41	– 41	–	–168	–168	–	–	–	–	–245	–285	–346
42	– 12	–	–129	–	–	–172	–189	–	–198	–243	–324
43	– 54	–	–174	–203	–	–203	–222	–	–238	–266	–348
44	–108	–	–263	–273	–	–295	–329	–	–350	–372	–458
45	–245	–	–438	–483	–	–497	–527	–	–540	–593	–673
'X25'	0										
'X29'			0	+ 6	+ 7	+ 9		– 15	– 12	– 28	– 58
II	0										+ 64

TABLE 5.4.3:1c
Readings of net balance stakes during 1980, in cm

Station No.	Jan 4–5	Feb 1	Feb 2	March 1	May 24	July 19–20	Sept 1	Sept 27	Oct 3	Nov 1	Dec 1	Dec 28–31
1	–112	–	–	–	–	–	–	–	–	–	–	new 0
2	– 5	–	–	–	–	–	–	–	new 0	– 7	+ 22	+ 14
3	+181	+163	+161	–	–	+ 89	– 41	–	+ 2	+ 9	+ 62	+ 59
4	– 7	– 52	– 54	–	–	–	–	–	–	–	–	new 0
5	– 14	– 34	– 36	–	–	– 98	–160	–	–124	–118	–	new 0
6	– 20	– 53	– 54	–	–	– 68	(–140)	(–156)	(–)	(–)	– 39	– 57
7	+ 76	+ 30	+ 29	–	–	– 58	–	– 57	– 59	– 51	– 31	– 51
8	– 9	– 62	– 65	–	–	new 0	+ 11	–	+ 8	+ 11	–	–
9	– 44	– 87	– 88	–	–	–	–	–	–	–	–	–
10	+196	+146	+147	–	–	+124	+116	–	+120	+125	+155	+133
11	+ 43	+ 20	+ 20	–	– 16	– 32	– 25	– 26	– 26	– 10	+ 13	–
12	– 5	– 24	– 29	–	–	– 71	– 66	–	– 79	– 64	– 32	– 34
13	+ 61	+ 33	+ 32	–	–	– 78	– 77	–	– 99	– 77	– 59	– 68
14	– 26	– 59	– 62	–	–	–	–	–	–	–	–	new 0
15	–110	–162	–162	–	–	–	–	–	–	–	–	–
21	–103	–164	–165	–	–235	new 0	+ 5	+ 1	– 1	+ 10	+ 15	– 1
22	–155	–196	–187	–276	–	–290	–	–313	–	–	–300	–331
23	–187	–225	–227	–259	–	–	–	–	–	–	–	–
24	–188	–233	–225	–300	–285	–304	–	–310	–	–	–	–295
25	–176	–220	–222	–267	–	–303	–320	–327	–341	–318	–314	–330
26	–216	–259	–258	–323	–	–334	–375	–365	–	–	–	–
31	–211	–261	–261	–321	–	–370	–387	–402	–413	–428	–376	–403
32	–245	–254	–297	–323	–	–336	–349	–356	–360	–350	–343	–373
33	–292	–337	–	–361	–448	–472	(–493)	–499	–507	–505	–501	–526
34	–364	–394	–394	–447	–	–437	(–432)	–462	–466	–462	–478	–549
35	–302	–352	–360	–441	–	–447	–	–455	–463	–462	–457	–484
41	–368	–416	–422	–463	–508	–526	–	–546	–552	–557	–578	–631
42	–345	–396	–400	–459	–	–514	–518	–515	–534	–506	–508	–578
43	–340	–382	–387	–436	–	–496	(–521)	–510	–517	–508	–519	–565
44	–477	–525	–520	–563	–	–611	–633	–	–638	–652	–658	–694
45	–689	–738	–	< –775	–	–	–	–	–	–	–	–
'X25'												
'X29'	– 74	–125	–130									
II	+ 56					–121						
M						new 0						– 7

TABLE 5.4.3:1d
Readings of net balance stakes during 1981, in cm

Station No.	Feb 1	Feb 7	March 1	Apr 1	May 2	June 1	June 7	July 2	Aug 1	Sept 1	Oct 3–4	Nov 5	Nov 23	Dec 27–29
1	+ 18	–	– 46	+ 3	+ 15	– 6	–	– 45	+ 9	+ 28	–	–	– 34	–
2	+ 12	–	– 42	+ 6	+ 37	+ 13	–	– 25	+ 18	+ 25	–	+ 10	+ 8	+ 9
3	+ 57	+ 27	+ 27	+ 70	+110	+ 76	–	+ 52	+107	+123	–	+157	(+166)	+173
4	–	– 79	– 66	– 31	– 12	– 36	–	– 70	– 13	+ 8	–	+ 31	+ 30	+ 34
5	– 3	–	–100	– 48	– 32	– 24	–	– 56	– 34	– 12	–	+ 25	+ 25	–
6	– 45	–101	–116	– 81	– 53	– 69	–	– 52	– 85	– 73	–	– 63	– 65	–
7	–	– 83	–109	– 57	– 52	– 73	– 71	– 87	– 49	– 22	– 12	– 3	0	– 10
8	–	–	–	–	–	–	–	–	–	–	–	–	–	–
9	–	–	–	–	–	–	–	–	–	–	–	–	–	–
10	+131	+104	+ 88	+117	+138	+117	+112	+ 79	+156	+183	+191	+197	+198	+189
11	+ 18	– 47	– 55	– 7	– 4	– 22	– 33	– 69	– 3	+ 23	+ 26	+ 38	+ 44	+ 31
12	– 39	– 61	– 57	– 9	+ 12	– 6	–	– 23	–	+ 23	–	+ 54	+ 58	+ 56
13	– 71	–	–138	– 84	–	–	–	–	–	–	–	–	–	–
14	– 8	–	–103	– 50	–	– 17	– 58	–	–	–	–	–	–	–
15	–	–	–	–	–	–	–	–	–	–	–	–	–	–
21	– 3	–	– 47	+ 3	–	–	–	–	–	–	–	–	–	–
22	(–324)	–	–	(–394)	(–404)	(–420)	–	(–455)	(–435)	(–395)	–	(–404)	(–404)	(–407)
23	–	–	–	–	–	–	–	–	–	–	–	–	–	–
24	–300	–	–	–	–	–	–	–	–	–	–	–	–	–
25	–362	–410	–431	–397	–487	(–426)	–	(–447)	(–421)	(–416)	–	(–416)	(–412)	(–413)
26	–	–	–	–	–	–	–	–	–	–	–	–	–	–
31	–408	–473	–498	–448	–484	–508	–503	–547	–504	–486	–	–485	–482	–499
32	–368	–476	–493	–446	–426	–443	–	–477	–	–	–	–	–	–
33	–585	–609	–634	–589	–	–	–	–685	–654	–643	–	–653	–645	–
34	–	–	–	–	–	–	–	–	–	–	–	–	–	–
35	–	–	–	–	–	–	–	–	–	–	–	–	–	–
41	–	–508	–725	–692	–688	–704	–	–743	–713	–693	–	–673	–665	–
42	–623	–678	–709	–687	–684	–701	–	–730	–689	–664	–	–674	–678	–
43	–567	–651	–672	–642	–662	–688	–	–709	–687	–663	–	–680	–611	–738
44	–697	–	(–708)	–	–	–	–	–	–	–	–	–	–	–
45	–	–	–	–	–	–	–	–	–	–	–	–	–	–
‘X25’														
‘X29’														
II														
M	+ 9	– 72	– 89	–	–122	–133	–	–157	–130	–110	– 81	– 79	– 75	–141

TABLE 5.4.3:1e

Readings of net balance stakes during 1982, in cm

Station No.	Dec 81 27–29	Jan 82 25–26	Feb 6	Feb 27	March 31	April 30	June 1	July 1	July 30	Sept 1	Oct 7	Oct 10	Nov 1	Nov 30	Dec 22–23
51	0	– 13	– 22	– 61	– 80	– 28	+ 9	+ 24	– 10	– 3	– 54	– 53	– 27	+ 26	+ 7
1	0	– 10	– 21	– 50	– 75	– 27	+ 9	+ 10	+ 3	+ 8	– 39	– 31	– 23	+ 48	+ 13
2	+ 9	–	–	–	–	–	–	–	–	–	–	–	–	–	–
3	+173	+162	–	+142	+109	+137	+199	+217	+211	+220	+162	–	+178	+266	+253
4B(4)	+ 34	+ 23	–	+ 3	– 10	+ 25	+ 43	+ 52	+ 40	+ 44	+ 10	+ 58	+ 37	–	–
4A			–	–	–	–	–	–	–	–	–	–	–	–	–
6	0	– 15	–	– 64	– 76	– 47	– 7	– 18	– 22	– 13	– 52	– 42	– 13	+ 38	+ 13
7	– 10	– 29	– 69	– 76	(– 4)?	(+ 28)?	(– 33)	(– 10)	(– 20)	–	(– 39)	(– 27)	– 5	+ 30	+ 42
10	+189	+169	–	+121	+101	+126	+151	+172	+166	+171	+178	+208	+202	+213	+263
11	+ 31	+ 7	–	– 4	– 58	– 25	– 11	+ 12	+ 6	+ 13	– 27	– 5	+ 2	+ 33	+ 40
12	+ 56	+ 44	–	+ 24	– 13	+ 44	+ 76	+ 87	+ 79	+ 80	(+ 25)	(+ 74)	0	+ 37	+ 96
13	0	– 8	–	– 35	– 66	– 31	– 13	+ 3	– 2	+ 3	(– 58)	(– 42)	– 41	+ 31	+ 42
71 (M)	–144	–181	–	–	–	–	–	–	–	–	–				
81	0	– 49	– 89	– 91	– 92	–	–	–	–	–	–				
22	–407	–450	–507	–525	–556	–531	–505	–492	–495	–491	–545	–	–	–	–
25	–413	–443	–488	–488	–500	–478	–467	–457	–481	–478	–510	–488	–468	–460	–453
31	–499	–544	–	–605	–664	–637	–610	–577	–587	–	–600	–	–578	–568	–545
32	0	– 52	– 95	–107	–143	–117	– 79	– 96	–109	–102	–147	(–133)	(–122)	–110	–119
33	0	– 41	– 69	– 72	–153	–128	– 98	– 82	– 91	– 87	–138	–	–109	–	–
41 (old 42)	0	– 35	–	– 89	– 97	– 73	– 49	– 35	– 51	– 44	– 87	–	– 89	– 73	– 93
42		0	–	– 60	– 71	– 46	– 27	– 16	– 28	– 27	–106	–	–	–	–
43	–738	–771	–	–823	–867	–835	–803	–788	–799	–763	–866	–	–840	–821	–835

TABLE 5.4.3:1f

Readings of net balance stakes during 1983, in cm

Station No.	Jan 83 31	Feb 6	Feb 28	Mar 15	Apr 4	May 1	June 1	July 1	Aug 1	Sept 1	Oct 1	Oct 31	Dec 1	Dec 83 19–22	Jan 84 1
51	– 52	– 56	– 77	–123	–157	–171	–181	–190	–196	–183	–177	–165	–159	–159	–149
1	– 43	– 56	– 79	– 93	–105	–118	–134	–130	–153	–139	–	–	–	0 (new)	–
2	–	–	–	–	–	– 9	– 16	– 27	– 41	– 39	– 30	– 16	– 2	–	–
3	+189	+179	+154	+119	+ 93	+ 65	+ 59	+ 45	+ 42	+ 63	+ 77	+ 99	+117	+155	+159
4B(4)	–	–	+ 23	– 38	– 66	–			–	–	–	–	–	–	–
4A		–							–						
6	– 64	–	–126	–	–138	–132	–144	–	–	–		–	+118	+130	+141
7	– 56	– 88	–118	–133	–	–	–	–	–	–	–	–	–	–	–
10	+183	+ 74	+142	+ 74	+ 50	+ 59	+ 38	+ 33	+ 31	+ 34	+ 52	+ 79	+101	+111	+121
11	– 47	– 58	– 80	–127	–	–	–	–	–	–	–	–	–	0 (new)	+ 18
12	+ 27	+ 15	+ 4	– 61	– 74	– 72	– 86	– 98	–101	– 95	– 59	– 39	– 32	– 35	– 26
13	– 31	– 41	– 65	–110	–131	–130	–185	–193	–199	–186	–168	–143	–131	–131	–103
71(M)	–	–	–	–	–	–	–	–	–	–	–	–			
81	–	–	–	–	–	–	–	–	–	–	–	–			
22	–	–	–	–714	–729	–735	–754	–	–	–	–	–			
25	–520	–551	–576	–596	–617	–625	–634	–	–	–	–	–	–616	–616	–604
31	–655	–669	–698	–743	–761	–768	–793	–827	–841	–838	–827	–813	–796	–775	–776
32	–224	–235	–259	–315	–333	–339	–351	–366	–392	–	–	–		0 (new)	+ 8
33	–235	–	–281	–302	–	–	–	(–230)	(–247)	(–241)	(–240)	(–228)	(–220)	(–224)	(–215)
41(old 42)	–127	–207	–235	–256	–327	–323	–339	–376	–384	–376	–354	–343	–335	0 (new)	+ 8
42	–	–178	–208	–289	–309	–303	–311	–286	–	–	–	–	–	–	–
43	–935	(–990?)	–	–1,291	–1,315	–1,309	–1,315	(–1,331?)	(–1,327)	(–1,258)	(–1,249)	(–1,233)	(–1,132)	(–1,132)	(–1,126)

TABLE 5.4.3:2a

Stake heights and height changes by budget your in cm

Station No.	1978 Jan 25–29 h [cm]	1978 March 4–5 h [cm]	March 78 to March 79 Δh [cm]	1979 March h [cm]	March 79 to March 80 Δh [cm]	1980 March h [cm]	March 80 to March 81 Δh [cm]		1981 March h [cm]	March 81 to March 82 Δh [cm]	1982 March h [cm]
1	0	–25	+ 64	(+ 39)	–219	(–180)	–	new	– 46	(– 37)	(– 83)
2	0	– 4	+192	(+188)	–279	(– 91)	–	new	– 42	(+ 11)	(– 31)
3	0	+29	+304	(+333)	–188	(+145)	–118		+ 27	+115	+142
4	0	– 9	+147	(+138)	–235	(– 97)	–	new	– 66	+ 69	+ 3
5	0	+37	+101	(+138)	–192	(– 54)	–	new	–100	(+ 74)	(– 26)
6	0	+28	+135	(+163)	–249	(– 86)	– 30		–116	(– 13)	(–129)
7	0	–14	+117	+103	–123	(– 20)	– 89		–109	+ 33	– 76
8	0	–20	+145	+125	–242	(–117)	–		–	–	–
9	0	–17	+ 80	(+ 63)	–193	(–130)	–		–	–	–
10	0	–18	+105	+ 97	+ 1	(+ 98)	– 10		+ 88	+ 33	+121
11	0	–14	+141	+137	–130	(– 3)	– 53		– 55	+ 51	– 4
12	0	+11	+220	(+209)	–276	(– 76)	+ 10		– 57	+ 81	+ 24
13	0	– 6	+165	(+159)	–154	(+ 5)	–143		–138	+103	– 35
14	0	–28	+137	(+109)	–201	(– 92)	–	new	–103	–	–
15	0	–61	+113	(+ 52)	–266	(–214)	–		–	–	–
21	0	–11	0	– 11	–214	(–215)	–	new	– 47	–	–
22	0	– 6	– 39	– 45	–251	–296	–156		(–452)	– 81	(–533)
23	0	–11	– 31	– 42	–217	–259	–		–	–	–
24	0	– 8	– 46	– 54	–246	–300	– 67		(–367)	–	–
25	0	–21	+ 18	– 3	–264	–267	–164		–431	– 57	–488
26	0	–18	– 59	– 77	–246	–323	–		–	–	–
31	0	–12	– 57	– 69	–257	–326	–172		–498	–107	–605
32	0	–21	– 77	– 98	–261	–359	–134		–493	(– 59)	(–552)
33	0	–12	–123	–135	–226	–361	–273		–634	(– 95)	(–729)
34	0	–22	–147	–169	–298	–467	–165		(–632)	–	–
35	0	–38	–118	–156	–285	–441	–182		(–623)	–	–
41	0	–29	–139	–168	–295	–463	–262		–725	(– 81)	(–806)
42	0	–39	– 90	–129	–330	–459	–250		–709	(–191)	(–900)
43	0	–41	–133	–174	–286	–460	–212		–672	–151	–823
44	0	–32	–231	–263	–300	–563	–217		–780	–	–
45	0	–66	–372	–438	–349	–787	–		–	–	–
'X29'		0		0	–176	(–176)					

1981 to late January 1982 was filled based on observations at stations 4 and 10. For station 6, the missing value from late November to late December 1981 was interpolated from stations 4, 7, and 10. For station 32 the gap from early July to late December 1981 was recovered from stations 31 and 33. For station 33, the missing value for late November to late December 1981 was approximated by observations at station 31. For station 41, the gap from late November to late December 1981 was filled from stations 31 and 43. For station 42, the missing record from late November 1981 to late January 1982 was recovered from stations 41 and 43.

Height changes obtained from Table 5.4.3:1 and the aforementioned data reductions

TABLE 5.4.3:2b
Stake heights and height changes by budget year in cm

Station No.	1982 March h [cm]	March 82 to March 83 Δh [cm]	1983 March h [cm]	March 83 to March 84 Δh [cm]	1984 March h [cm]
1	(−83)	−29	(−112)		
51		−16	(−77)	−63	(−140)
2	(−31)	–	–		
3	+142	+12	+154	−21	+133
4	+3	+20	+23		
6	(−129)	−62		+93	
7	−76	−42	−118		
10	+121	+21	+142	−80	+62
11	−4	−76	−80		
12	+24	−20	+4	−76	−72
13	−35	−30	−65	−83	−148
71	0	–	–		
81	0	–	–		
22	(−533)				
25	−488	−88	−576	−40	−616
31	−605	−93	−698	−186	−884
32	(−552)	−152	(−704)	(−165)	(−869)
33	(−729)	−209	(−938)	−57	(−995)
41	(−806)	−146	(−952)	(−175)	(−1,127)
42	(−900)	−148	(−1048)	−136	(−1,184)
43	−823	−197	(−1020)		

TABLE 5.4.3:3
Densities in pits dug on 30/31 December 1978. The two half-years are defined by ice horizons, in g cm^{-3}

Pit station	Density	
	Sept 77–March 78	March–Sept 78
0	0.625	0.601
3	0.677	0.550
13	0.595	0.578
21	0.770	0.770

TABLE 5.4.3:4

Budget year March 1978 to March 1979. Height changes in cm, density in g cm^{-3}, and net balance in cm of liquid water equivalent

Station No.	Height change (cm)	Density (g cm^{-3})	Net balance (cm)
1	+ 64	0.60	+ 38
2	+192	0.60	+115
3	+304	0.60	+182
4	+147	0.60	+ 88
5	+101	0.60	+ 60
6	+135	0.65	+ 88
7	+117	0.70	+ 82
8	+145	0.72	+104
9	+ 80	0.64	+ 51
10	+105	0.69	+ 72
11	+141	0.74	+104
12	+220	0.60	+132
13	+165	0.60	+ 99
14	+137	0.60	+ 82
15	+113	0.60	+ 68
21	0	0.77	0
22	− 39	0.90	− 35
23	− 31	0.90	− 28
24	− 46	0.90	− 41
25	+ 18	0.90	− 16
26	− 59	0.90	− 53
31	− 57	0.90	− 51
32	− 77	0.90	− 69
33	−123	0.90	−111
34	−147	0.90	−132
35	−118	0.90	−106
41	−139	0.90	−125
42	− 90	0.90	− 81
43	−133	0.90	−120
44	−231	0.90	−208
45	−372	0.90	−335

were multiplied by an appropriate density to give water-equivalent net balance. For stations in the accumulation region, densities representative of the two accumulation seasons were derived from pit profiles at stations II, 3, 13, and 21 (Table 5.4.3:3). The former three stations provide a spatial sampling of the upper part of the accumulation region. From the tolerance in the subsequent analysis, and the similar magnitude of numbers in Table 5.4.3:2, differentiation of density values between the two half-years and various locations in the upper accumulation region is not found warranted. Accordingly, a density of 0.60 g cm^{-3} is assigned to the upper accumulation region, that is to stations 1 through 5 and 12 through 15 for both semesters and the total year. For stations 6 through 11 and 15, in the transition from accumulation to ablation region, density

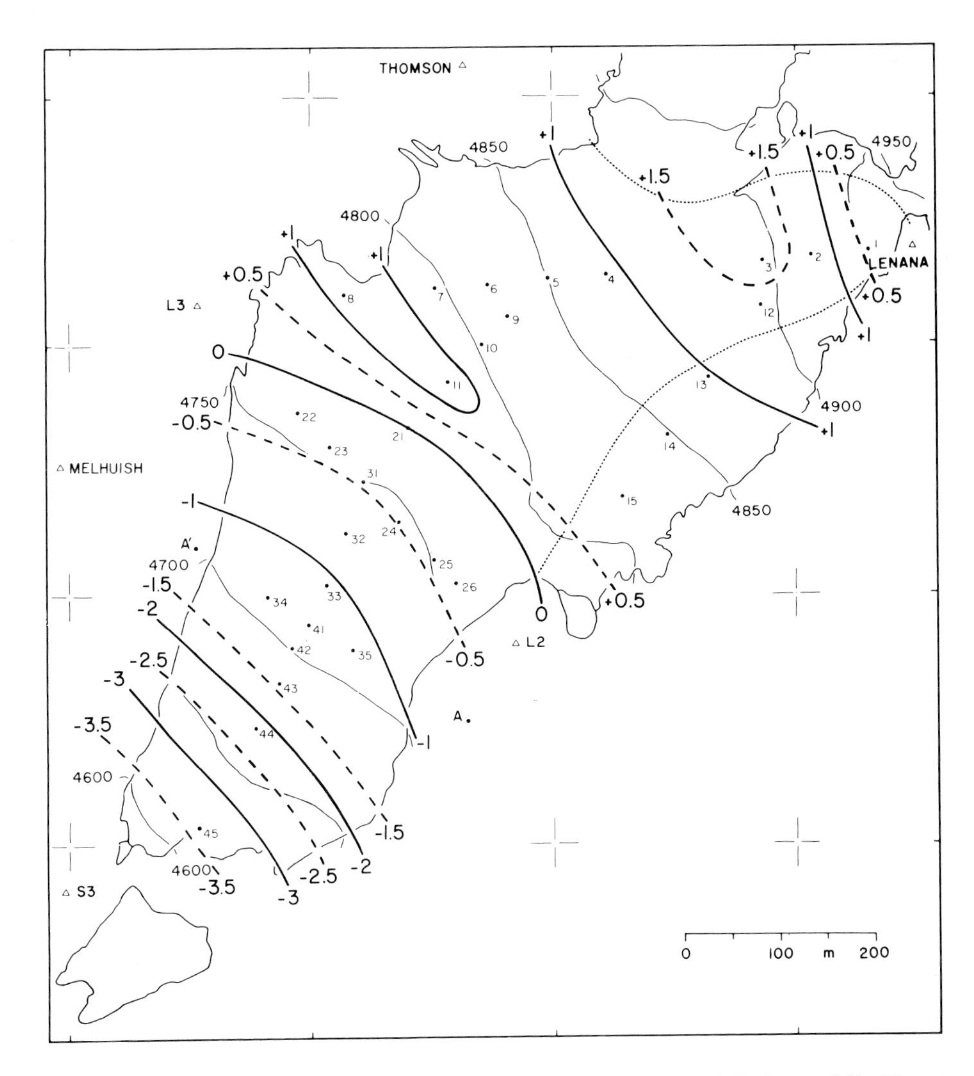

Map 5.4.3:1. Net balance during budget year March 1978 to March 1979, in m of liquid water equivalent.

was interpolated linearly by elevation between stations 5 and 21. At station 25 a value of 0.80 g m^{-3} was used for both semesters and the total year, being intermediate between the density for station 21 and the maximum firn density of 0.85 g m^{-3}. Finally, for the stations in the ablation area, namely stations 22 through 24 and 26 through 45, annual height changes were multiplied by 0.9 g m^{-3}, the density of ice.

Annual net balance thus obtained for all stations is listed in Tables 5.4.3:4, 5.4.3:6, 5.4.3:8, and 5.4.3:10, and plotted in Maps 5.4.3:1–5.4.3:4. Tables 5.4.3:5, 5.4.3:7, 5.4.3:9, and 5.4.3:11 show the evaluation of net balance from Maps 5.4.3:1–5.4.3:4 by 50 m elevation bands. The corresponding vertical profiles of net balance are furthermore illustrated in Figs. 5.4.3:1–5.4.3:4.

TABLE 5.4.3:5

Budget year March 1978 to March 1979. Net balance by 50 m elevation bands as obtained from Map 5.4.3:1 by planimetering. Area in m^2, liquid water equivalent net balance in cm, and volume equivalent in m^3

m	Area in 1978 in 10^2 m^2	Net balance cm	Volume m^3
5 000			
	37	+ 25	+ 925
4 950			
	156	+110	+17 130
4 900			
	592	+104	+61 290
4 850			
	500	+ 70	+34 860
4 800			
	705	+ 28	+19 790
4 750			
	442	− 91	−40 085
4 700			
	320	−169	−54 190
4 650			
	184	−300	−55 205
4 600			
	14	−370	− 5 180
4 550			
total glacier	2 950	− 7	−20 665

Consistent with the precipitation records, Table 5.4.2:2, the Maps 5.4.3:1–5.4.3:4 and the vertical profiles plotted in Fig. 5.4.3:5 illustrate the strongly negative net balance during 1979/80 and 1980/1, as compared to the more nearly balanced conditions in 1978/9 and 1981/2.

It seems plausible to expect a relation between the spatial patterns of net balance and of the surface heat budget, especially ablation. Considering the mass redistribution associated with the flow dynamics, maps of ice thickness differences obtained from successive topographic mappings (Maps 5.2.7:2, 5.2.7:3, 5.2.7:4) are less pertinent than the results from a stake network (Maps 5.4.3:5, 5.4.3:6). Comparison of the 1978–82 average net balance, Map 5.4.3:5, with the annual mean clear-sky downward-directed shortwave radiation, Map 5.4.5:1 shows little pattern similarity. However, the latter quantity is not representative of the ablation pattern, given the spatial distribution of surface albedo, among other factors. In fact, the prominent altitude dependence of net balance indicated by the Map 5.4.3:5 is qualitatively consistent with that of albedo, Map 5.4.6:1.

For the four year period 1978–82 as a whole, two independent estimates of the mass continuity are available from (a) the aero-photogrammetric mappings of the surface topography (Maps 4.3:10*, 4.3:11*, 5.2.7:3), and (b) the stake network (Tables 5.4.3:5,

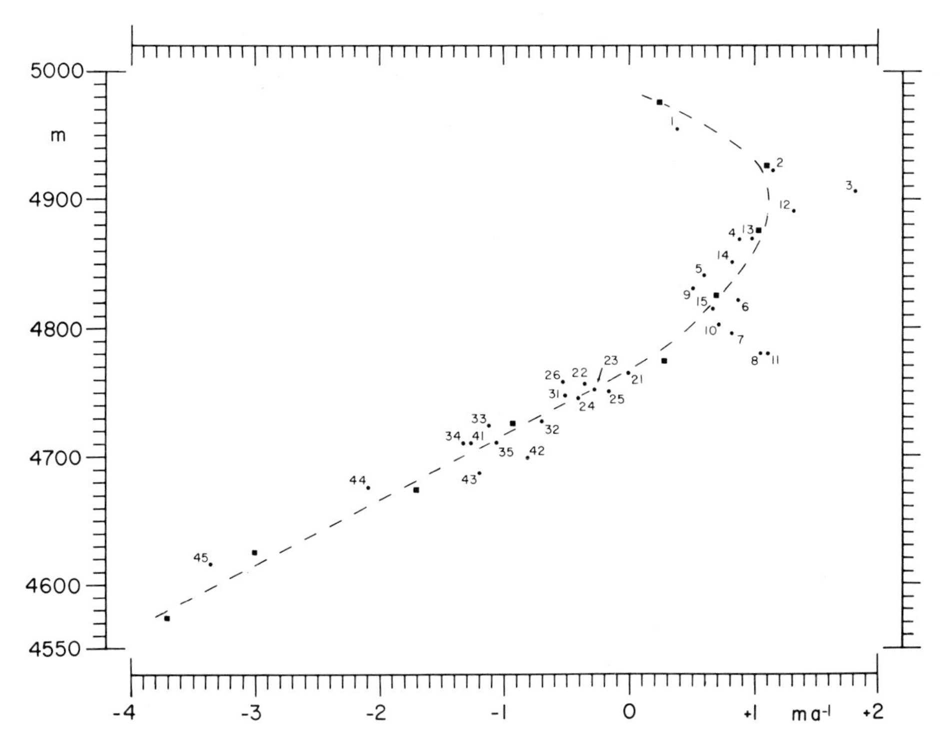

Fig. 5.4.3:1. Vertical net balance profile during budget year March 1978 to March 1979, in m of liquid water equivalent.

5.4.3:7, 5.4.3:9, 5.4.3:11, Maps 5.4.3:5, 5.4.3:6). Maps 5.2.7:3 and 5.4.3:6 are dimensionally comparable. Moreover, results by 50 m elevation bands are summarized in Table 5.4.3:12.

Maps 5.2.7:3 and 5.4.3:6 show changes of similar magnitude and largest negative values in the lower glacier, but the patterns differ in detail. This is in part due to the spatial resolution of the stake network, but in part to the redistribution of mass related to the flow dynamics. The latter factor is also indicated in Table 5.4.3:12, in that for most of the upper glacier the height decrease derived from the 1978 and 1982 mappings is larger than that obtained from the stake network, while the reverse is true for the lower glacier. For the glacier as a whole the two independent techniques yield thickness or volume decreases that agree closely within plausible error tolerances. Referring to the discussion in Section 5.4.1, the close agreement between the two independent methods indicates that melting at depth greater than the annual layer is of subordinate importance (in fact $\Delta_s \approx \Delta_{sa}$ and $Q_{mm} \approx Q_{ma}$).

The relation of surface lowering, net balance, and flow dynamics suggested from the intercomparison of Maps 5.2.7:3, 5.4.3:6, and Table 5.4.3:12, is in terms of the pertinent processes substantiated in Fig. 5.3.2:6, especially part B. The vertical component of ice flow at the surface is downward in the upper, and upward in the lower glacier. This corresponds to surface velocity vectors plotted in the longitudinal–vertical cross-section in part A of Fig. 5.3.2:6. Returning to part B of Fig. 5.3.2:6, net balance over the four

TABLE 5.4.3:6

Budget year March 1979 to March 1980. Height changes in cm, density in g cm^{-3}, and net balance in cm of liquid water equivalent

Station No.	Height change (cm)	Density (g cm^{-3})	Net balance (cm)
1	−219	0.60	−131
2	−279	0.60	−167
3	−188	0.60	−113
4	−235	0.60	−141
5	−192	0.60	−115
6	−249	0.65	−161
7	−123	0.70	− 86
8	−242	0.72	−174
9	−193	0.64	−124
10	+ 1	0.69	+ 1
11	−130	0.74	− 96
12	−276	0.60	−166
13	−154	0.60	− 92
14	−201	0.60	−120
15	−266	0.60	−159
21	−214	0.77	−165
22	−251	0.90	−226
23	−217	0.90	−195
24	−246	0.90	−222
25	−264	0.90	−237
26	−246	0.90	−222
31	−257	0.90	−231
32	−261	0.90	−235
33	−226	0.90	−302
34	−298	0.90	−268
35	−285	0.90	−257
41	−295	0.90	−266
42	−330	0.90	−297
43	−286	0.90	−257
44	−300	0.90	−270
45	−349	0.90	−314
'X29'	−176		−158

year interval 1978–82 is negative at all elevations along the longitudinal axis, with losses increasing towards the lower glacier. The sum of these two quantities, both expressed as geometric distance rather than liquid water equivalent units, is expected to represent the change in surface topography over the four year interval 1978–82. The available data allowed this evaluation for five of the stations. These calculations should be compared with the observed surface lowering, as documented by the 1978 and 1982 mappings of the glacier topography. The two independent approaches agree in major respects: (a) negative values are obtained at all stations, with surface lowering increasing from the upper towards the lower glacier; (b) the surface lowering exceeds the negative net balance in the upper, but is less than the net balance effect in the lower glacier, thus demonstrating the role of vertical velocity, and mass redistribution through flow dynamics. The values

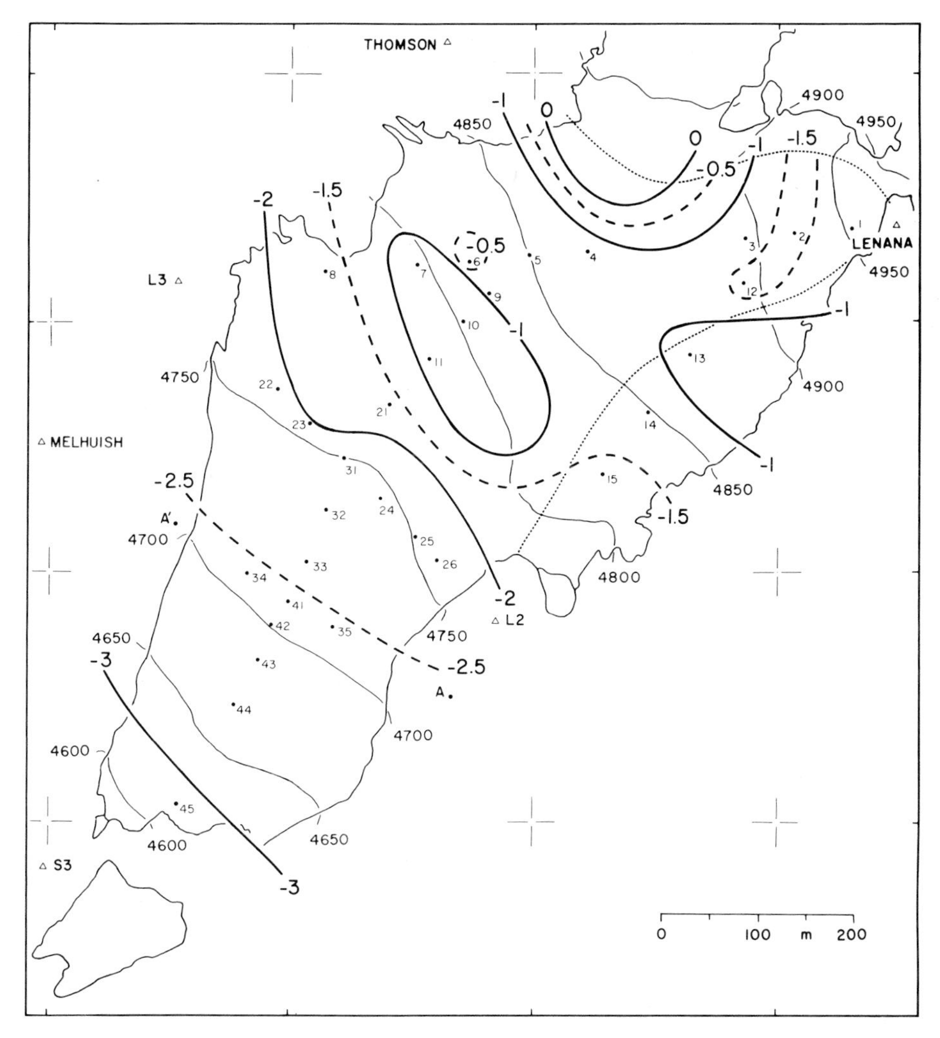

Map 5.4.3:2. Net balance during budget year March 1979 to March 1980, in m of liquid water equivalent.

of surface lowering obtained by the two independent methods differ, which may in part be due to the error tolerances in field measurements and evaluation procedures. However, it is apparent from part B of Fig. 5.3.2:7 that in the upper glacier the surface lowering calculated from vertical flow velocity and net balance slightly exceeds the value documented by the 1978 and 1982 mappings, while the reverse is the case in the lower glacier. These systematic departures may be due to the fact that the flow is confluent in the upper glacier where its bed narrows downstream, while in the lower glacier difference is favored both by the widening of the valley and the upward convex bedrock topography apparent in the central portion of the snout region. Part B of Fig. 5.3.2:7 thus underlines the important role of flow dynamics and mass redistribution in modulating the effect of net balance on surface topography.

TABLE 5.4.3:7

Budget year March 1979 to March 1980. Net balance by 50 m elevation bands as obtained from Map 5.4.3:2 by planimetering. Area in m^2, liquid water equivalent net balance in cm, and volume equivalent in m^3

m	Area in 1978 in 10^2 m^2	Net balance cm	Volume m^3
5 000			
	37	−140	− 5 180
4 950			
	156	−136	− 21 160
4 900			
	592	− 97	− 57 700
4 850			
	500	−126	− 63 050
4 800			
	705	−167	−117 520
4 750			
	442	−239	−105 820
4 700			
	320	−270	− 86 400
4 650			
	184	−302	− 55 520
4 600			
	14	−340	− 4 760
4 550			
total glacier	2 950	−175	−517 110

While this monograph continued in press, the mass balance year March 1982 to March 1983 was completed. The precipitation measurements are listed in Table 5.4.2:1 and the net balance stake readings in Tables 5.4.3:1e to f and 5.4.3:2b. Measurements were evaluated as described above for the earlier years. As is apparent from Table 5.4.3:1e to f observations were complete for all stations, except stake 22 which was lost early in the mass budget year, and stakes 13 and 43 for which the interval 6–28 February 1983 was missing. This was recovered for stake 13 with reference to station 3, and for station 43 with reference to station 41. The evaluation of net balance by stations is summarized in Table 5.4.3:13. The station values of net balance are plotted in the vertical profile Fig. 5.4.3:6. The net balance at the stations was further plotted in a map and contour analyzed as shown in Map 5.4.3:7. Net balance by 50 m elevation bands is obtained from Map 5.4.3:7 by planimetering. Results are listed in Table 5.4.3:14 and plotted in the vertical net balance profile Fig. 5.4.3:6. For the glacier as a whole a volume loss of 188×10^3 m^3 is obtained, corresponding to a glacier average net balance of −0.72 m. This should be viewed in context with the annual precipitation of about 940 mm (Table 5.4.2:2.). Tables 5.4.2:2 and 5.4.3:1e to f, show the continuation of the observation program into the sixth consecutive mass budget year.

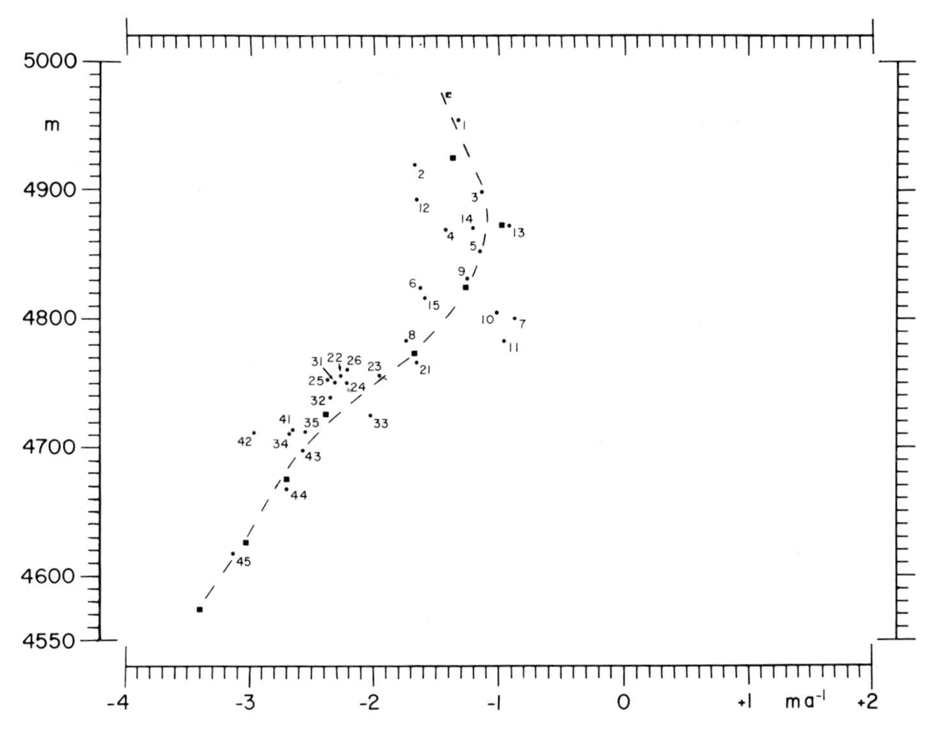

Fig. 5.4.3:2. Vertical net balance profile during budget year March 1979 to March 1980, in m of liquid water equivalent.

TABLE 5.4.3:8

Budget year March 1980 to March 1981. Height changes in cm, density in g cm^{-3}, and net balance in cm of liquid water equivalent

Station No.	Height change (cm)	Density (g cm^{-3})	Net balance (cm)
1	–	0.60	–
2	–	0.60	–
3	–118	0.60	– 71
4	–	0.60	–
5	–	0.60	–
6	– 30	0.65	– 20
7	– 89	0.70	– 62
8	–	0.72	–
9	–	0.64	–
10	– 10	0.69	– 7
11	– 53	0.74	– 40
12	+ 10	0.60	+ 6
13	–143	0.60	– 86
14	–	0.60	–
15	–	0.60	–
21	–	0.77	–
22	–156	0.90	–140
23	–	0.90	–
24	– 67	0.90	– 60
25	–164	0.90	–148
26	–	0.90	–
31	–172	0.90	–155
32	–134	0.90	–121
33	–273	0.90	–246
34	–165	0.90	–148
35	–182	0.90	–164
41	–262	0.90	–236
42	–250	0.90	–225
43	–212	0.90	–191
44	–217	0.90	–195
45	–	0.90	–

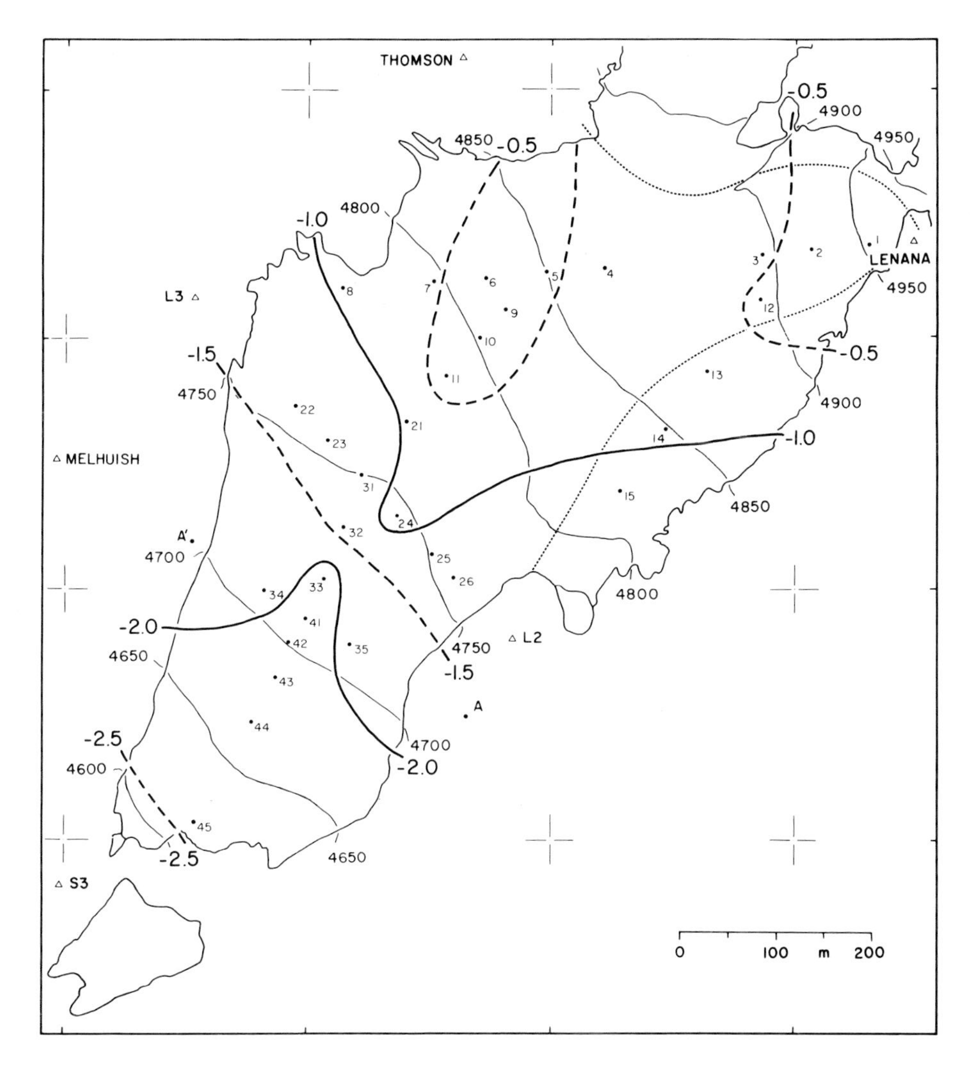

Map 5.4.3:3. Net balance during budget year March 1980 to March 1981, in m of liquid water equivalent.

TABLE 5.4.3:9

Budget year March 1980 to March 1981. Net balance by 50 m elevation bands as obtained from Map 5.4.3:3 by planimetering. Area in m^2, liquid water equivalent net balance in cm, and volume equivalent in m^3

m	Area in 1978 in 10^2 m^2	Net balance cm	Volume m^3
5000			
	37	– 40	– 1480
4950			
	156	– 40	– 6240
4900			
	592	– 72	– 43050
4850			
	500	– 78	– 39100
4800			
	705	–103	– 73167
4750			
	442	–171	– 75601
4700			
	320	–224	– 71698
4650			
	184	–250	– 46000
4600			
	14	–250	– 3500
4550			
total glacier	2950	–121	–359836

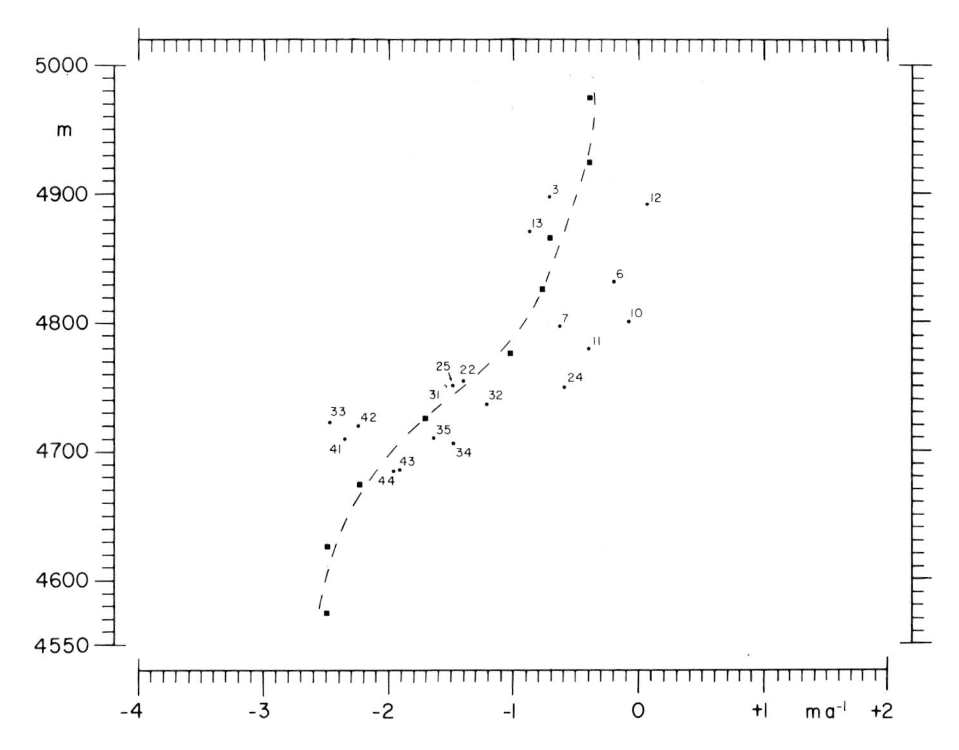

Fig. 5.4.3:3. Vertical net balance profile during budget year March 1980 to March 1981, in m of liquid water equivalent.

TABLE 5.4.3:10

Budget year March 1981 to March 1982. Height changes in cm, density in g cm^{-3}, and net balance in cm of liquid water equivalent

Station No.	Height change (cm)	Density (g cm^{-3})	Net balance (cm)
1	(– 37)	0.60	– 22
2	(+ 11)	0.60	+ 7
3	+115	0.60	+ 69
4	+ 69	0.60	+ 41
5	(+ 74)	0.60	+ 44
6	(– 13)	0.65	– 8
7	+ 33	0.70	+ 23
8	–	0.72	–
9	–	0.64	–
10	+ 33	0.69	+ 23
11	+ 51	0.74	+ 38
12	+ 81	0.60	+ 48
13	+103	0.60	+ 62
14	–	0.60	–
15	–	0.60	–
21	–	0.77	–
22	– 81	0.90	– 73
23	–	0.90	–
24	–	0.90	–
25	– 57	0.90	– 51
26	–	0.90	–
31	–107	0.90	– 96
32	(– 59)	0.90	– 53
33	(– 95)	0.90	– 85
34	–	0.90	–
35	–	0.90	–
41	(– 81)	0.90	– 73
42	(–191)	0.90	–172
43	–151	0.90	–136
44	–	0.90	–
45	–	0.90	–

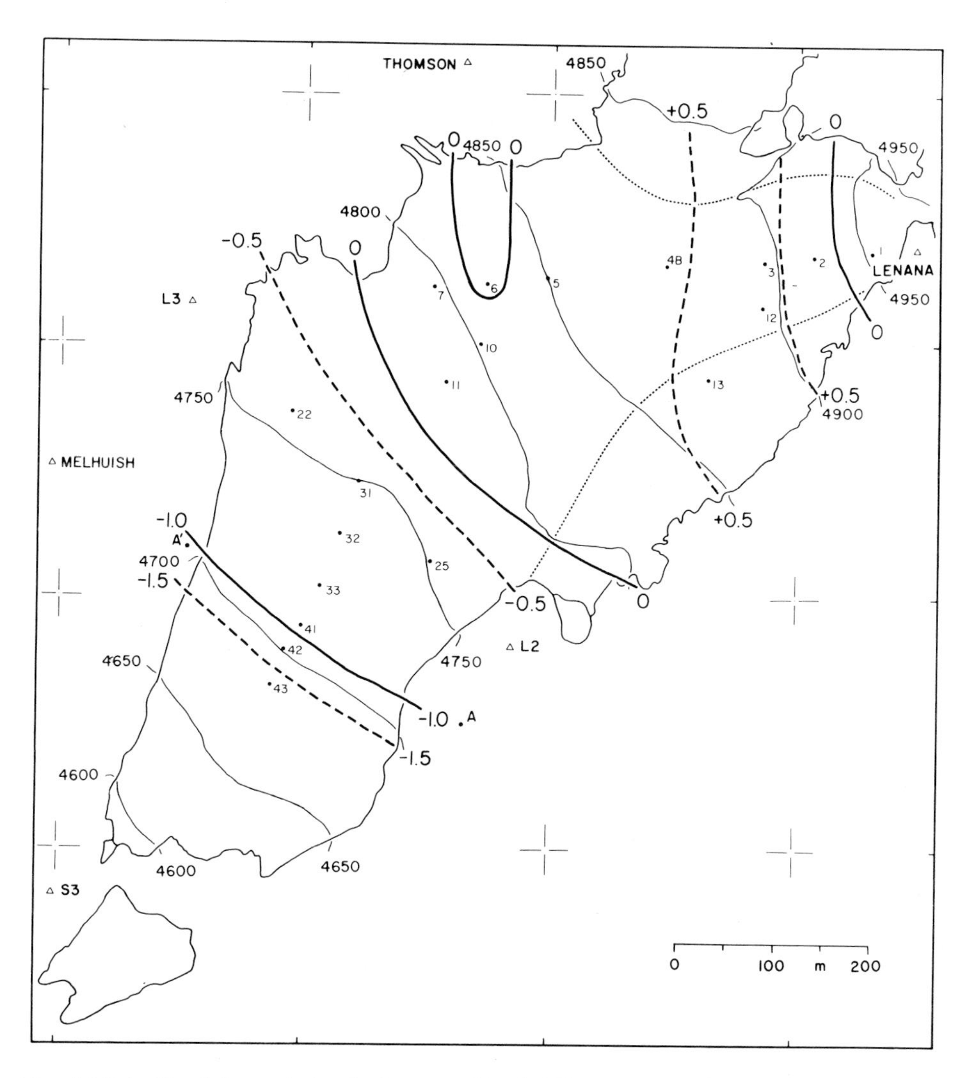

Map 5.4.3:4. Net balance during budget year March 1981 to March 1982, in m of liquid water equivalent.

TABLE 5.4.3:11

Budget year March 1981 to March 1982. Net balance by 50 m elevation bands as obtained from Map 5.4.3:4 by planimetering. Area in m^2, liquid water equivalent net balance in cm, and volume equivalent in m^3

m	Area in 1978 in 10^2 m^2	Net balance cm	Volume m^3
5000			
	37	− 21	− 765
4950			
	156	+ 18	+ 2771
4900			
	592	+ 44	+ 26121
4850			
	500	+ 16	+ 7850
4800			
	705	− 22	− 15857
4750			
	442	− 83	− 36852
4700			
	320	−170	− 52893
4650			
	184	−200	− 36800
4600			
	14	−200	− 2800
4550			
total glacier	2950	− 37	−109225

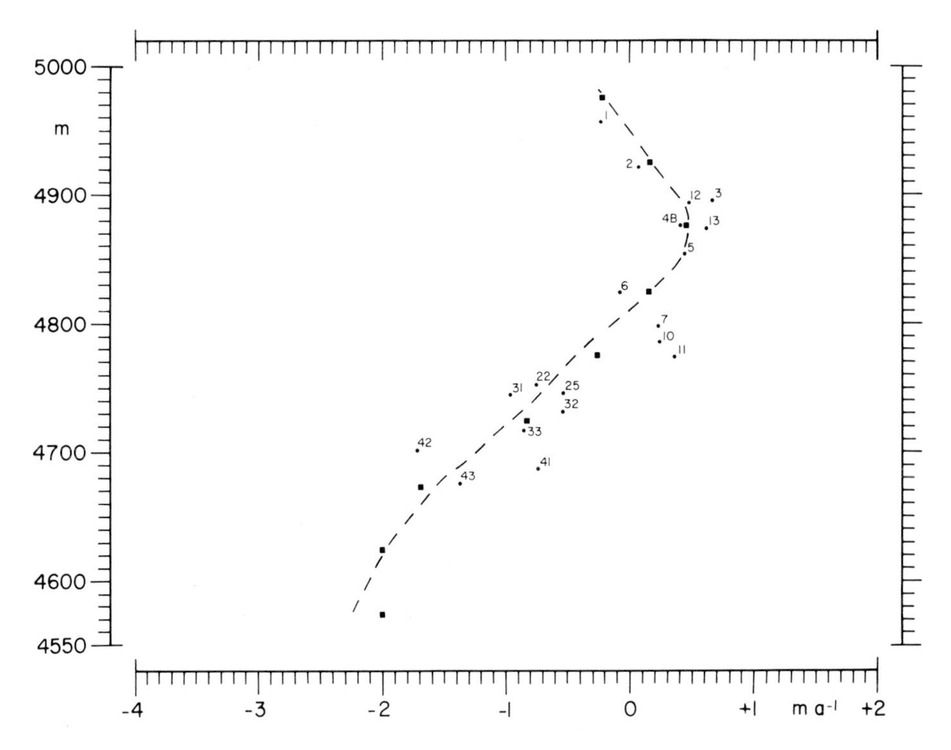

Fig. 5.4.3:4. Vertical net balance profile during budget year March 1981 to March 1982 in m of liquid water equivalent.

TABLE 5.4.3:12

Net balance and height changes in cm from March 1978 to March 1982, by 50 m elevation bands (of 1978 mapping). (a) H_m, height change from 1978 to 1982 mapping (source: Maps 4.3:10*, 4.3:11*, 5.2.7:3); (b) H_{sa}, height change from stake network (source: Map 5.4.3:6); (c) Δ_{sa}, net balance from stake network in liquid water equivalent (source: Tables 5.4.3:5, 5.4.3:7, 5.4.3:9, 5.4.3:11, and Map 5.4.3:5)

m	H_m	H_{sa}	Δ_{sa}
5 000			
	– 150	– 292	– 176
4 950			
	– 180	– 80	– 48
4 900			
	– 263	– 35	– 21
4 850			
	– 154	– 173	– 118
4 800			
	– 259	– 325	– 264
4 750			
	– 455	– 650	– 584
4 700			
	– 719	– 925	– 833
4 650			
	– 946	–1 170	–1 052
4 600			
	– 900	–1 290	–1 160
4 550			
glacier average	– 362	– 396	– 340
volume equivalent in 10^3 m^3	–1 067	–1 169	–1 007

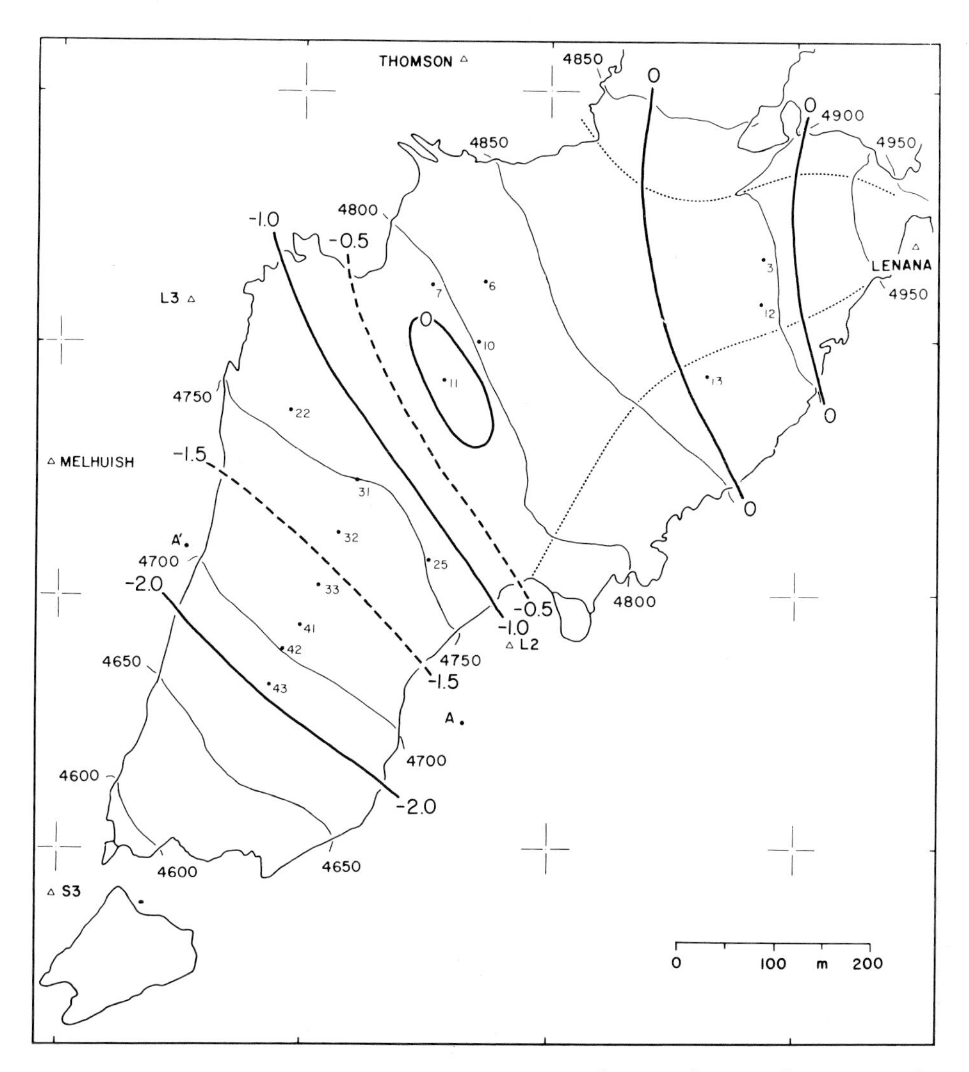

Map 5.4.3:5. Net balance averaged for the budget years 1978/79, 1979/80, 1980/81, and 1981/82, in m of liquid water equivalent; from Tables 5.4.3:4, 5.4.3:6, 5.4.3:8, 5.4.3:10, and Maps 5.4.3:1–5.4.3:4.

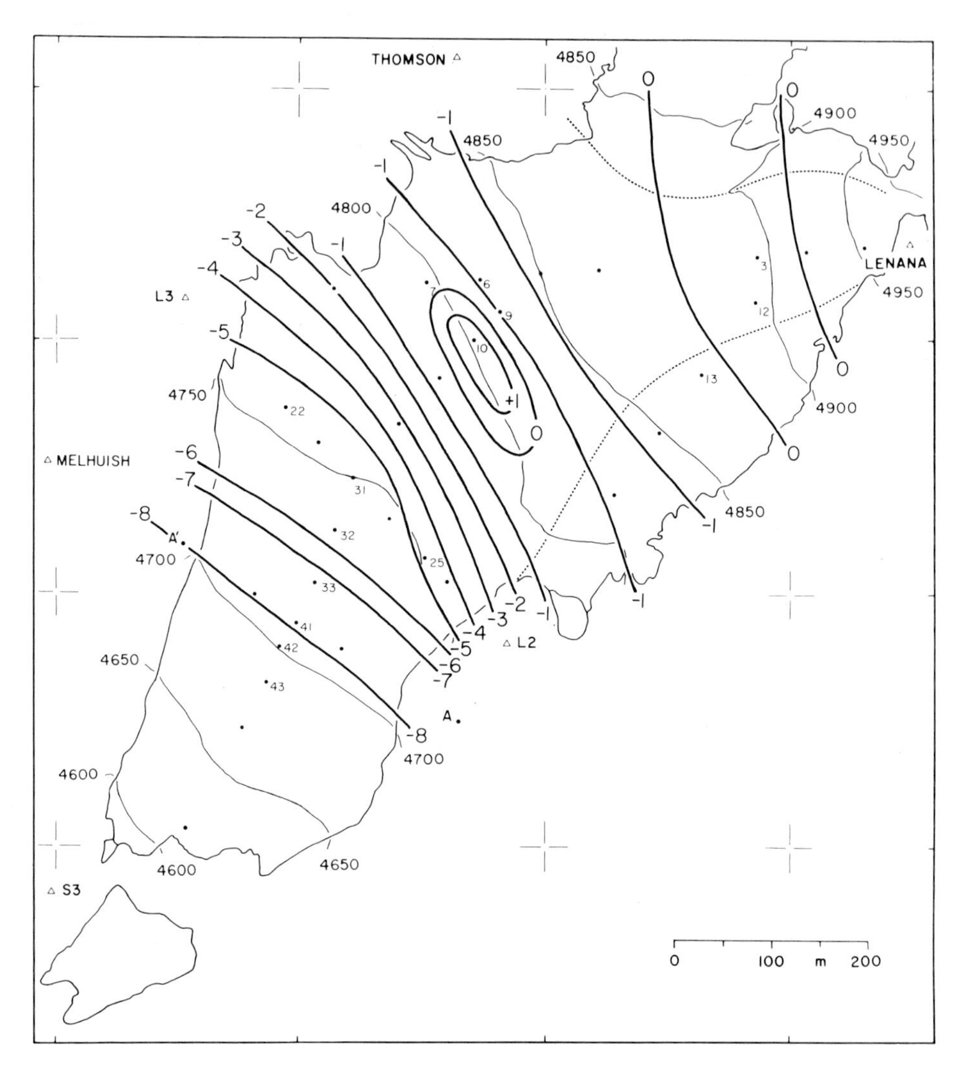

Map 5.4.3:6. Height change at stake network summed over the budget years 1978/79, 1979/80, 1980/81, 1981/82, in m; from Tables 5.4.3:4, 5.4.3:6, 5.4.3:8, 5.4.3:10.

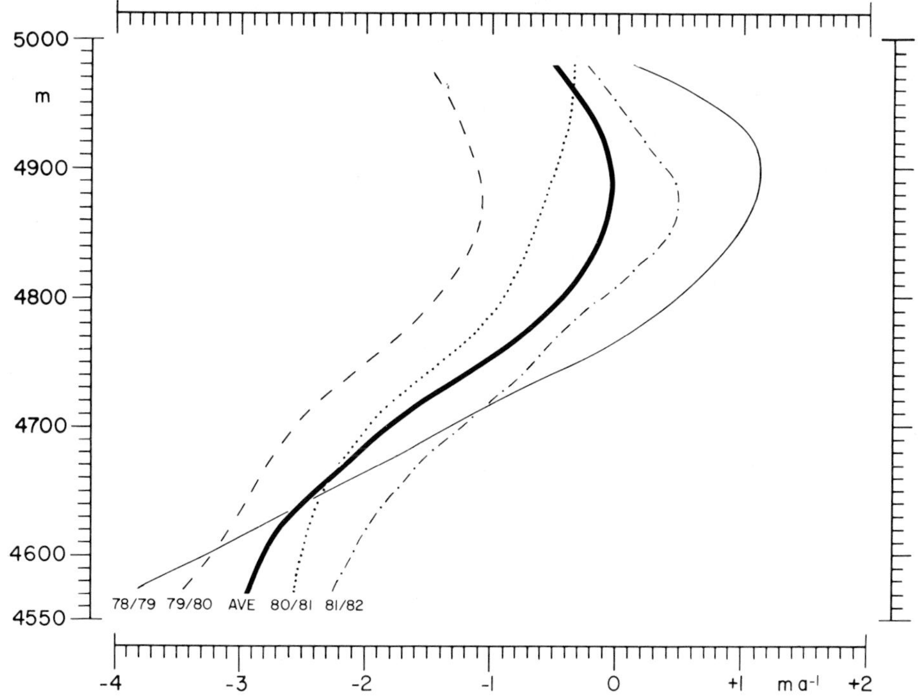

Fig. 5.4.3:5. Vertical net balance profiles, in m of liquid water equivalent, for the budget years 1978/79 thin solid, 1979/80 broken, 1980/81 dotted, 1981/82 dash-dotted, and the four years combined heavy solid lines; from Figs. 5.4.3:1–5.4.3:4.

TABLE 5.4.3:13

Budget year March 1982 to March 1983. Height changes in cm, density in g cm^{-3}, and net balance in cm of liquid water equivalent

Station No.	Height change (cm)	Density (g cm^{-3})	Net balance (cm)
51	− 16	0.60	− 10
1	− 29	0.60	− 17
–	–	0.60	–
3	+ 12	0.60	+ 7
4B	+ 20	0.60	+ 12
6	− 62	0.65	− 40
7	− 42	0.70	− 29
10	+ 21	0.69	+ 15
11	− 76	0.74	− 56
12	− 20	0.60	− 12
13	− 41	0.60	− 24
22	–	0.90	–
25	− 88	0.90	− 80
31	− 93	0.90	− 84
32	−152	0.90	−137
33	−209	0.90	−188
41	−146	0.90	−131
42	−148	0.90	−133
43	−197	0.90	−177

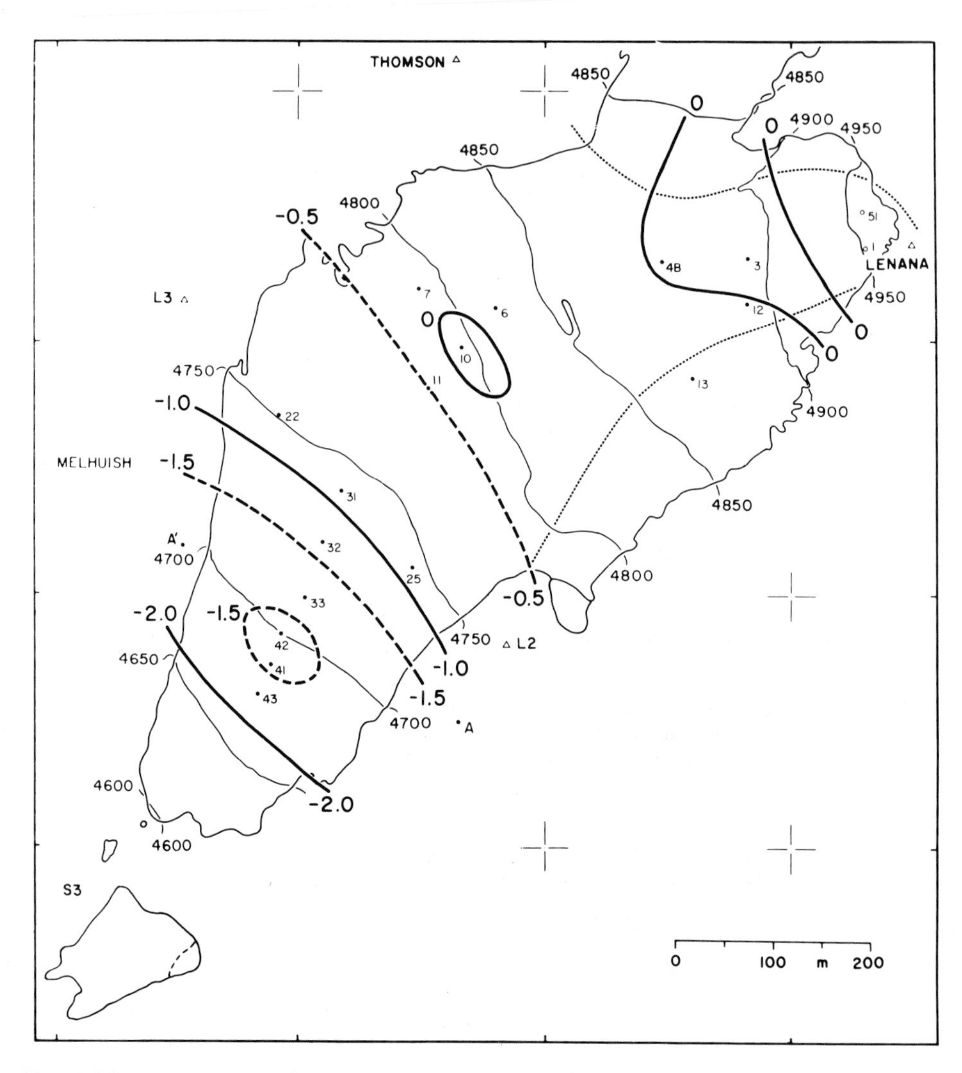

Map 5.4.3:7. Net balance during budget year March 1982 to March 1983, in m of liquid water equivalent.

TABLE 5.4.3:14

Budget year March 1982 to March 1983. Net balance by 50 m elevation bands as obtained from Map 5.4.3:7 by planimetering. Area in m^2, liquid water equivalent net balance in cm, and volume equivalent in m^3

m	Area in 1982 in 10^2 m^2	Net balance (cm)	Volume (m^3)
5000			
	28	− 10	− 280
4950			
	140	− 2	− 375
4900			
	568	− 14	− 7790
4850			
	438	− 37	− 16085
4800			
	628	− 52	− 32800
4750			
	430	−135	− 58000
4700			
	247	−173	− 42970
4650			
	129	−230	− 29670
4600			
	1	−250	− 250
4550			
total glacier	1609	− 72	−188220

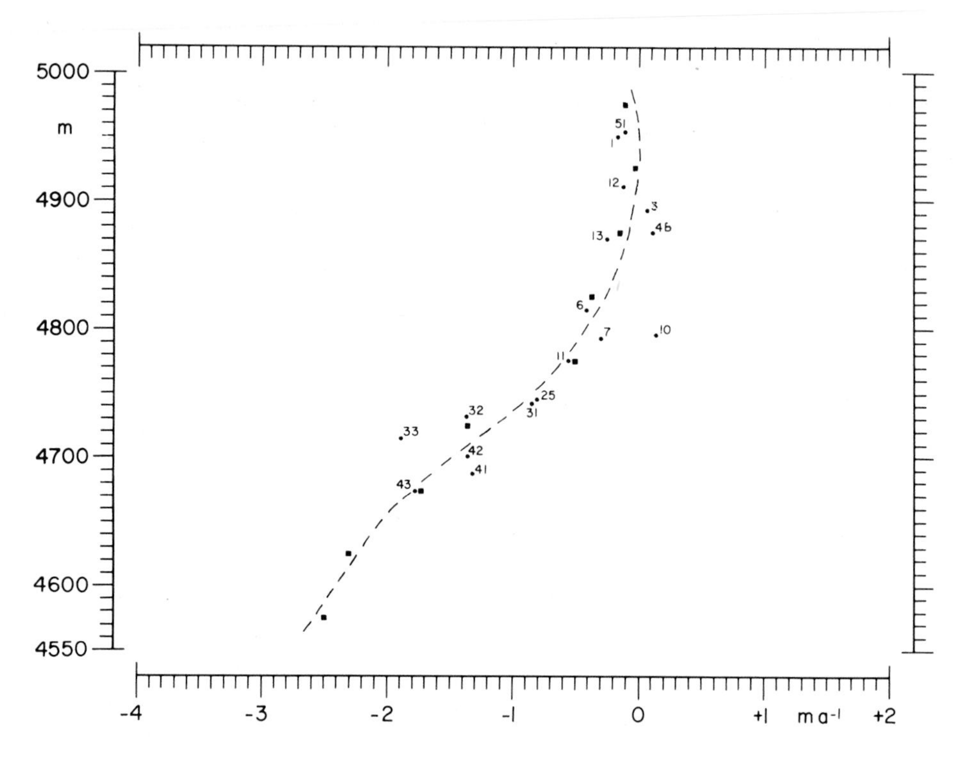

Fig. 5.4.3:6. Vertical net balance profile during the budget year March 1982 to March 1983, in m of liquid water equivalent.

5.4.4. WATER RUNOFF

The first measurements of water runoff from the glacier were performed by the IGY Mount Kenya Expedition during 2–13 January 1958, by means of a V-notch weir and a water level recorder installed at the exit of Lewis Tarn. A facsimile of the chart record was obtained from Frank Charnley, Nairobi. Ice formation and frost heaving affect the recorder mechanism during the night and morning while the trace is good for the greater part of the diurnal cycle. Information on representative average amounts and characteristics of the daily march of discharge is summarized in Table 5.4.4:1 and Fig. 5.4.4:1.

Discharge is calculated from the measured water levels at the weir by an empirical formula. For a 90 degree notch, King and Brater (1963, pp. 5–16) give

$$Q = 2.52 \times H^{2.47} \qquad 5.4.4:(1)$$

where H is the water level above the apex of the V-notch in feet, and Q the discharge in cubic feet per second.

In January 1974 we installed a V-notch weir of KWD at the exit of Lewis Tarn. The daily march of water level was monitored through several diurnal cycles. Readings at various times of day were obtained during occasional visits in the course of about a half-year period. Seepage through the small dam piled up at the weir later precluded meaningful measurements. Water level readings were converted to discharge using formula 5.4.4:(1). Results of this measurement series are presented *in extenso* in Table 5.4.4:2 and Fig. 5.4.4:1.

During the 1977–8 expedition, a current meter was available to measure velocities in two vertical transects, a few meters apart, across the rocky outlet below Lewis Tarn. Water levels were measured concurrently with the flow velocities, at three points for the upper and at four for the lower transect. On this basis, water level-discharge relationships were established for each transect, as illustrated in Fig. 5.4.4:2. Results from the two gauging sites agree closely, the lower transect yielding only slightly larger discharge values over most of the range of interest. Fig. 5.4.4:2 reduces the task of discharge determination to the measurement of water levels. Measurements through various diurnal

TABLE 5.4.4:1

Water discharge at the exist of Lewis Tarn during 4–12 January 1958, in 10^{-3} m^3 s^{-1} (= ℓ s^{-1})

	Amount	hr LT
mean	17	–
average maximum	29	17:50
average minimum	9	12:10
average on hour	21	15:00
average on hour	21	21:00
absolute maximum	42	17:30 (12 Jan)
absolute minimum	4	12:30 (5 Jan)

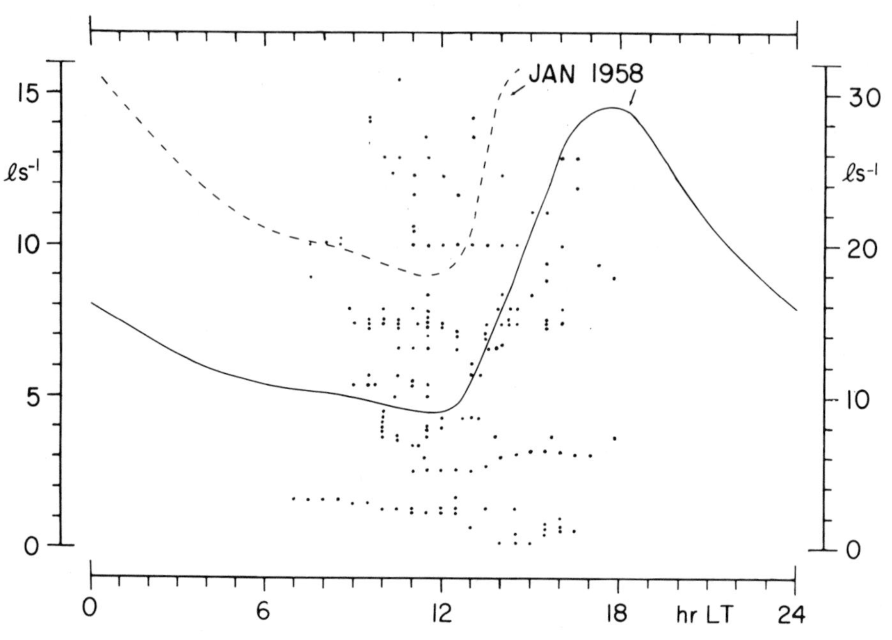

Fig. 5.4.4:1. Water discharge measured at the exit of Lewis Tarn. The average diurnal cycle during 4–12 January 1958 is depicted by solid line referring to right-hand scale. The 0 to 14:30 hr LT portion of this cycle is re-plotted as broken-line referring to left-hand scale (source: Table 5.4.4:1). Dots referring to left-hand scale denote measurements during January–June 1974 (source: Table 5.4.4:2). Units in 10^{-3} m^3 s^{-1} (= ℓ s^{-1}).

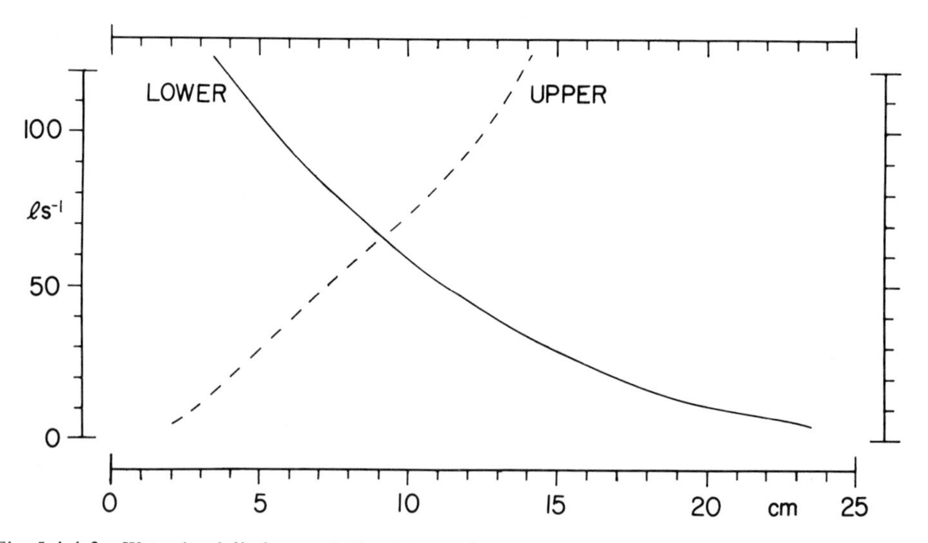

Fig. 5.4.4:2. Water level-discharge relationship, ℓ s^{-1} *vs* cm, for two transects below the exit of Lewis Tarn. Solid and broken lines refer to lower and upper transects, respectively.

TABLE 5.4.4:2

Water discharge at the exit of Lewis Tarn during 1974, in 10^{-3} m^3 s^{-1} (= ℓ s^{-1})

January									February		
day	hr LT	amount	day	hr LT	amount	day	hr LT	amount	day	hr LT	amount
3	11:00	2	4	07:00	2	14	10:00	4	3	14:00	7
	11:30	3		07:30	2	15	11:24	4	5	12:00	4
	12:00	3		08:00	2	17	17:50	4	6	13:15	6
	12:30	3		08:30	2	19	15:45	4	7	14:15	8
	13:00	3		09:00	2	20	11:25	4	10	11:30	5
	13:30	3		09:30	2	22	11:15	4	11	13:00	4
	14:00	3		10:00	1	24	14:15	7	12	11:00	5
	14:30	3		10:30	1	25	13:30	7	13	12:00	4
	15:00	3		11:00	1	27	16:00	8	15	11:30	7
	15:30	3		11:30	1	29	15:30	7	16	10:30	7
	16:00	3		12:00	1	31	13:50	4	17	9:30	6
	16:30	3		12:30	1				18	13:15	4
	17:00	3	5	15:30	1				21	13:30	7
				16:00	1				22	14:00	7
				16:30	1				23	11:30	7
			6	14:30	1				25	11:30	4
				15:30	1				26	13:00	6
									27	13:50	7
									28	11:30	5

March														
day	hr LT	amount	day	hr LT	amount	day	hr LT	amount	day	hr LT	amount	day	hr LT	amount
2	9:30	7	3	9:00	5	15	10:00	7	16	7:30	10	18	12:30	7
	10:00	8		9:30	7		10:30	7		8:00	10	19	11:30	7
	10:30	7		10:00	7		11:00	7		8:30	10	20	10:30	6
	11:00	5		10:30	4		11:30	10		9:00	7	24	11:00	3
	11:30	8		11:00	1		12:00	10		9:30	7	26	11:30	7
	12:00	7		12:00	1		12:30	10		10:00	7	27	10:30	4
	12:30	7		12:30	1		13:00	10		10:30	7	28	10:00	4
	13:00	6		13:00	1		13:30	7		11:00	10	30	10:30	7
	13:30	7		13:30	1		14:00	7		11:30	10	31	11:00	11
	14:00	7		14:00	0		14:30	7		12:00	7			
	14:30	8		14:30	0		15:00	7		12:30	10			
	15:00	8		15:00	0		15:30	8		13:00	10			
	15:30	9		15:30	1		16:00	8		13:30	10			
	16:00	10		16:00	1		16:30	8		14:00	10			
	16:30	13	5	12:45	4		17:00	8		14:30	10			
			6	16:00	7		17:30	8		15:00	11			
			8	13:50	8					15:30	11			
			9	13:50	7					16:00	13			
			10	12:30	7					16:30	13			
			11	11:30	7									
			13	11:00	7									
			14	9:45	5									

Table 5.4.4:2, continued

April day	hr LT	amount	May day	hr LT	amount	June day	hr LT	amount
1	14:00	8	1	10:30	16	1	11:00	8
13	12:00	12	2	8:30	10	2	15:30	7
15	10:00	13	3	11:30	13	3	10:05	4
16	8:50	8	4	9:30	5	11	10:05	4
20	12:00	10	5	11:30	4	12	15:30	9
21	10:15	12	6	12:30	2	13	9:30	14
22	11:25	14	7	14:30	1	14	10:00	4
			9	12:30	12	16	11:15	7
			10	10:30	13			
			11	11:30	7			
			12	9:30	14			
			16	10:00	4			
			17	11:00	12			
			18	13:00	14			
			20	10:00	4			
			21	14:30	7			
			22	13:00	14			
			25	7:30	9			
			26	10:25	5			
			27	11:00	12			
			28	15:30	7			
			29	17:15	9			
			30	11:00	11			
			31	14:00	12			

cycles performed during IGY in January 1958 (Table 5.4.4:1) and during the 1977–8 expedition (Fig. 5.4.4:3) showed the timing of minimum and maximum to be around 11:30–12:00 and 17:30–18:00 hours LT. Evaluation of complete diurnal cycles indicates that the arithmetic mean of values at these two times of day provides a good estimate of the daily average discharge. Accordingly, water level measurements were performed by a local collaborator at the beginning of each month. Results of this monitoring effort are summarized in Table 5.4.4:3.

The measurements during January 1958, January–June 1974, and January 1978–June 1980, as summarized in Tables 5.4.4:1–5.4.4:3, and Figs. 5.4.4:1 and 5.4.4:3 provide an orientation on the characteristics of water runoff from Lewis Glacier. There is a marked diurnal cycle, with a minimum towards 12:00 and a maximum towards 18:00 hours LT, reflecting the phase lag against the diurnal cycle of melting. As with the latter, the water runoff varies appreciably between days and with season. The average annual discharge rate is estimate to be of the order of 15 $\ell\,s^{-1}$ throughout the day. This seems a characteristic magnitude of the 'background' runoff throughout long periods with conditions not particularly favorable for melting. By contrast, on a few days during the ablation season, such as on 5–6 January 1978 and 1 January 1980 (Table 5.4.4:3, Fig. 5.4.4:3), maximum discharge in the evening was found to approach 100 $\ell\,s^{-1}$, with a corresponding minimum of only 10 $\ell\,s^{-1}$. These are rare events, but in view of the large water volume involved, they are not unimportant in the annual mass budget.

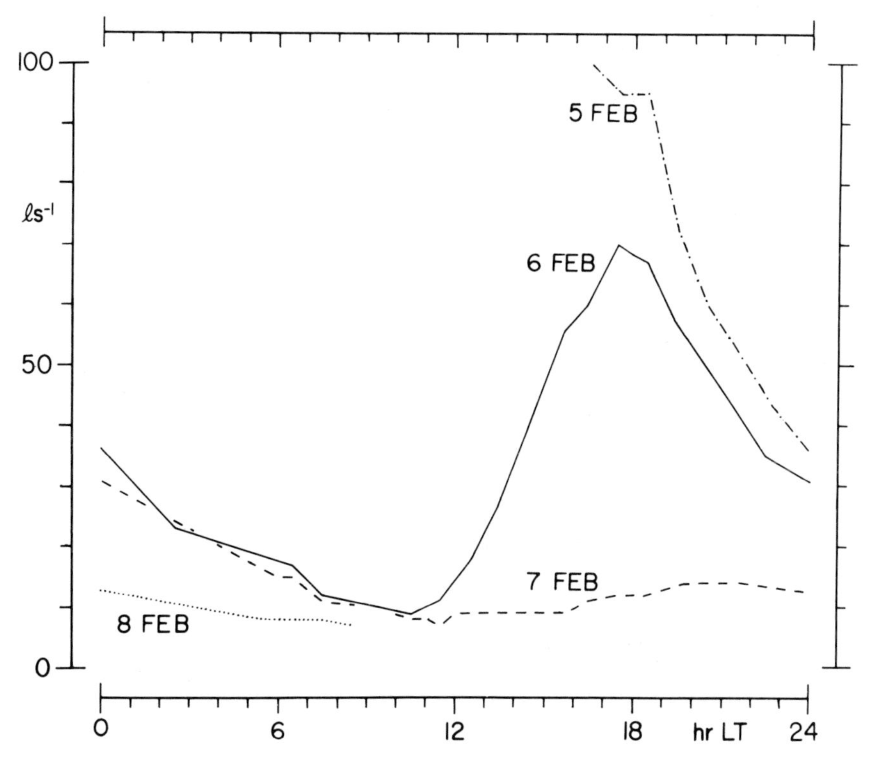

Fig. 5.4.4:3. Diurnal march of water discharge measured at the exit of Lewis Tarn in February 1978, in 10^{-3} m^3 s^{-1} (= ℓ s^{-1}). Source: Table 5.4.4:3.

TABLE 5.4.4:3

Water discharge at the exit of Lewis Tarn during 1978, 1979, 1980, in 10^{-3} m^3 s^{-1} (= ℓ s^{-1})

February 1978

day	hr LT	amount	day	hr LT	amount	day	hr LT	amount	day	hr LT	amount	day	hr LT	amount
5	16:35	100	6	16:30	60	7	13:30	9	26	12:00	8	28	13:30	7
	17:35	95		17:30	70		14:05	9		12:30	8		14:30	8
	18:35	95		18:00	68		15:45	9		13:30	8		15:30	8
	19:30	72		18:30	67		16:30	11		14:30	8		16:30	8
	20:30	60		19:30	57		17:30	12		15:30	10		17:00	11
	22:30	44		22:30	35		18:00	12		16:30	13		17:30	12
6	2:30	23	7	6:00	15		18:30	12		17:00	14		18:00	14
	6:30	17		6:30	15		19:45	14		17:30	14			
	7:30	12		7:30	11		21:35	14		18:00	14			
	8:30	11		8:30	11	8	5:30	8	28	8:30	8			
	9:30	10		9:40	10		6:30	8		9:30	8			
	10:30	9		10:00	9		7:30	8		10:00	8			
	11:30	11		10:30	8		8:30	7		10:30	8			
	12:40	18		11:00	8	9	11:10	6		11:00	8			
	13:30	27		11:30	7	12	17:40	14		11:30	8			
	14:30	40		12:00	9	26	11:05	10		12:00	8			
	15:40	56		12:30	9		11:30	10		12:30	8			

Table 5.4.4:3, continued

March														
day	hr LT	amount	day	hr LT	amount	day	hr LT	amount	day	hr LT	amount	day	hr LT	amount
1	9:00	7	2	8:30	7	3	11:30	8	4	8:30	8	5	8:30	7
	9:30	7		9:30	7		12:00	8		9:30	8		9:30	7
	10:00	7		10:00	7		12:30	8		10:00	8		10:00	7
	10:30	7		10:30	7		13:30	8		10:30	8		10:30	7
	11:00	7		11:00	7		14:30	8		11:00	8		11:00	7
	11:30	7		11:30	7		15:30	8		11:30	8		11:30	7
	12:00	7		12:00	7		16:30	8		12:00	8		12:00	7
	12:30	7		12:30	7		17:00	12		12:30	8		12:30	7
	13:30	7		13:30	7		17:30	12		13:30	8		13:30	8
	14:30	7		14:30	8		18:00	12		14:30	8		14:30	8
	15:30	7		15:30	8					15:30	8		15:30	8
	16:30	8		16:30	11					16:30	8		16:30	8
	17:00	8		17:00	13					17:00	8		17:00	8
	17:30	8		17:30	14					17:30	8		17:30	9
	18:00	8		18:00	14					18:00	8		18:00	9

April-Dec 1978			Jan-Nov 1979			January 1980					
day	hr LT	amount	day	hr LT	amount	day	hr LT	amount	day	hr LT	amount
Apr 1	17:30	34	Jan 1	13:15	5	1	10:25	10	5	12:30	10
May 1	10:30	7		17:30	6		11:00	10		13:00	10
	17:30	6	March 1	10:30	8		11:30	10		13:30	10
June 1	10:30	1	May 1	10:30	8		12:00	10		14:30	10
	17:30	1		18:00	8		12:30	13		15:30	19
July 1	12:15	1	June 1	10:30	9		13:30	24		16:00	23
	17:30	1		17:30	9		14:30	52		16:30	23
Aug 1	10:30	1	Sept 1	10:30	7		15:30	68		17:30	28
	17:30	1	Oct 1	10:30	7		16:30	80		18:00	27
Sept 1	10:30	4		18:00	7		17:00	85	Feb 1	10:30	8
	17:30	4	Nov 1	10:30	9		18:00	85		18:00	13
Oct 1	10:30	2		17:30	15	4	14:00	49	July 20	13:30	8
	17:30	3					14:30	57			
Nov 1	10:30	3					15:00	65			
	17:30	6					15:30	65			
Dec 1	10:30	6					16:30	64			
	17:30	14					17:30	60			
							18:00	58			

5.4.5. GLOBAL RADIATION

In the context of the glacier heat budget, representative monthly mean values of global radiation are desired. To this end, calendar month estimates are first derived of the downward directed direct shortwave radiation on a horizontal surface under clear sky conditions. Actual measurements provide an idea of the ratio of diffuse to direct radiation. The effect of cloudiness shall be accounted for by empirical formula.

As Lewis Glacier is located above a substantial part of atmospheric mass and most of the tropospheric mosture and dust, available tabulations of monthly clear sky global radiation (Budyko 1974, pp. 46–7) are not applicable. Instead, use is made of the theoretical framework presented by Bernhardt and Philipps (1958). The clear-sky solar radiation on a horizontal surface is given by Bernhardt and Philipps (1958, p. 10) as

$$SW{\downarrow}(\text{dir}) = \frac{I_0}{\rho^2(\delta)} \cos\theta \left[\frac{0.907}{(\cos\theta)^{0.018}}\right]^{T/\cos\theta} \qquad 5.4.5{:}(1)$$

The zenith angle of the sun, θ, is determined by the declination δ, the latitude ϕ, and the hour angle ω:

$$\cos\theta = \sin\delta \sin\phi + \cos\delta \cos\phi \cos\omega \qquad 5.4.5{:}(2)$$

The value for the solar constant $I_0 = 1352 \text{ W m}^{-2}$, the ratio of true to mean distance of the sun ρ, and the solar declination δ, for the middle of each calendar month, were taken from List (1968, pp. 414–17, 495–6).

T is the so-called 'old' Linke turbidity factor. Representative of the environment of Lewis Glacier are atmospheric pressure of around 590 mbar, vapor pressure of about 4.3 mbar corresponding to 70% relative humidity at 0°C, and very little dust. The atmospheric pressure describes the mass of the air column above the location, and the surface vapor pressure is related to the precipitable water. On this basis and from the information given by Bernhardt and Philipps (1958, pp. 17–19), an effective T of about unity was chosen here.

Calculations with Eq. 5.4.5:(1) were performed for the middle of each of the 12 one-hour intervals from 6 am to 6 pm, and the middle of each calendar month. Results were then combined into a daily figure for each calendar month as presented in Table 5.4.5:1. These values are essentially identical to the Rayleigh radiation as presented by Bernhardt and Philipps (1958, p. 158). Sample calculations with Eq. 5.4.5:(1) were also carried out for turbidity factors of $T = 0.9$ and 1.1. Values obtained differ from Table 5.4.5:1 by less than 2%. Even with some uncertainty in the turbidity factor T, the values listed in Table 5.4.5:1 are therefore regarded as reasonable estimates of clear-sky downward directed direct shortwave radiation on a horizontal surface at Lewis Glacier.

At high altitudes, the diffuse component is very small compared to the intensity of the direct solar beam under clear sky conditions. Actual measurements show that the diffuse component accounts only for 5–10% of the clear-sky global radiation at Mount Kenya, while even smaller percentages were measured at the Quelccaya Ice Cap in the Peruvian Andes (Hastenrath, 1978). However, under cloudy or overcast conditions, rather larger contributions are obtained for the diffuse component, in both relative and absolute terms. This is commensurate with the fact that the ratio of global radiation at overcast to that at completely clear sky is about as large at high altitude as at sea level

TABLE 5.4.5:1

Monthly means of (a) direct solar radiation on a horizontal surface under clear-sky conditions calculated for Lewis Glacier from Bernhardt and Philipps (1958); (b) clear-sky global radiation calculated from (a) and the assumption of 5 percent for the ratio of diffuse to global radiation (c) global radiation at overcast sky, calculated from (b) and Eq. 5.4.5:(3); in cal cm^{-2} day^{-1} and W m^{-2}

	J	F	M	A	M	J	J	A	S	O	N	D	Year
	in cal cm^{-2} day^{-1}												
(a)	750	783	790	765	720	688	700	740	776	779	755	739	749
(b)	788	823	830	803	758	722	735	772	815	818	795	776	788
(c)	250	258	263	255	240	229	233	247	259	260	252	246	250
	in W m^{-2}												
(a)	364	379	383	371	349	334	340	358	376	377	366	358	363
(b)	382	398	402	390	367	351	357	376	385	396	384	376	381
(c)	121	126	128	124	116	111	113	120	125	126	122	119	121

TABLE 5.4.5:2

Daily global radiation measured in January–March 1978 on rock ridge overlooking Lewis Glacier, in W m^{-2}

Date	Amount	Date	Amount	Date	Amount	Date	Amount
Jan 31	238	Feb 6	291	Feb 13	250	March 2	316
Feb 1	246	7	166	25	208	3	215
2	287	8+9	190	26	262	4	162
3	304	10	148	27	225	5	159
4	342	11	139	28	200		
5	330	12	257	March 1	139		

(Hastenrath, 1978). Clear-sky global radiation at Lewis Glacier is thus estimated at 5–10% larger than the values of direct radiation (Table 5.4.5:1). Global radiation under cloudy sky can be calculated from (Budyko, 1974, pp.45–8)

$$SW{\downarrow} = (SW{\downarrow})_0 \ (1 - k\,C) \qquad 5.4.5{:}(3)$$

where the subscript o refers to a completely cloudless day, C signifies total cloudiness, and k is an empirical coefficient for which a numerical value of 0.65 is given at the Equator.

Formula 5.4.5:(1) together with Table 5.4.5:1 and the estimate of the diffuse component amounting to 5–10% of the clear-sky global radiation permit estimates of global radiation at Lewis Glacier for various fractions of cloud cover.

Daily global radiation measured in Janunary–March 1978 by a RIMCO counting radiometer installed on a rock ridge overlooking the glacier is listed in Table 5.4.5:2 for comparison with Table 5.4.5:1. Measurements during an expedition in July–August 1975 are included in Table 5.4.9:1. For January–March 1978 Table 5.4.5:2 and 5.4.9:1 show daily average, maximum, and minimum values of 240, 342, and 139 W m^{-2}, respectively. The latter two figures are consistent with the values listed in Table 5.4.5:1 for global

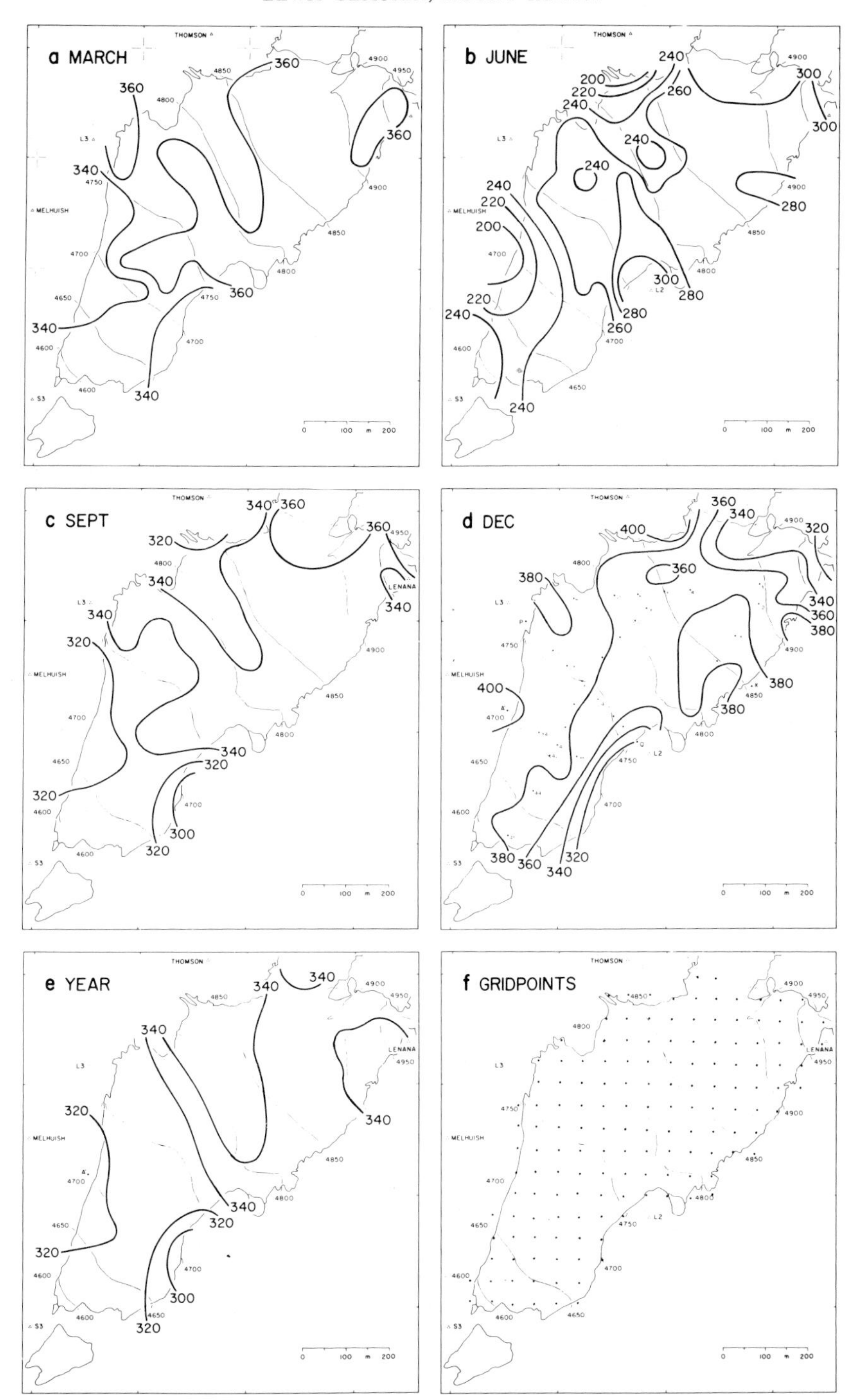

Map 5.4.5:1. Patterns of clear-sky downward directed direct shortwave radiation on sloping surface at Lewis Glacier. (a) March, (b) June, (c) September, (d) December, (e) year, (f) location of grid points. In W m^{-2}.

radiation in February under completely clear and completely overcast sky of 398 and 126 W m^{-2}, respectively. Complete absence of clouds, including high cirrus, is not common. Our measurements thus support the model calculations exemplified in Table 5.4.5:1. Rather larger uncertainties are related to the estimate of cloud amount.

Discussion was here limited to radiation fluxes with reference to an extended horizontal surface. The spatial distribution, especially of the direct shortwave radiation is complicated, however, by the varying slope and orographic obstruction of the different portions of Lewis Glacier. P. Kruss performed model calculations of the direct beam radiation at discrete grid points accounting for slope and horizon, with a value of $T = 1.0$. Spatial patterns for the equinox months March and September, the solstice months June and December, and the year as a whole, are illustrated in the Maps 5.4.5:1.

The diffuse component of solar radiation on a horizontal surface can be approximated as amounting to about 5 and 33% (Eq. 5.4.5:(3)) of the daily clear-sky direct radiation on a horizontal surface, for completely clear and overcast sky, respectively. Linear interpolation is considered appropriate for intermediate fractional cloud cover. It can further be assumed that the diffuse radiation varies comparatively little with the orientation of the receiving surface.

Accordingly, the total downward directed shortwave radiation at any location on the glacier can be approximated from the Maps 5.4.5:1 and Table 5.4.5:1. Thus, the direct component is assumed to decrease from the values shown in Maps 5.4.5:1 for clear sky linearly to zero at overcast. Concomitantly, the diffuse component is postulated to increase from 5% of the value listed in lines (a) of Table 5.4.5:1 at clear sky linearly to 33% at overcast. The daily total (direct plus diffuse) downward directed shortwave radiation at any location on the glacier can thus be estimated from the Maps 5.4.5:1 and Table 5.4.5:1 for prescribed values of cloudiness.

5.4.6. ALBEDO

The upward directed shortwave radiative flux is obtained from the downward shortwave radiation as discussed in Section 5.4.5, and the albedo of the glacier surface. The latter was repeatedly sampled by measurements near the various net balance stakes, as described in Section 5.1.7.

At all elevations, albedo is particularly large immediately after fresh snowfall, but decreases in subsequent days with the deterioration of the snow surface especially at the lower elevations. However, a separation by calendar months or individual days was not found warranted in view of sampling errors of various kind.

At each stake location, measurements were averaged separately for the accumulation (March to June and September to December) and the ablation seasons (January to early March and July to early September). Averages are a few percent larger for the accumulation than for the ablation seasons. In the computation of representative annual values, the aforementioned seasonal averages of albedo were weighted according to the approximate duration of the accumulation and ablation seasons of about 8 and 4 months, respectively. The resulting annual pattern is shown in Map 5.4.6:1. Albedo broadly decreases from the upper towards the lower glacier. However, comparatively low albedo is also characteristic of the approximately westward facing steep slopes below Point Lenana and in a band extending from Point Thomson towards L2.

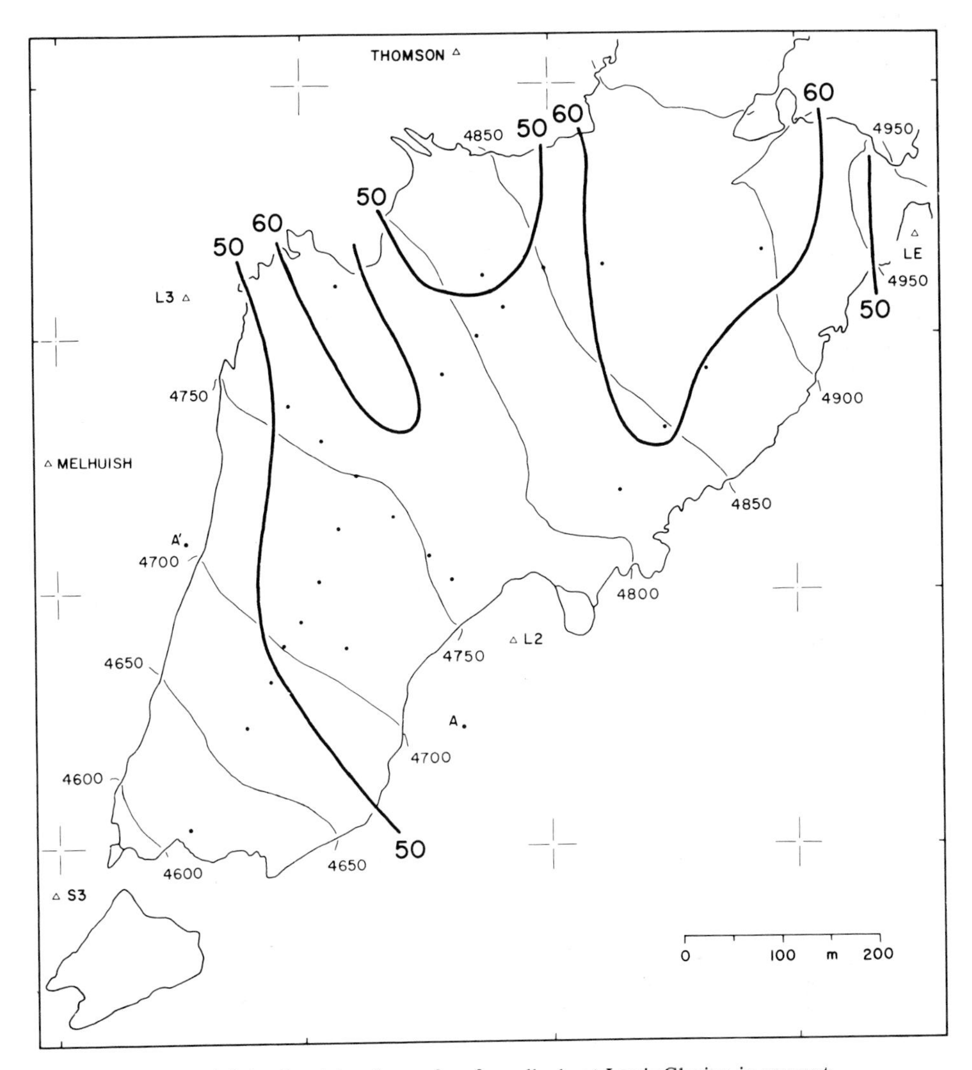

Map 5.4.6:1. Spatial pattern of surface albedo at Lewis Glacier, in percent.

5.4.7. NET LONGWAVE RADIATION

As the upward generally exceeds the downward flux, the net longwave radiation amounts to a heat loss to the glacier. Component fluxes were measured during the 1977–8 field experiment concurrent with temperature and humidity. Net longwave radiation can also be calculated from an empirical formula proposed by Budyko (1974, pp. 57–60)

$$LW\uparrow\downarrow = \epsilon\sigma T_a^{\,4}\ (0.39 - 0.05\ \sqrt{e})\ (1 - 0.53C^2) + 4\epsilon\sigma T^3\ (T_0 - T_a) \qquad 5.4.7{:}(1)$$

where ϵ signifies emissivity, here taken as 0.95 (Sellers, 1965, p. 41; Budyko, 1974, p. 58), $\sigma = 567 \times 10^{-10}$ W m^{-2} K^{-4} (= 826×10^{-13} cal cm^{-2} min^{-1} K^{-4}) is the

TABLE 5.4.7:1

Longwave radiation at station 21 on Lewis Glacier during February 1978. Surface and instrument temperature T_{sfc} and T_i in °C, vapor pressure e in mbar, measured and calculated (Eq. 5.4.7:(1)) net longwave radiation $LW\uparrow\downarrow_{meas}$, $LW\uparrow\downarrow_{calc}$, (a) 10^{-3} in cal cm^{-2} min^{-1}, and (b) in W m^{-2}

Day	hr LT	T_{sfc}	T_i	e	(a) $LW\uparrow\downarrow_{meas}$	$LW\uparrow\downarrow_{calc}$	(b) $LW\uparrow\downarrow_{meas}$	$LW\uparrow\downarrow_{calc}$	Cloudiness
5	19:01	–	1.2	(5.4)	150	121	105	84	clear
7	19:10	−1.2	−1.0	(5.5)	120	–	84	–	?
26	18:21	−4.0	−2.5	(5.0)	120	–	84	–	?
	20:01	−5.5	−3.8	4.4	93	102	65	71	clear
	20:21	−6.5	−3.0	4.3	93	91	65	63	clear
	20:51	−6.1	−2.8	4.2	90	94	63	65	clear
27	04:56	−8.5	−6.0	3.8	66	–	46	–	some ac
	05:16	−6.5	−4.5	2.8	37	43	26	30	overcast at station
	05:39	−6.5	−4.0	(4.2)	70	68	49	47	7/10 ac
	20:21	−6.1	−3.5	2.9	103	103	72	72	clear
	20:31	−7.0	−3.2	3.1	107	95	75	66	clear
28	05:21	−8.8	−3.5	5.6	87	71	61	49	clear
	05:44	−8.4	−2.5	2.0	76	85	53	59	clear

Stefan–Boltzmann constant, e vapor pressure in mbar, T_a and T_0 air and surface temperature in K, and C fractional cloudiness. The second bracket in the first right-hand term reflects the reduction of net longwave radiation with cloudiness, and the second term accounts for the effect of thermal stratification near the surface. Net longwave radiation at clear sky can also be obtained graphically from the 'Integrated Elsasser Chart' (Sellers, 1965, p. 55), instead of Eq. 5.4.7:(1).

Measured values of longwave radiation fluxes, temperature, and humidity, and net longwave radiation calculated from Eq. 5.4.7:(1) are summarized in Table 5.4.7:1. Radiation measurements were performed with a UPYR Universal Pyranometer (Rosenhagen, Hamburg) installed at station 21 in the middle portion of the glacier. Nighttime measurements are regarded as most reliable, since shortwave radiative effects on the instrument are eliminated. Under clear sky conditions, the net longwave radiation is found to be of the order of 70 W m^{-2} (= 10^{-1} cal cm^{-2} min^{-1}), agreement between measured and calculated values being reasonable.

5.4.8. NET ALLWAVE RADIATION

Of the various radiative flux components, the downward directed shortwave radiation $SW\downarrow$ (Section 5.4.5) is by far the largest; its spatial pattern being primarily determined by slope and orographic obstruction. The upward directed shortwave flux $SW\uparrow$ amounts to more than half of $SW\downarrow$, with an increase of this fraction from the lower towards the upper glacier, according to the spatial pattern of albedo (Section 5.4.6). The pattern of net shortwave radiation $SW\uparrow\downarrow$ is influenced by that of albedo, as well as slope and surrounding orography. The net longwave radiation $LW\uparrow\downarrow$ (Section 5.4.7) is rather smaller than the shortwave flux components, but of a magnitude approaching that of $SW\uparrow\downarrow$. Some spatial variation can be expected from the complex topographic configuration of Lewis Glacier, but the typical magnitude is indicated by Table 5.4.7:1.

As a resultant of the various flux components considered above, the net allwave radiation $SWLW\uparrow\downarrow$ is a comparatively small positive quantity, the estimation of which is fraught with appreciable relative errors. The spatial pattern may be characterized by an increase from the lower towards the upper glacier in consequence of the albedo distribution, but even the relative spatial pattern is uncertain. Assuming a cloudiness of 5 tenths, and an albedo of 60%, representative annual estimates of $SW\downarrow$, $SW\uparrow$, $LW\uparrow\downarrow$, and $SWLW\uparrow\downarrow$, for the glacier as a whole are 228, 137, 61, and 30 W m^{-2}. The figure suggested for net allwave radiation is plausible, but cannot be safeguarded from rigorous error analysis.

5.4.9. SENSIBLE HEAT FLUX AND STORAGE

The basics of sensible Q_s, and latent heat transfer, Q_{ei}, across the glacier-air interface can be understood from the bulk-aerodynamic formulations

$$Q_s = c_p \, \rho \, C_D \, (T_0 - T_a) \, V \qquad 5.4.9{:}(1)$$

and

$$Q_{ei} = L_s \, \rho \, C_D \, (q_0 - q_a) \, V \qquad 5.4.9{:}(2)$$

The specific heat at constant pressure $c_p = 10^3$ J kg^{-1} K^{-1} = 0.24 cal g^{-1} K^{-1}, the latent heat of sublimation $L_s = 284 \times 10^4$ J kg^{-1} = 680 cal g^{-1}, air density $\rho = 0.740$ kg m^{-3} at 580 mbar and 0°C. C_D is the drag coefficient describing the roughness of the surface. For the surface of Lewis Glacier this is chosen as $C_D = 2 \times 10^{-3}$ (Priestley, 1959, p. 21). V is wind speed, T temperature, and q specific humidity, the subscripts o and a referring to ice surface and air, respectively.

Eqs. 5.4.9:(1) and 5.4.9:(2) are of analogous structure. The latter will be used mainly in Section 5.4.10. Eq. 5.4.9:(1) shows that the interface temperature difference determines the direction of the sensible heat flux and in large part also its magnitude variations, in that relative changes of V tend to be smaller than those of $(T_0 - T_a)$; the other three factors in Eq. 5.4.9:(1) being constant for all practical purposes.

Measurements performed on Lewis Glacier in December 1957–January 1958 (Brinkman *et al.*, 1968), in April 1960 (Platt, 1966; Brinkman *et al.*, 1968), in July–August 1975 (Davies *et al.*, 1977), and during our expedition in January–February 1978, offer an idea of the sensible heat flux Q_s. During three of these experiments, stations were installed at essentially the same location, in the flat area near our station 21 (Map 4.3:10*).

TABLE 5.4.9:1

Summary of meteorological observations on Lewis Glacier. $\overline{T}$, T_x, T_n, T_{xa}, T_{na}, T_s, T_{sx}, T_{sn} denote daily average, average maximum and minimum, absolute maximum and minimum air temperature, and average, and average maximum and nighttime minimum surface temperature in °C; RH, RH_x, RH_n are daily average, and average maximum and minimum relative humidity in percent; V, V_d, V_n, is average daily, daytime and nighttime wind speed in cm s^{-1}. S_A, Sp is sunshine duration before and after noon, in hours. R, R_x, R_n, R_{xa}, denote daily average, maximum and minimum, and hourly maximum global radiation in W m^{-2}. Brackets contain approximate time in hours LT. (a) 19 December 1957–17 January 1958 (source: Brinkman *et al.*, 1968); (b) 2–5 April 1960 (source: Platt, 1966; Brinkman *et al.*, 1968); (c) 23 July–5 August 1975 (source: Davies *et al.*, 1977); (d) 30 January–13 February and 25 February–4 March 1978

	(a)		(b)		(c)		(d)	
$\overline{T}$	– 1.9		– 0.7		– 0.8			
T_x	– 0.2	(16:00)	+ 0.4	(13:30)	+ 0.5	(11:30)	+ 2.8	
T_n	– 3.5	(5:30)	– 1.5	(6:00)	– 1.5	(23:30)	– 2.3	
T_{xa}	+ 5.0	(15:30)	+ 1.7	(15:30)	+ 1.2		+ 5.0	(13:00)
T_{na}	– 8.3	(4:30)	– 2.3	(5:30)	– 1.6		– 3.5	(6:30)
T_s	– 8.5		–		(– 3.0)?		–	
T_{sx}	– 3.7		–		–		0.0	
T_{sn}	– 12.9		–		–		– 3.7	
RH	59		73		–		–	
RH_x	77	(14:00)	82	(14:00)	–		(80)	(14–17)
RH_n	49	(0:30)	64	(06:30)	–		(40)	(05–08)
V	400				430		310	
V_d							270	
V_n							400	
S_A			3.0				–	
Sp			1.9					
R					203		240	
R_x					223	(12:00)	342	
R_n					178		139	
R_{xa}					780	(12:00)		

TABLE 5.4.9:2

Minimum T_n and maximum temperature T_x at various depths, in °C. Approximate hour in LT is given in parentheses. (a) in January 1960 (source: Platt, 1966); (b) in July–August 1975 (source: Davies *et al.*, 1977); (c) in February–March 1978

	(a)				(b)				(c)			
	T_n		T_x		T_n		T_x		T_n		T_x	
cm												
1	−4.7	(07:30)	0.0		–		–		−5.1	(6:00)	0.0	(11–18)
5	–		–		−2.4	(08:00)	−0.5	(16:00)	–		–	
9	–		–		−2.2	(09:00)	−0.4	(20:00)	–		–	
10	–		–		–		–		−2.0	(6:00)	0.0	(11–18)
15	−4.0	(07:30)	−1.1	(11:00)	−2.0	(09:00)	−0.7	(19:00)	–		–	
25	–		–		−1.7	(13:00)	−0.7	(07:00)	−0.9	(8:40)	0.0	(11–18)
29	−2.0	(08:00)	−1.1	(11:00)	–		–		–		–	

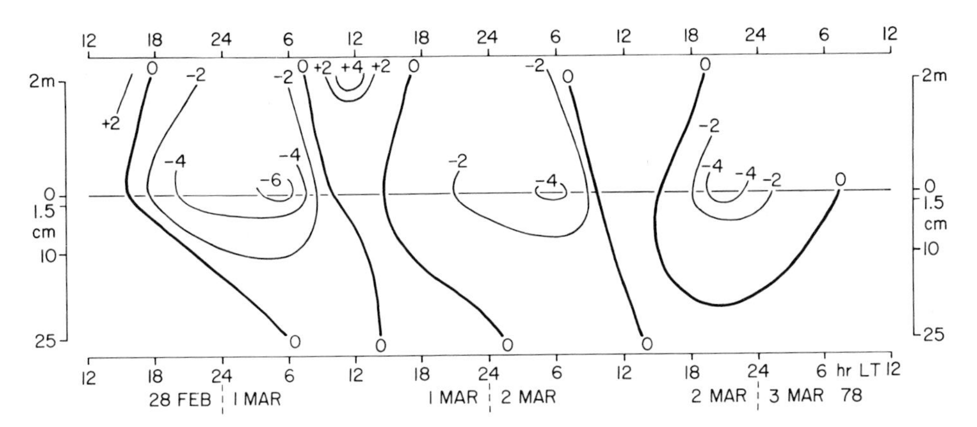

Fig. 5.4.9:1. Variation of temperature in the vertical and with time at station 21 (see Map 4.3.10*) during 28 February to 3 March 1978, in °C. Dots denote measuring times in 2 m above the surface, at the surface, and at depths of 1.5, 10, and 25 cm. Vertical scale for air is one tenth of that for ice.

The installation of Davies *et al.* (1977) seems to have been located further towards the Southeast, about half way to the ice rim.

Characteristic parameters obtained in the various experiments are summarized in Table 5.4.9:1. Information on temperature variation at various depths in the snow is given in Table 5.4.9:2. Fig. 5.4.9:1 is a time section plot of temperature variations in the air and at various depth at station 21 (see Map 4.3:4) during 28 February–3 March 1978. Fig. 5.4.9:1 shows largest temperature variations near the surface, a phase lag with depth, and the vanishing of the diurnal temperature wave near a depth of 25 cm. Presumably radiationally controlled surface inversions prevail from the afternoon throughout the night into the morning hours, while lapse conditions are limited to the middle of the day. Vertical temperature gradients are strongest during the nighttime inversions.

As representative values of the drag coefficient C_D as well as of wind speed and temperature are only approximately known, a detailed evaluation of the sensible heat flux Q_s through Eq. 5.4.9:1 is not warranted. However, information in Table 5.4.9:1 and Fig. 5.4.9:1 allows an estimate of the general magnitude and the sign of Q_s. A combination of $C_D = 2 \times 10^{-3}$, $V = 5$ m s^{-1}, $(T_0 - T_a) = 1$°C would, according to Eq. 5.4.9:(1), correspond to a sensible heat flux $Q_s = 7.4$ W m^{-2} = 15.2 cal cm^{-2} day^{-1}. When the ice is colder/warmer than the air, the flux is directed downward/upward. As is illustrated by Table 5.4.9:1 and Fig. 5.4.9:1, interface temperature differences during most hours of the nighttime inversion are much larger than 1°C. During daytime, the glacier surface reaches close to 0°C, while the air temperature may rise above freezing only during a few hours, if at all. Interface temperature differences during daytime lapse conditions are much smaller than with the nighttime inversions. Accordingly, the upward directed sensible heat flux Q_s during the day would be much smaller than the downward flux at night, so that Q_s may provide a heat gain to the glacier of the order of 7.4 W m^{-2} = 15.2 cal cm^{-2} day^{-1} for the day as a whole. In view of the general decrease of air temperature from the lower to the upper glacier, the daily average sensible heat gain may be largest in the lower, and smaller in the upper portions of Lewis.

The general magnitude of heat storage within the ice in the course of the diurnal cycle can be estimated from Table 5.4.9:2 and Fig. 5.4.9:1 as follows. The top 25 cm are of particular interest, since the diurnal temperature wave vanishes at greater depth. The lowest temperature of the layer surface to 25 cm of about –2°C is found in the early morning. The layer mean temperature is highest with the 0°C isothermy in the afternoon. With a snow density of 500 kg m^{-3}, and a specific heat of ice of 2.1×10^3 g kg^{-1} K^{-1} = 0.5 cal g^{-1} K^{-1} (List, 1968, p. 343), the layer-mean temperature difference of 2°C corresponds to a change of heat content in the course of the diurnal cycle of 52×10^4 J m^{-2} = 13 cal cm^{-2}. As this storage/depletion takes place over intervals of about 12 hours, the corresponding rate of change of heat content is of the order of 12 W m^{-2} = 3×10^{-4} cal cm^{-2} s^{-1}. Heat storage is vanishingly small for the day as a whole and over longer periods.

The estimates of sensible heat flux across the ice-air interface and of heat storage within the top layer of the glacier arrived at here are further of interest for the diurnal forcing-response relationships to be discussed in Section 5.4.12.

5.4.10. MELTING AND SUBLIMATION

Ablation can take place through two processes: sublimation is energetically expensive (284×10^4 J kg^{-1} = 680 cal g^{-1}), melting is not (33×10^4 J kg^{-1} = 80 cal g^{-1}) but is precluded at low temperatures. Commensurate with the thermal conditions at Lewis Glacier, melting is the ablation mode prevalent for the mass budget, while sublimation may play a comparably important role in the energetics. Possible approaches for estimating ablation are as follows.

The sum of melting plus sublimation can be inferred from the difference between solid precipitation and net balance measured at a stake network. On similar principles, the corresponding energies involved in the phase change can be obtained as residual when all other components of the surface heat budget (Eq. 5.4.1:(1)) are determined independently. These two lines shall be pursued in a closure of budgets in Section 5.4.11.

Melting by itself can be estimated by lysimeter measurements (Section 5.1.7), and water runoff (Sections 5.1.6 and 5.4.4). Sublimation by itself can also be estimated by lysimeter measurements (Section 5.1.7), and independently by the bulk-aerodynamic method, Eq. 5.4.9:(2), requiring as inputs wind speed and humidity in the air, surface temperature, and a measure of surface roughness.

Results of lysimeter measurements during an intensive observation program in January–February 1978 are summarized in Table 5.4.10:1. Melting rates are smallest at the high station 3, and largest at the lowest station 44, and are of the general order of 1 g cm^{-2} day^{-1} = 10 kg m^{-2} day^{-1}. Melting becomes particularly large in the afternoon and vanishes with the temperature decrease at night. Sublimation is generally a much smaller quantity, but the fractional mass change is not negligible. Nighttime mass gain shown by the lysimeter measurements reflects a downward directed water vapor flux.

The available data allow only a broad qualitative discussion of the vertical water vapor and latent heat flux in terms of the bulk-aerodynamic Equation 5.4.9:(2). Consider a representative wind speed of about 5 m s^{-1}, a drag coefficient of $C_D = 2 \times 10^{-3}$, and relative humidity of the air of 70%. The magnitude and direction of the latent heat flux then depend essentially on surface and air temperature. An interface difference in specific humidity of 1 g kg^{-1} would correspond to a latent heat flux of 21 W m^{-2} = 43.5 cal cm^{-2} day^{-1} or 6.4×10^{-2} g H_2O cm^{-2} day^{-1}. Table 5.4.10:2 shows interface

TABLE 5.4.10:1

Lysimeter measurements during January–March 1978, at stations 3 (4896 m), 21 (4765 m), and 44 (4668 m). Amounts are given in 10^{-3} g cm^{-2} = 10^{-2} kg m^{-2}. A minus sign denotes a mass *gain*

Day and hr LT	Sublimation	Melting	Remarks
Station 3			
3 Feb			
07:41			
	176	410	
14:01			
	– 98	– 295	
16:36			
	317	?	all melted, but frozen overnight
5 Feb			
09:41			
	141	0	
16:00			time approx
	0	0	
6 Feb			
05:31			
	0	0	
08:21			
	48	286	
11:06			
	12	1280	
14:21			
	20	545	
17:11			
	57	0	
7 Feb			
06:01			
	0	44	
11:16			
	9	295	
14:01			
	0	361	
17:11			
	– 26	0	
8 Feb			
05:46			
	9	0	
08:31			
Station 21			
31 Jan			
10:06			
	53	805	
16:41			
	132	760	

Table 5.4.10:1, continued

Day and hr LT	Sublimation	Melting	Remarks
1 Feb 13:49			
	432	300	
2 Feb 12:38			
	440	0?	
3 Feb 07:16			
	220	560	
13:19			
	26	360	
15:41			
	562	1280	
4 Feb 14:31			
	356	1495	
5 Feb 17:39			
	36	0	
6 Feb 06:26			
	40	0	
08:41			
	26	1040	
11:33			
	22	1370	
14:31			
	9	686	
17:35			
	22	0	
7 Feb 06:29			
	0	0	
08:32			
	12	382	
11:44			
	0	355	
14:34			
	13	364	
17:41			
	− 44	0	
8 Feb 06:30			
	22	0	
09:01			

Table 5.4.10:1, continued

Day and hr LT	Sublimation	Melting	Remarks
26 Feb			
12:01			
	35	520	
15:31			
	219	153	
17:46			
	−280	0	
27 Feb			
05:41			
	101	0	
09:01			
	−285	537	
14:41			
	− 22	88	
17:31			
	31	0	
20:16			
	− 53	0	
28 Feb			
05.44			
	61	0	
08:36			
	44	0	
11:39			
	–	–	snowfall
14:31			
	–	–	snowdrift
1 March			
09:19			
	44	0	
11:23			
	–	–	snowfall
17:41			
	61	0	
2 March			
08:47			
	40	0	
11:21			
	127	105	
14:31			
	57	0	
16:26			
	–	–	snowfall
3 March			
14:46			
	− 9	0	
16:31			
	−145	0	snowfall ?

Table 5.4.10:1, continued

Day and hr LT	Sublimation	Melting	Remarks
4 March			
07:26			
	79	0	
08:36			
	26	0	
09:36			
	4	0	
10:38			
	9	0	
12:06			
	− 9	0	
12:36			
	–	–	snowfall
14:11			
	0	0	
15:01			
	66	0	
15:31			
	− 35	0	
15:56			
	− 18	0	
16:31			
	40	0	
17:31			
Station 44			
31 Jan			
10:46			
	22	715	
16:26			
	397	360	
3 Feb			
12:36			
	97	600	
16:03			
	885	890	
4 Feb			
15:26			
	620	1225	
5 Feb			
17:41			
	158	0	
6 Feb			
06:46			
	22	0	
09:06			
	39	765	

Table 5.4.10:1, continued

Day and hr LT	Sublimation	Melting	Remarks
6 Feb			
12:01			
	26	730	
15:01			
	3	590	
18:11			
	0	0	
7 Feb			
06:51			
	44	0	
09:11			
	0	366	
12:16			
	31	225	
15:06			
	0	264	
18:06			
	− 22	0	
8 Feb			
06:43			
	9	0	
09:26			

TABLE 5.4.10:2

Specific humidity difference surface minus air Δq in g kg^{-1} for various combinations of surface T_0 and air temperature T_a (°C) and relative humidities of the air of 60, 70, and 80%

T_0	T_a	Δq (60%)	Δq (70%)	Δq (80%)
−12	−8	+0.3	−0	−0.3
− 8	−4	+0.5	+0	−0.5
− 4	−2	+1.4	+0.9	+0.3
− 2	−4	+2.7	+2.2	+1.7
− 2	−3	+2.4	+1.9	+1.4
− 2	−2	+2.2	+1.7	+2.0
− 2	−1	+1.8	+1.2	+0.6
− 2	0	+1.6	+1.0	+0.3
0	−4	+3.7	+3.2	+2.7
0	−3	+3.4	+2.9	+2.4
0	−2	+3.2	+2.6	+2.1
0	−1	+2.9	+2.3	+1.7
0	0	+2.6	+2.0	+1.3
0	+1	+2.3	+1.6	+0.9
0	+2	+1.9	+1.2	+0.4
0	+3	+1.6	+0.8	+0
0	+4	+0.9	+0.4	−0.5

differences of specific humidity corresponding to plausible combinations of surface and air temperature. At low temperature associated with the characteristic nighttime surface inversion, the latent heat flux would be directed from air to glacier. This is consistent with the nighttime mass gain frequently observed in the lysimeters. Under all other circumstances the latent heat flux would be directed upward. Negative air temperatures with the surface at melting point, a common combination during daytime, allow particularly large sublimation rates. From the examples in Table 5.4.10:2 sublimation rates for the day as a whole can be estimated to be of the order of 0.1 g H_2O cm^{-2} day^{-1} = 1 kg H_2O m^{-2} day^{-1}, as compared to a value an order of magnitude larger for melting. These figures are broadly consistent with the lysimeter measurements, and illustrate that sublimation is of subordinate importance in the mass, but not in the heat budget of the glacier.

5.4.11. CLOSURE OF ANNUAL MASS AND HEAT BUDGETS

Estimates of various budget components have been presented in the preceding Sections 5.4.2–5.4.10. These are fraught with differing degrees of uncertainty. The task here is to combine the component estimates into internally consistent mean annual mass and heat budgets, within plausible error tolerances.

Table 5.4.11:1 offers models of the annual mean radiation budget for clear sky, 5 tenths cloudiness, and overcast, respectively. Models were constructed as follows.

TABLE 5.4.11:1

Models of annual mean radiation budget, constructed from Maps 5.4.5:1e and 5.4.6:1, Tables 5.4.5:1 and 5.4.7:1, and Eqs. 5.4.5:(3), and 5.4.7:(1). Upper and lower glacier refers to areas above and below 4800 m, respectively. Albedos of 60 and 55% are assumed for these two domains. Symbols are as defined in Section 5.4.1. Heat gain for the glacier is counted positive, in W m^{-2}

	Upper	Lower	All
(a) clear sky			
$SW\downarrow$	+399	+384	+390
$SW\uparrow\downarrow$	+159	+172	+166
$LW\uparrow\downarrow$	− 70	− 70	− 70
$SWLW\uparrow\downarrow$	+ 89	+102	+ 96
(b) 5 tenths cloudiness			
$SW\downarrow$	+266	+256	+260
$SW\uparrow\downarrow$	+106	+116	+112
$LW\uparrow\downarrow$	− 60	− 60	− 60
$SWLW\uparrow\downarrow$	+ 46	+ 56	+ 52
(c) overcast			
$SW\downarrow$	+133	+128	+130
$SW\uparrow\downarrow$	+ 53	+ 58	+ 56
$LW\uparrow\downarrow$	− 35	− 35	− 35
$SWLW\uparrow\downarrow$	+ 18	+ 23	+ 21

Annual mean downward directed direct shortwave radiation $SW\downarrow$ on the upper (> 4800 m) and lower (< 4800 m) glacier were obtained from Map 5.4.5:1. These values were referred to the horizontal, under consideration of the surface slope of the upper and lower glacier as obtained from Map 4.3:10*. In accordance with Table 5.4.5:1, a constant value of 39 W m^{-2} representing the downward directed diffuse radiation at clear sky was added, so as to yield the total (direct plus diffuse) downward directed shortwave radiation $SW\downarrow$ at clear sky, as listed in part (a) of Table 5.4.11:1. Using Eq. 5.4.5:(3), the total downward directed shortwave radiation $SW\downarrow$ at overcast in part (c) of Table 5.4.11:1 is obtained as 33% of the above clear-sky values. The corresponding figure for 5 tenths cloudiness in part (b) of Table 5.4.11:1 results by linear interpolation from the respective values in parts (a) and (c).

Surface albedo is from Map 5.4.6:1 estimated at 60 and 55% for the upper and lower glacier, respectively. This in conjunction with the values of $SW\downarrow$ permits calculation of the net shortwave radiation $SW\uparrow\downarrow$ in parts (a), (b), and (c) of Table 5.4.11:1.

In accordance with Section 5.4.7 and Table 5.4.7:1, net longwave radiaton $LW\uparrow\downarrow$ with clear sky is estimated at about 70 W m^{-2}, as given in part (a) of Table 5.4.11:1. From this value and Eq. 5.4.7:(1), the figures of $LW\uparrow\downarrow$ in parts (b) and (c) of Table 5.4.11:1 are calculated.

The difference between $SW\uparrow\downarrow$ and $LW\uparrow\downarrow$ yields the net allwave radiation $SWLW\uparrow\downarrow$, as listed in the three parts of Table 5.4.11:1. The dependence of $SWLW\uparrow\downarrow$ on cloudiness and albedo is further illustrated in Fig. 5.4.11:1, constructed from Eqs. 5.4.5:(3) and 5.4.7:(1), and the clear-sky values of $SW\uparrow\downarrow$ and $LW\uparrow\downarrow$ presented in Table 5.4.11:1. Fig. 5.4.11:1 shows the strong decrease of net allwave radiation $SWLW\uparrow\downarrow$ with increasing cloudiness and albedo. It is furthermore noted from Table 5.4.11:1 and Fig. 5.4.11:1 that a one percent change in surface albedo is commensurate with alterations in the net allwave radiation $SWLW\uparrow\downarrow$ of 4, 3, and 1 W m^{-2}, for the clear sky, 5 tenths cloudiness, and overcast models, respectively.

The net allwave radiation $SWLW\uparrow\downarrow$ is an important component of the surface heat budget, Table 5.4.11:2. Estimates of the sensible and latent heat fluxes, Q_s and Q_{ei}, are entered from Sections 5.4.9 and 5.4.10. These terms are considered as approximately invariant with cloudiness. In the light of the order of magnitude estimates in Section 5.4.1, all other terms in the heat budget Equation 5.4.1:(1) are disregarded. The heat used in melting, as listed in Table 5.4.11:2, is then obtained as residual.

The dependence of the sensible and latent heat fluxes, Q_s and Q_{ei}, on temperature and specific humidity differences at the glacier–air interface, $\Delta T = (T_0 - T_a)$ and $\Delta q = (q_0 - q_a)$, can be appreciated from the bulk-aerodynamic Equations 5.4.9:(1) and 5.4.9:(2). With a representative windspeed of about 5 m s^{-1} and values of c_p, L_S, C_D, and ρ as given in Section 5.4.9, we obtain $Q_s = 7.4 \times \Delta T$ and $Q_{ei} = 21 \times \Delta q$; where ΔT is in K, Δq in g kg^{-1}, Q_s and Q_{ei} in W m^{-2}, and positive values of Q_s and Q_{ei} represent a heat loss for the glacier. Consider temperatures around 0°C and refer to Table 5.4.10:2. Higher/lower air temperature would correspond to an increase/decrease of sensible heat (Q_s) gain by the glacier, but without drastic change in relative humidity this would be approximately offset by an increase/decrease of the latent heat (Q_{ei}) loss. Increased/decreased relative humidity without change in air temperature would entail decreased/increased latent and sensible heat ($Q_{ei} + Q_s$) expenditure by the glacier. An appropriately different energy amount is then available for melting.

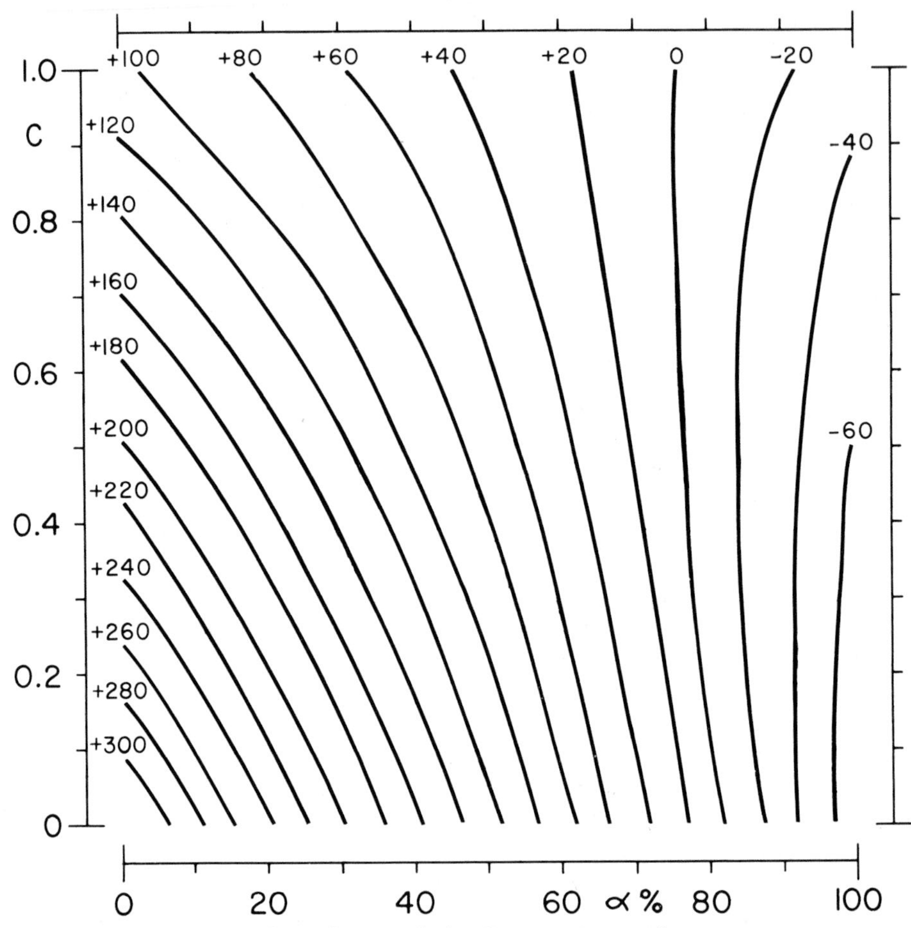

Fig. 5.4.11:1. Dependence of net allwave radiation $SWLW\uparrow\downarrow$ (in W m^{-2}) on cloudiness C (in tenths) and surface albedo a (in percent). Calculations are based on clear-sky values of downward directed shortwave $SW\downarrow$ and net longwave radiation $LW\uparrow\downarrow$ in Table 5.4.11:1 and Eqs. 5.4.5:(3) and 5.4.7:(1).

Based on calculations analogous to those used in the compilation of Table 5.4.11:1–5.4.11:3, the approximate annual heat and mass budgets displayed in Figs. 5.4.11:2 and 5.4.11:3 were constructed. These are patterned after the schemes in Figs. 5.4.1:1 and 5.4.1:2. A glacier average albedo of 57% and cloudiness of about 6 tenths are adopted. These interdependent heat and mass budgets are consistent with an annual average water discharge at the exit of Lewis Tarn of about 15 ℓ s^{-1}, as estimated in Section 5.4.4. Fig. 5.4.11:2 and 5.4.11:3 are intended to illustrate the gross characteristics of the multi-annual mean heat and mass budgets.

Table 5.4.11:3 lists solid precipitation adopted from Section 5.4.2, and sublimation and melting taken from Table 5.4.11:2. Contrary to the presentation in Tables 5.4.11:2 and 5.4.11:3, sublimation is expected to decrease with increasing cloudiness and precipitation in response to larger atmospheric humidity. Accordingly, melting would depend somewhat less strongly on cloudiness than suggested by Tables 5.4.11:2 and 5.4.11:3. In addition to sublimation, solid precipitation is here also taken as invariant with cloudiness,

TABLE 5.4.11:2

Models of annual mean surface heat budget, constructed from Tables 5.4.11:1 and 5.4.10:1. Domains and sign convention are as in Table 5.4.11:1, and symbols as defined in Section 5.4.1, in W m^{-2}

	Upper	Lower	All
(a) clear sky			
$SWLW\uparrow\downarrow$	+89	+102	+96
Q_s	+ 7	+ 7	+ 7
Q_{ei}	−30	− 25	−27
Q_m	−66	− 84	−76
(b) 5 tenths cloudiness			
$SWLW\uparrow\downarrow$	+46	+ 56	+52
Q_m	+23	− 38	−32
(c) overcast			
$SWLW\uparrow\downarrow$	+18	+ 23	+21
Q_m	+ 5	− 5	0

TABLE 5.4.11:3

Models of annual mean mass budget constructed from Tables 5.4.11:2, 5.4.4:1, 5.4.4:2, and 5.4.4:3. Refer to Section 5.4.1 for symbols: Solid precipitation P_s, sublimation Q_{ei}/L_s, melting Q_{mm}/L_m, and difference $D = P_s - Q_{ei}/L_s - Q_{mm}/L_m$. Domains and sign convention are as in Tables 5.4.11:1 and 5.4.11:2. Values in parts a, b, c, are in cm of liquid water equivalent or g H_2O cm^{-2} = 10 kg m^{-2}. Figures in part d are in ℓ s^{-1}

	Upper	Lower	All
(a) clear sky			
P_s	+ 85	+ 85	+ 85
Q_{ei}/L_s	− 34	− 28	− 30
Q_{mm}/L_m	−629	−808	−730
D	−578	−751	−675
(b) 5 tenths cloudiness			
Q_{mm}/L_m	−217	−369	−295
D	−166	−270	−240
(c) overcast			
Q_{mm}/L_m	+ 50	− 48	0
D	+101	+ 9	+ 55
(d) melting (Q_{mm}/L_m times area)			
clear sky	− 26	− 42	− 68
5 tenths cloudiness	− 9	− 19	− 28
overcast	+ 2	− 2	0

for purposes of calculation. From order of magnitude estimation (Section 5.4.1), the important terms in Eq. 5.4.1:(3) are solid precipitation P_s, sublimation Q_{ei}/L_s, melting Q_{mm}/L_m, and lateral ice flow divergence $(F_{is} - F_{os})$; the latter quantity vanishing for the glacier as a whole. The difference $D = P_s - Q_{ei}/L_s - Q_{mm}/L_m$ thus represents the contribution of portions of the glacier to the net balance of the glacier as a whole, for which $D = \Delta_s$. Similarly, Q_{mm}/L_m integrated over the glacier as a whole corresponds to

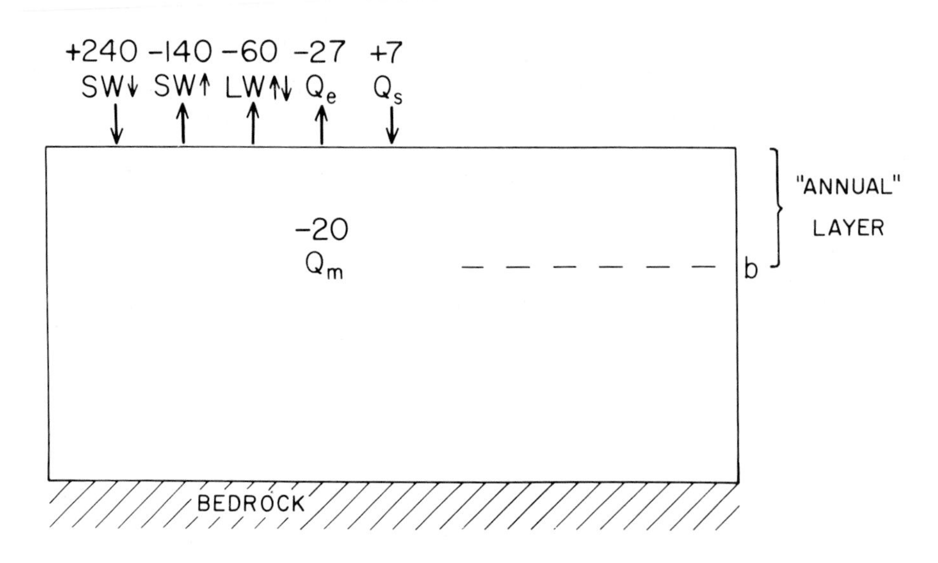

Fig. 5.4.11:2. Scheme of average annual heat budget of Lewis Glacier. $SW\downarrow$ downward directed shortwave radiation; $SW\uparrow$ upward directed shortwave radiation; $LW\uparrow\downarrow$ net longwave radiation; Q_s sensible heat flux at glacier-air interface; Q_e latent heat flux at glacier-air interface; Q_m heat used in melting throughout a column except at base. A glacier average albedo of 57 percent, and cloudiness of about 6 tenths are adopted. Numbers are in W m^{-2}.

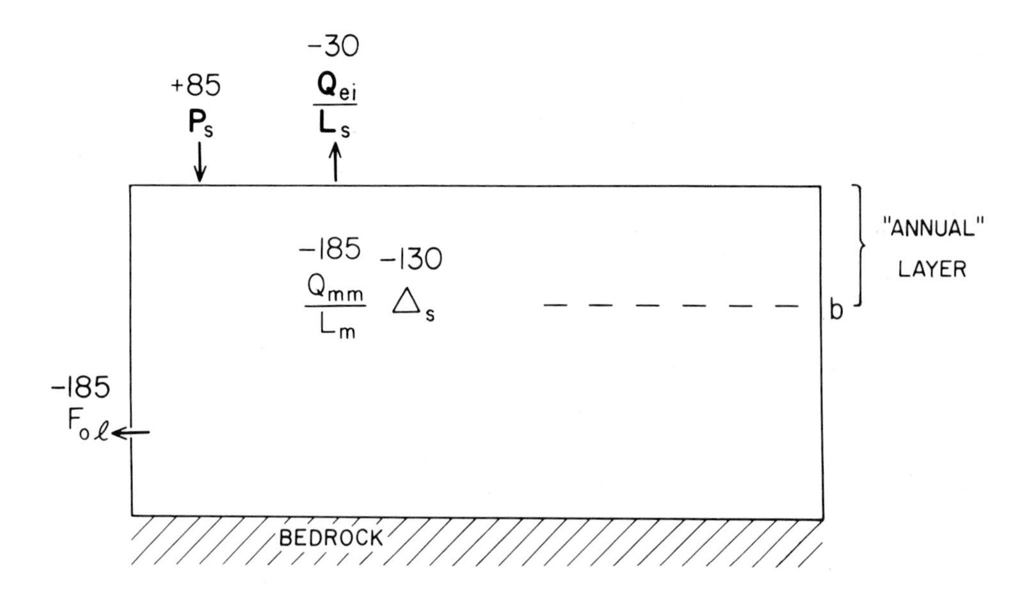

Fig. 5.4.11:3. Scheme of averge annual mass budget of Lewis Glacier, commensurate with heat budget illustrated in Fig. 5.4.11:2. P_s solid precipitation; Q_{ei}/L_s sublimation; Q_{mm}/L_m melting; Δ_s net balance; F_{ol} water discharge. Numbers are in cm of liquid water equivalent or g H_2O cm^{-2} = 10 kg m^{-2} (F_{ol} = 1850 kg m^{-2} corresponds to 17 ℓ s^{-1}).

the total water discharge from the glacier, barring seepage. Table 5.4.11:3 illustrates the strong dependence of melting and discharge on cloudiness.

Table 5.4.11:3 may shed some light on the cause of interannual variations in melting, but it is less informative concerning net balance, in that both precipitation and surface albedo are presumably positively correlated with cloudiness. The latter effect would enhance the inverse dependence of melting on cloudiness, and both factors would contribute towards a stronger positive coupling of net balance with cloudiness than indicated in Table 5.4.11:3.

It is desirable to estimate the interdependence of cloudiness, precipitation, albedo, and net balance empirically. For evaluation of the cloudiness-precipitation relation, mean monthly values were read from the maps of Atkinson and Sadler (1970) and Jackson (1961) for grid points from 1 °S to 8 °N along the 36 °E meridian. These data suggest a rainfall increase of the order of 40 mm a^{-1} per increase of 0.1 tenths of cloudiness. Independently, individual monthly values of sunshine duration and rainfall for Kabete and Dagoretti in Central Kenya and Kisumu at Lake Victoria (Kenya Meteorological Department, unpublished data) were evaluated, yielding similar results.

Regarding the relation of albedo to melting and net balance, Maps 5.4.3:1–5.4.3:4 and 5.4.6:1, Figs. 5.4.3:1–5.4.3:4, and Tables 5.4.3:5, 5.4.3:7, 5.4.3:9, 5.4.3:11, and 5.4.11:3, are of interest, in that they depict the increase of albedo and net balance, and the decrease of melting, with elevation. These data suggest an albedo increase of 1% per roughly 3×10^2 kg m^{-2} = 30 g cm^{-2} increase in net balance or decrease in melting.

Such numbers in conjunction with Tables 5.4.11:1–5.4.11:3 permit us to estimate the sensitivity of net balance to cloudiness and albedo changes. Consider an increase of one tenth in cloudiness from initial conditions of 5 tenths cloudiness and precipitation of 850 mm a^{-1}. Under consideration of the precipitation and albedo change associated with that in cloudiness, the glacier average net allwave radiation would decrease by about 12 W m^{-2} and the net balance would increase by a liquid water equivalent of the order of 200 cm a^{-1}, as compared to only about a third of that value if precipitation and albedo were left invariant of cloudiness.

These sensitivity calculations shall now be compared with observation. Table 5.4.11:4 summarizes conditions during the contrastingly different net balance years 1978/9, 1979/80, 1980/1, 1981/2. Comparison is made particularly between the most diverse years 1978/9 and 1979/80. Precipitation differed between the 1978/9 and 1979/80 years by 400 mm a^{-1}, and glacier average net balance by as much as 1680 mm a^{-1} liquid water equivalent. The interannual difference in ablation is considered to reflect mainly differences in melting. The interannual difference of the corresponding latent heat is, from Table 5.4.11:4, seen to be about 13 W m^{-2}. The above sensitivity calculations for concomitant increases of one tenth in cloudiness and 400 mm a^{-1} in solid precipitation are thus in good agreement with the inferred difference in net allwave radiation and the observed contrast in net balance between the 1978/9 and 1979/80 years.

The sensitivity analysis supported by Tables 5.4.11:1–5.4.11:3 indicates the way in which precipitation–net balance relationships are to be understood quantitatively. In good agreement, sensitivity analysis and comparison of the 1978/9 versus the 1979/80 observations indicate, respectively, increments of 500 and 420 mm a^{-1} liquid water equivalent net balance per 100 mm a^{-1} increase in solid precipitation.

TABLE 5.4.11:4

Comparison of mass and heat budget during the budget years 1978/79, 1979/80, 1980/81, and 1981/82, based on Figs. 5.4.3:1–5.4.3:4, Maps 5.4.3:1–5.4.3:4, and Tables 5.4.2:2, 5.4.3:5, 5.4.3:7, 5.4.3:9, 5.4.3:11. Refer to Section 5.4.1 for symbols: solid precipitation P_s from daily and accumulating gauges, Δ_{sa} net balance from stake network, difference $M = -(P_s - \Delta_{sa})$. Domains and sign convention are as in Tables 5.4.11:1–5.4.11:3, and symbols as defined in Section 5.4.1. Units in cm liquid water equivalent, except fourth line of (e), which is in W m^{-2}

	Upper	Lower	All
(a) 1978/79			
P_s	+105	+105	+105
Δ_{sa}	+ 89	− 81	− 7
M	− 16	−186	−112
(b) 1979/80			
P_s	+ 65	+ 65	+ 65
Δ_{sa}	−114	−222	−175
M	−179	−287	−240
(c) 1980/81			
P_s	+ 65	+ 65	+ 65
Δ_{sa}	− 70	−162	−121
M	−135	−227	−186
(d) 1981/82			
P_s	+120	+120	+120
Δ_{sa}	+ 28	− 87	− 37
M	− 92	−207	−157
(e) difference $(a) - (b)$			
$P_s(a) - P_s(b)$	+ 40	+ 40	+ 40
$\Delta_{sa}(a) - \Delta_{sa}(b)$	+203	+141	+168
$M(a) - M(b)$	+163	+101	+128
melting heat equivalent of $M(a) - M(b)$	+ 17	+ 10	+ 13

The observed dependence of net balance on precipitation is further illustrated in Fig. 5.4.11:4. In addition to the observations from the stake network and precipitation gauges for the budget years 1978/9, 1979/80, 1980/1, 1981/2 and 1982/3, and the 1978/82 period as a whole, data based on the topographic mappings in 1963, 1974, 1978, and 1982, as displayed in Fig. 5.2.7:1, are also plotted. Precipitation values for the corresponding intervals 1963/74 and 1974/8, are approximated by the observations over the periods 1964/72 and 1972/9 at the Klarwill Hut and Austrian Hut gauges as available from Table 5.4.2:1.

The observations from the stake network and precipitation gauges indicate an increment of about 300 mm a^{-1} liquid water equivalent net balance per 100 mm a^{-1} increase in solid precipitation. For the 1978/82 interval, the results from the stake network and the topographic mappings agree closely. By contrast, comparatively large negative net balance values are found for the 1963/74 and 1974/78 periods based on the mappings. Among the possible factors for this apparent discrepancy are the mismatch in available observation periods for precipitation versus net balance, and observation errors. Regarding

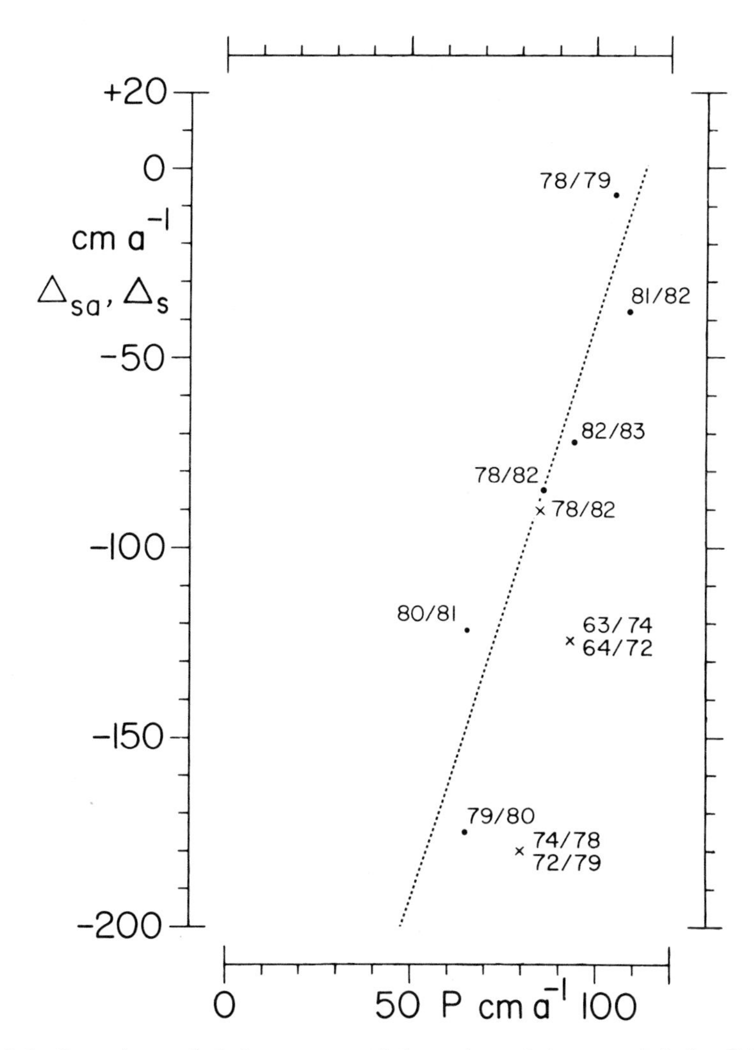

Fig. 5.4.11:4. Dependence of glacier-average net balance Δ_{sa} and Δ_s on precipitation P. Dots refer to net balance Δ_{sa} obtained from the stake network for the budget years 1978/79, 1979/80, 1980/81, 1981/82, 1982/83, and the period 1978/82 as a whole (source: Tables 5.4.3:5, 5.4.3:7, 5.4.3:9, 5.4.3:11, 5.4.3:14, 5.4.2:2). Crosses indicate net balance Δ_s obtained from successive topographic mappings for the intervals 1963/74, 1974/78, and 1978/82, and precipitation for the periods 1964/72, 1972/79, and 1978/82 (source: Fig. 5.2.7:1 and Tables 5.4.2:1 and 5.4.2:2). Coordinates are in cm a^{-1} of liquid water equivalent, or g H_2O cm^{-2} = 10 kg m^{-2}.

the systematic observation program since 1978, Fig. 5.4.11:4 shows a clear dependence of glacier average net balance on precipitation. This offers the prospect of estimating glacier average net balance – and in conjunction with information such as that contained in Fig. 5.4.3:5 even the vertical net balance profile – from data as straightforward as precipitation gaugings at Teleki Ranger Camp.

5.4.12. DIURNAL FORCING-RESPONSE RELATIONSHIPS

The following presentation is based on Hastenrath (1983b). On prolonged sojourns at Lewis Glacier, one is impressed by the marked daily cycles of radiation, melting, and water runoff, as exemplified in part by Figs. 5.4.4:1, 5.4.4:3, 5.4.9:1 and Tables 5.4.4:3, 5.4.7:1, 5.4.9:1, 5.4.9:2, and 5.4.10:1. Moreover, it is striking how sequences of cloudless days without fresh snowfall and hence low albedo are associated with large water discharge, while in episodes with abundant cloudiness, snowfall, and thus high albedo, runoff decreases and may vanish altogether. The largest water discharge is found in the evening (Figs. 5.4.4:1, and 5.4.4:3, and Tables 5.4.4:1–5.4.4:3), that is well after the daily maximum of surface net radiation. The daily march of net radiation represents the major heat budget forcing, and the discharge at the exit of Lewis Tarn manifests the water budget response of the glacier. Intensive observation programs aimed at such forcing-response relationships were conducted on Lewis Glacier during limited periods in January–March 1978. Observations of this experiment are in the following evaluated with regard to the daily march of the surface heat budget, and are discussed in terms of the daily cycle of melting and runoff. Reference is made to section 5.4.10 for comparison of bulk-aerodynamic and lysimeter estimates. Discussion of melting and evaporation is here primarily based on lysimeter measurements.

Net allwave radiation exhibits a very prominent daily variation. Global radiation was measured near the glacier over intervals of up to a few hours, as mentioned in Section 5.4.5. In conjunction with estimates of albedo in early February 1978 at stations 3, 21, and 44, of 70, 40, and 30%, respectively, these measurements permit inference of the net shortwave radiation $SW\uparrow\downarrow$. Net longwave radiation $SW\uparrow\downarrow$ at clear sky is of the order of 70 W m^{-2} (Section 5.4.7). Since no monitoring of cloudiness as such was obtained, $LW\uparrow\downarrow$ at actual cloud cover was estimated as follows. Clear-sky global radiation at each full hour LT was calculated for early February and a turbidity factor of $T = 1$, using Eqs. 5.4.5:(1) and 5.4.5:(2), as discussed in Section 5.4.5. From the calculated clear-sky and the actually measured global radiation, an effective cloud cover can be computed from Eq. 5.4.5:(3). This value of cloudiness and the aforementioned figure of $LW\uparrow\downarrow$ for clear sky are then inserted in Eq. 5.4.7:(1), so as to yield an improved estimate of $LW\uparrow\downarrow$ with actual cloud cover. From the component fluxes thus estimated, the daily march of net allwave radiation $SWLW\uparrow\downarrow$ was constructed. $SWLW\uparrow\downarrow$ reaches the largest positive values around the middle of the day but before rather than after noon, in consequence of the afternoon maximum of cloudiness. Largest negative values occur at night. Estimates of $SWLW\uparrow\downarrow$ for 06:00–18:00 and 18:00–06:00 hr LT during the experiment days at stations 3, 21, and 44 are listed in Table 5.4.12:1.

Melting and sublimation as evaluated from Table 5.4.10:1 attain their maxima around to about an hour later than net allwave radiation $SWLW\uparrow\downarrow$. Melting ceases with temperature below freezing at night. Estimates for 06:00–18:00 and 18:00–06:00 hr LT during the experiment days at the same three stations are included in Table 5.4.12:1.

Net allwave radiation $SWLW\uparrow\downarrow$ and the sum of the latent heat equivalents of melting and sublimation ($Q_m + Q_{ei}$), listed in Table 5.4.12:1, are of comparable magnitude. Discussion of differences between these two quantities is not warranted in view of the large uncertainty of terms; estimates of $SWLW\uparrow\downarrow$, for example, being strongly affected by the choice of a representative albedo. However, Table 5.4.12:1 does illustrate the

TABLE 5.4.12:1

Daytime and nighttime values of net allwave radiation $SWLW\uparrow\downarrow$, and latent heat equivalents of sublimation Q_{ei}, and melting, Q_m, at stations 3, 21, and 44, for 5–8 February 1978, and at station 21 for 27 February to 2 March 1978; in W m^{-1}

	5		6		7		8 Feb 1978
	6	18	6	18	6	18	6 12 hr LT
sta 3							
$SWLW\uparrow\downarrow$	+153	−60	+145	−60	+ 74	−60	+159
Q_{ei}	176	82	73	35	26	16	26
Q_m	0	0	219	0	57	0	0
sta 21							
$SWLW\uparrow\downarrow$	+383	−60	+460	−60	+200	−60	+351
Q_{ei}	187	22	69	13	18	27	7
Q_m	91	0	255	0	90	0	0
sta 44							
$SWLW\uparrow\downarrow$	+444	−60	+414	–	+244	−60	+415
Q_{ei}	325	95	63	–	115	−13	26
Q_m	75	0	210	–	88	0	0
	27		28 Feb		1 Mar		2 Mar 1978
	6	18	6	18	6	18	6 18 hr LT
sta 21							
$SWLW\uparrow\downarrow$	+250	−60	+209	–	+129	−60	+447
Q_{ei}	272	89	141	–	168	32	231
Q_m	49	0	0	–	0	0	12

increase of net allwave radiation and of ablation from the higher to the lower elevations of the glacier, largely as a result of the altitudinal variation of surface albedo.

Table 5.4.12:1 illustrates a nighttime heat loss by net allwave radiation $SWLW\uparrow\downarrow$ of the order of 60 W m^{-2}. In accordance with Eqs. 5.4.1:(1) and 5.4.1:(2), various processes can be considered to provide the energy for this expenditure. Of particular interest are the following terms: sensible heat flux at the glacier-air interface Q_s, transient heat storage within the ice Q_t, vertical heat flux in the surface layer V, and freezing of liquid water within the column F; downward directed latent heat flux Q_{ei} at night being unimportant at Lewis, as shown by observation. According to the order of magnitude calculations in Sections 5.4.1 and 5.4.9, the sum $(Q_s + Q_t + V)$ may supply only about 20 W m^{-2}. By contrast, considerations in Section 5.4.1 underline the relative importance of F, in that freezing of 10 kg m^{-2} of water over 12 hours would amount to 76 W m^{-2}. The respective quantity of one gram of liquid water per cm^2 in a column of ice, say about 1 m deep, seems modest indeed. Accordingly, the nighttime refreezing of daytime melt is an important feature in the daily march of the surface heat budget.

TABLE 5.4.12:2

Melting and discharge during 5–8 February 1978. Volume of melting during 6–18 hr LT is estimated as product of lysimeter measurements at stations 3, 21, and 44 times area of the respective elevation bands. Values are detailed by bands for the Western portion of Lewis, while totals over all elevations are given for the Western and the Western plus Eastern portions of Lewis Glacier. Discharge measurements at the exit of Lewis Tarn refer to 6–18 hr LT of day indicated and the subsequent interval 18–6 hr LT; in 10^4 ℓ

Station	Elevation (m)	Area 10^2 m^2 W	Area 10^2 m^2 W + E	5	6	7 Feb
				melting 6–18 hr LT, in 10^4 ℓ		
				W glacier		
3	⩾ 4830	754	985	0	216	56
21	4830 – 4720	1150	1180	137	384	135
44	⩽ 4720	695	785	66	191	78
total		2599		203	791	269
				W + E glacier		
total			2950	216	891	300
				discharge in 10^4 ℓ		
6–18 hr LT				–	121	52
18–6				173	156	48

Melting is in the following considered in perspective with the water discharge at the exit of Lewis Tarn. Table 5.4.12:2 is based on Fig. 5.4.4:3, and Tables 5.4.4:3 and 5.4.10:1. The lysimeter stations 3, 21, and 44, are regarded as representative of three altitude domaines. Only the Western portion of the glacier is considered, which according to the surface topography drains to Lewis Tarn, although subsurface drainage may well link the Eastern to the Western glacier segment and thus increase the total melt volumes shown in Table 5.4.12:2. Deficiencies of the poor-man lysimeters may in part compensate, in that the tray is heated by solar radiation but radiation penetrates to depths greater than the lysimeter. Melt is given for 06:00–18:00, and the measured discharge at the exit of Lewis Tarn for 06:00–18:00 and 18:00–06:00 hr LT. Table 5.4.12:2 shows daytime melt volumes for the glacier as a whole greatly exceeding the measured discharge. In this context, consider a nighttime freezing of daytime melt in the modest amount of one gram of liquid water per cm^2, as contemplated above. Integrated over the Western glacier as a whole, this would correspond to 26×10^5 l. By comparison, a water discharge rate at the exit of Lewis Tarn of 15 l s^{-1} (ref. Section 5.4.4) amounts to a daily total of 13×10^5 l. These order of magnitude estimates indicate that a substantial portion of the daytime melt does not reach Lewis Tarn.

5.5. Ice Stratigraphy and Cores

During the various expeditions pits were dug at six locations of the glacier, as shown in Map 5.5:1. Also plotted in Map 5.5:1 are the 1978 network of net balance stakes and the approximate drill sites of the University of East Anglia expeditions in July–August of 1975 and 1977 (Davies *et al.*, 1977a, b; personal communication 1978–9). The stratigraphy at stations II, 3, 4, 13, 21, is illustrated in Figs. 5.5:1–5.5:5. Details of ice cores retrieved at sites I and II are presented in Fig. 5.5:6–5.5:9.

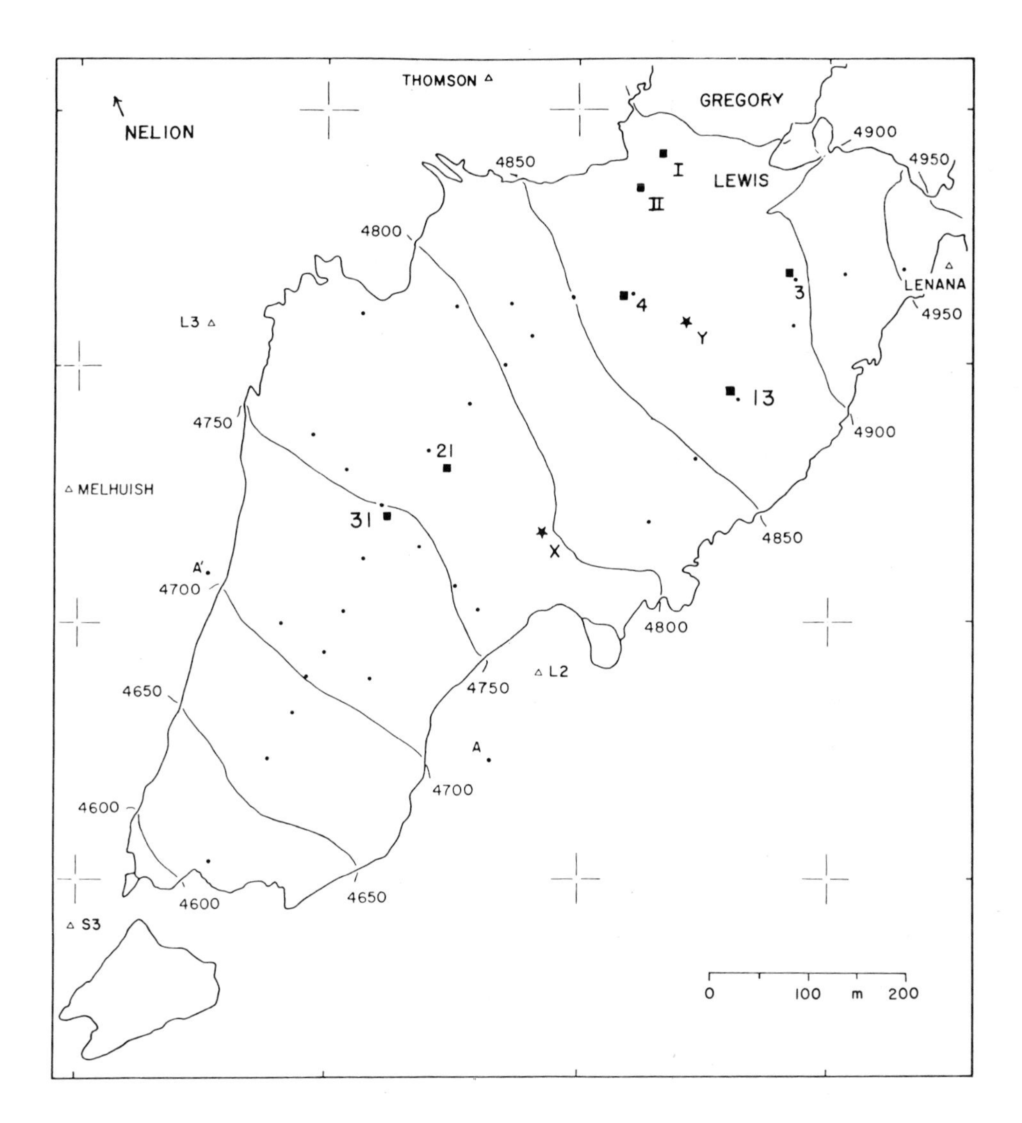

Map 5.5:1. Orientation map. Rectangles denote sites of ice pits 3, 4, 13, 21, 31 and of ice cores I and II; and dots refer to the network of net balance stakes. Asterisks denote drill sites by University of East Anglia expeditions in August 1975 (*X*) and July 1977 (*Y*).

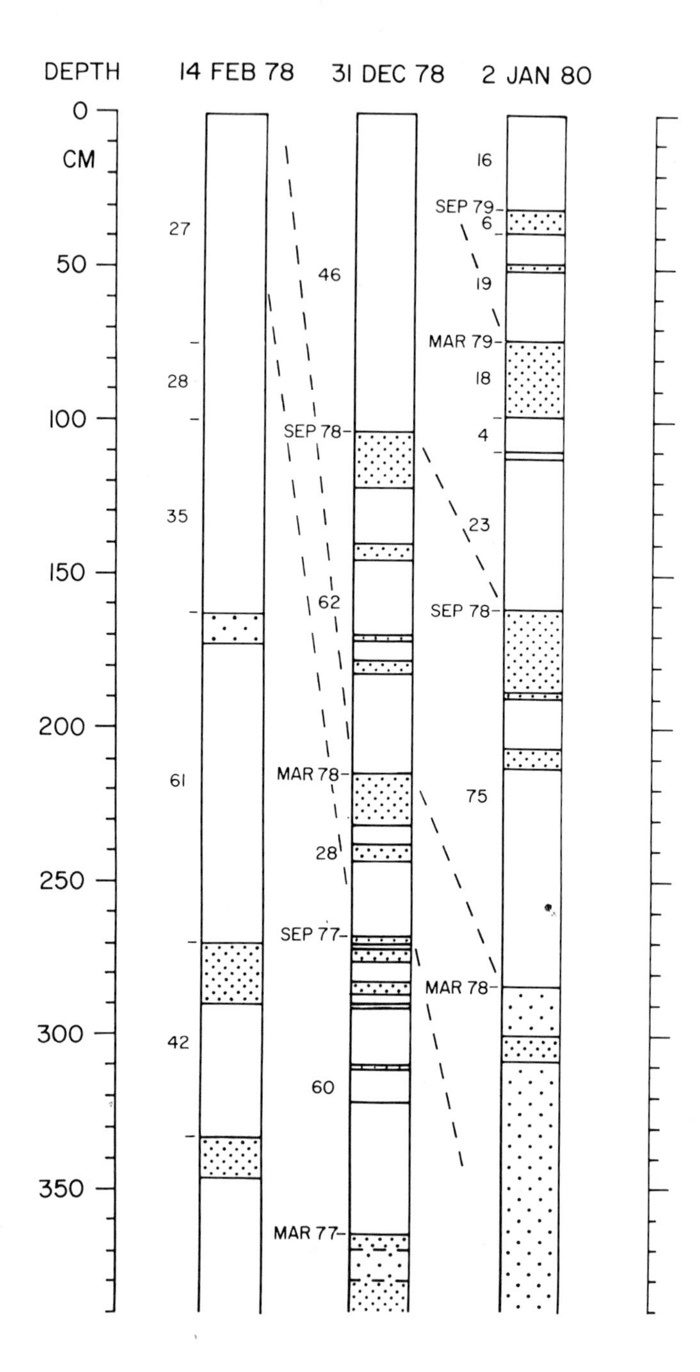

Fig. 5.5:1. Pit profiles at station II studied on 14 February 1978, 31 December 1978, and 2 January 1980. Dense and sparse dot raster denote strong and weaker ice horizons, and blank surfaces snow layers. Broken lines show suggested correlation of chronology. Numbers to the left of columns indicate mass content of layers in g cm^{-2} = 10 kg m^{-2}.

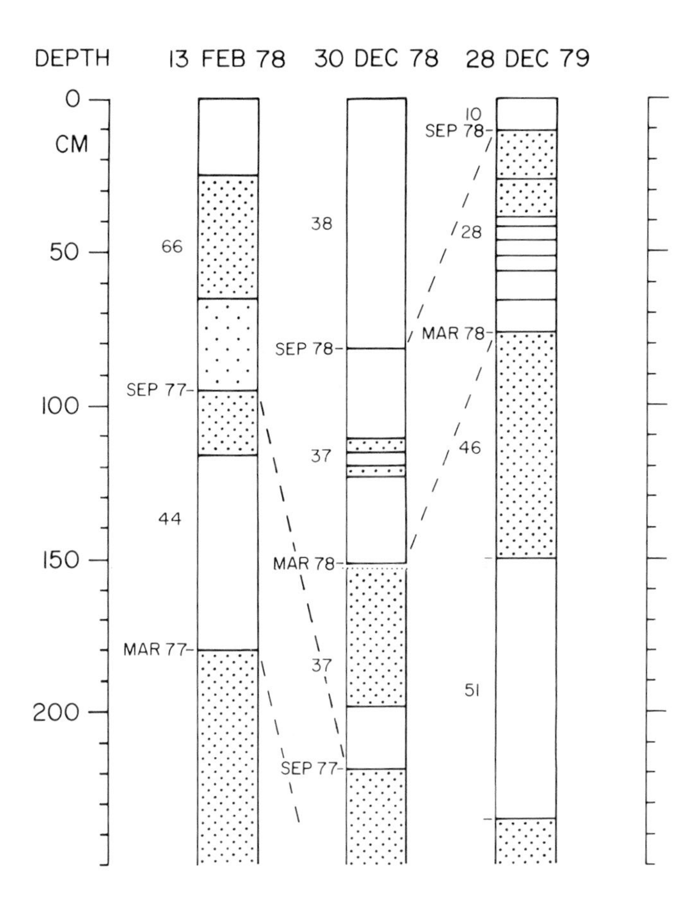

Fig. 5.5:2. Pit profiles at station 3, studied on 13 February 1978, 30 December 1978, and 28 December 1979. Symbols as for Figs. 5.5:1–5.5:5. Horizontal dotted line signifies cross bar marking March 1978 horizon.

Of particular interest in the profiles Figs. 5.5:1–5.5:5 and 5.5:7–5.5:9 are (a) the spatial continuity of prominent ice horizons, and (b) the consistency of mass content of identifiable layers with the net balance obtained from stake readings. Two major ice horizons per year correspond to the January–February and July–August ablation seasons. Proceeding from the upper portions of the accumulation region downward (Fig. 5.5:1, 5.5:2, and 5.5:4), individual seasonal layers thin out and disappear altogether near the equilibrium line (Fig. 5.5.5:5). Little wooden bars placed on the surface at the net balance stakes allowed a firm stratigraphic control for the period February to December 1978 at stations 3 and 21. Regarding other intervals and locations, the mass content of layers has a similar magnitude as the net balance obtained from stake readings. Complete agreement is not expected because of percolation effects especially in the lower portions of the accumulation area. The negative net balance during the budget year March 1979 to March 1980 over most of the glacier resulted in the annihilation of parts of the earlier profile at locations in the accumulation area (Figs. 5.5:2–5.5:4).

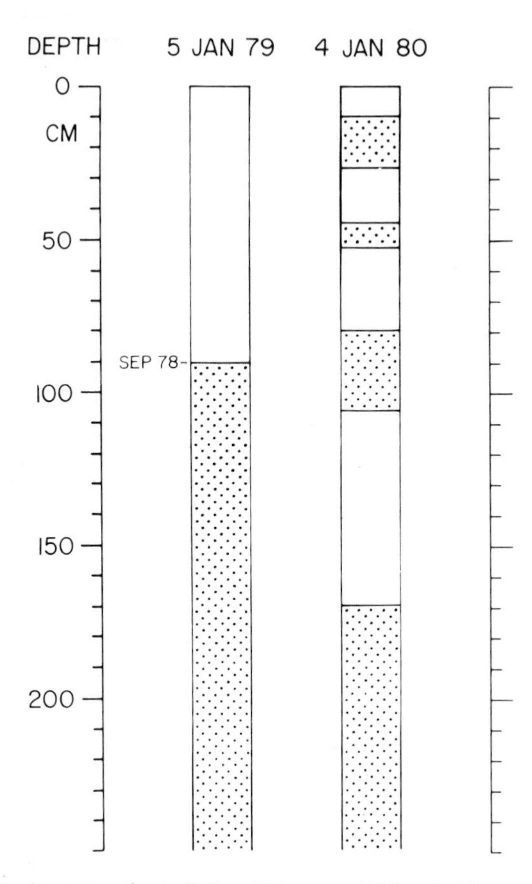

Fig. 5.5:3. Pit profiles at station 4, studied on 5 January 1979 and 4 January 1980. Symbols as for Figs. 5.5:1–5.5:5.

The numerous paleoclimatic reconstructions for high latitudes (Dansgaard, 1964; Dansgaard *et al.*, 1969) contrast with the information gap for the tropics. Therefore the possibility of applying the ice core paleoclimatic technique in the low latitudes merits particular attention (Hastenrath, 1975). In February 1978 ice cores were retrieved at Lewis Glacier for climate reconstruction through analysis of oxygen isotope, total β radioactivity, and microparticle content. Further samples were obtained in January 1980 from a pit in the same general location. The background provided by the overall project, and in particular the stratigraphic control available from pits, net balance, and precipitation measurements, are considered fundamental to the interpretation of ice cores. Results are discussed here in the context of our earlier ice core experience at the Quelccaya Ice Cap in Peru (Thompson *et al.*, 1979), largely following a previous publication (Thompson and Hastenrath, 1980).

Among the pits dug in February–March 1978 are two on the flat col between the Lewis and Gregory Glaciers around 4870 m (Map 5.5:1; Figs. 5.5:1). The more northerly of these (I) extended to 2.5 m, from where a SIPRE drill provided a core to a total depth

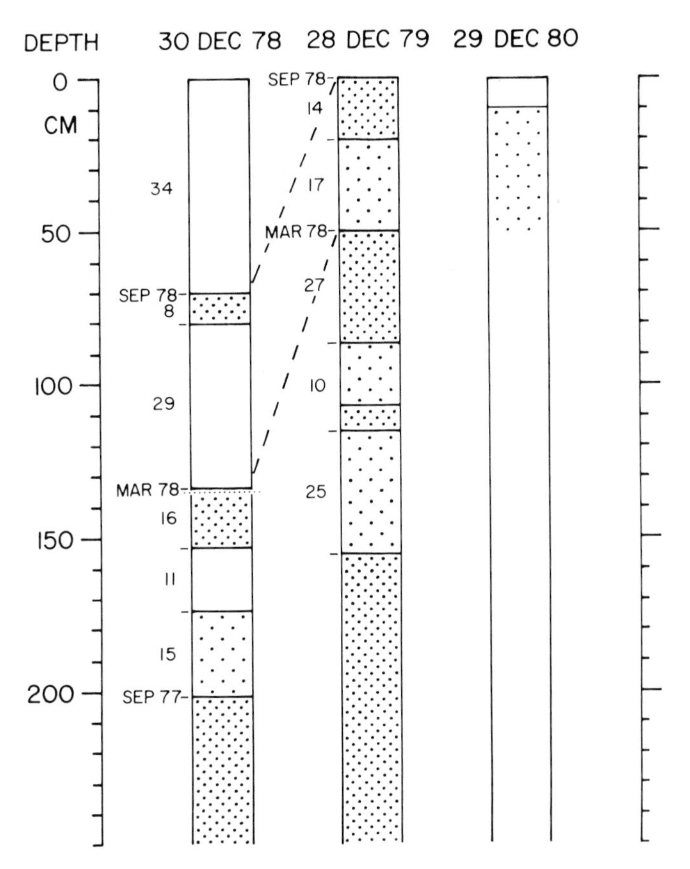

Fig. 5.5:4. Pit profiles at station 13, studied on 30 December 1978 and 28 December 1979. Symbols as for Figs. 5.5:1–5.5:5. Horizontal dotted line signifies cross bar marking March 1978 horizon.

of 13.4 m. The other pit (II) some 40 m further South was dug to 2.1 m, and an ice core was drilled to a total depth of 11.4 m. At site I drilling was unusual, in that voids were encountered between 8.3 and 8.6 m and between 9.4 and 12.9 m depth. The core at site II is continuous.

519 samples were cut, melted, and placed in clean containers while in the field. In the process, each core was split in half. One half was analyzed for oxygen isotope and total β radioactivity at the Geophysical Isotope Laboratory in Copenhagen, and the other for microparticles at the Ohio State University under Class 100 clean room conditions.

In December 1978 a pit was excavated to 3.8 m in the same general area (Fig. 5.5:1), for study of stratigraphy and density profile. In December 1979–January 1980, a pit was dug to 3.1 m, and 16 samples were retrieved for isotope and microparticle analysis (Fig. 5.5:1).

Temperature at the surface is below freezing during the night and early morning; however, firn temperatures increase to 0°C at 0.5 m depth, thus indicating that the ice is temperate. Fig. 5.5:6 illustrates a rapid increase of density with depth, presumably resulting from percolation.

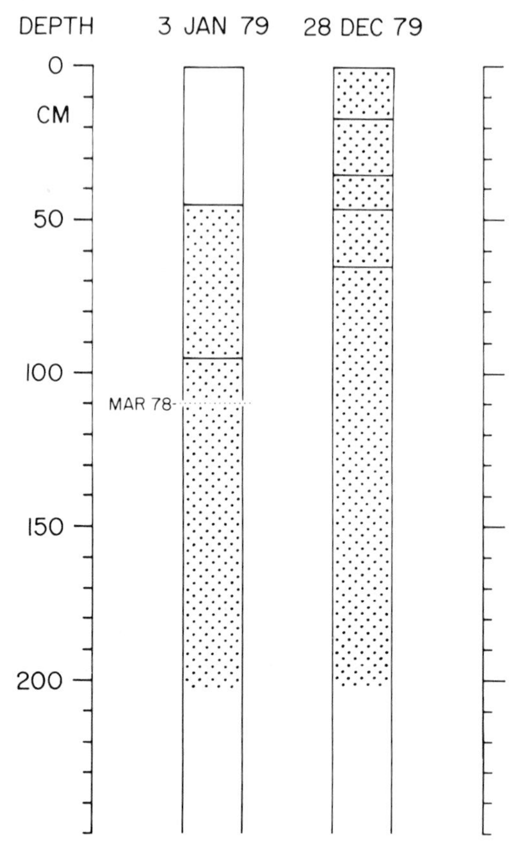

Fig. 5.5:5. Pit profiles at station 21, studied on 3 January 1979 and 28 December 1979. Symbols as for Figs. 5.5:1 and 5.5:5. Horizontal dotted line signifies cross bar marking March 1978 horizon.

Both cores I and II (Figs. 5.5:7 and 5.5:8) contain well developed ice horizons and visually detectable dirt bands. Throughout the cores large particle concentrations occur simultaneously with high β radioactivity. $\delta^{18}O$ ratios exhibit a range of 5.4 ‰ at the surface, but this is smoothed out below 1 m presumably due to meltwater percolation.

For core I, Fig. 5.5:7, a hiatus is indicated at 6 m by a 45 degree tilting of the strata, as well as by the voids at 8.3–8.6 m and 9.4–12.9 m. Most remarkable is the large range of $\delta^{18}O$ in the ice downward from 8.6 m, below the first void. As at the surface, the less negative $\delta^{18}O$ values are associated with high particle concentrations. Open crevasses could allow snowfall to become trapped at greater depth, thus contributing to the irregularity of the $\delta^{18}O$ profile. Although no crevasses were noted near the sampling site, some mobility in crack formation and closure cannot be ruled out even in this location of gentle topography. Based upon the aforementioned interruptions and irregularities core I appears little suited for the reconstruction of a continuous climatic time series.

Ice core II (Fig. 5.5:8) differs remarkably from core I, despite their close proximity. Microparticle concentrations and total β radioactivity are particularly large in the upper few meters, but greatly decrease with depth, broadly parallel with increasingly larger

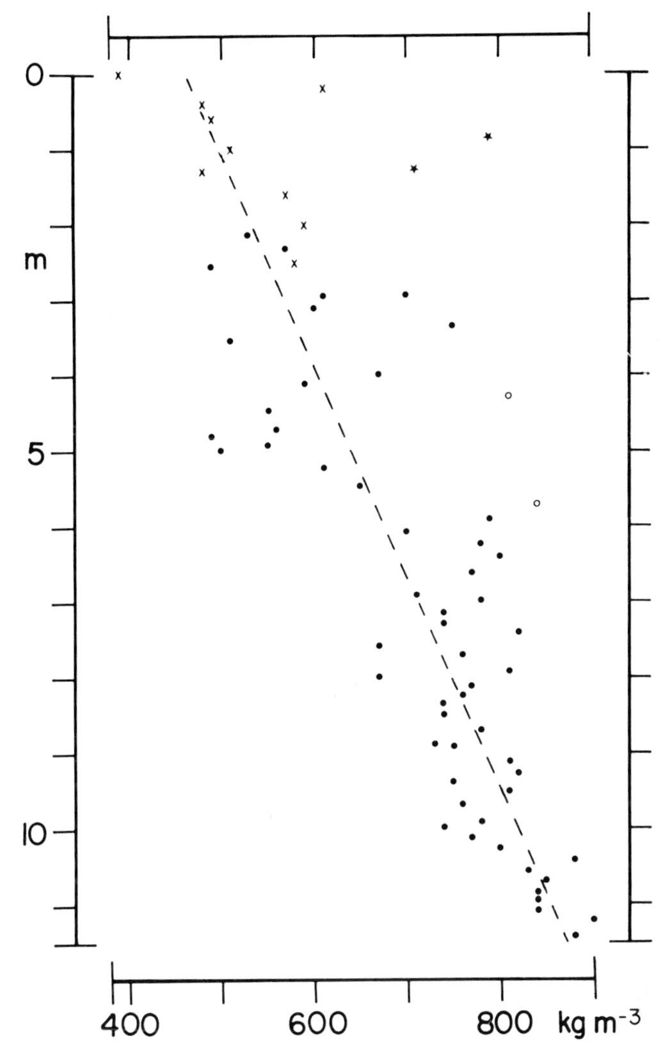

Fig. 5.5:6. Density profile. Crosses and asterisks refer to site I, and dots and circles to site II, asterisks and circles denoting ice horizons.

negative $\delta^{18}O$ values. The decrease with depth of visually apparent impurities in Lewis Glacier has been noted earlier (Hastenrath, 1975). The microparticle profile in Fig. 5.5.8(A) is consistent with the cleaning of snow by percolation as reported for temperate glaciers (Glen *et al.*, 1977). Whether the comparatively high radioactivity in the top few meters is related to the presence of algae, as suggested by the pink color of samples from the wall of pit II, is open to speculation.

The change toward more negative $\delta^{18}O$ values with depth in core II, Fig. 5.5:8, contrasts with conditions observed in most temperate glaciers (Sharp *et al.*, 1960; Ambach *et al.*, 1972; Thompson *et al.*, 1979). Refreezing of the melt and rain water percolating into cold underlying snow or firn is a viable mechanism for homogenization and less

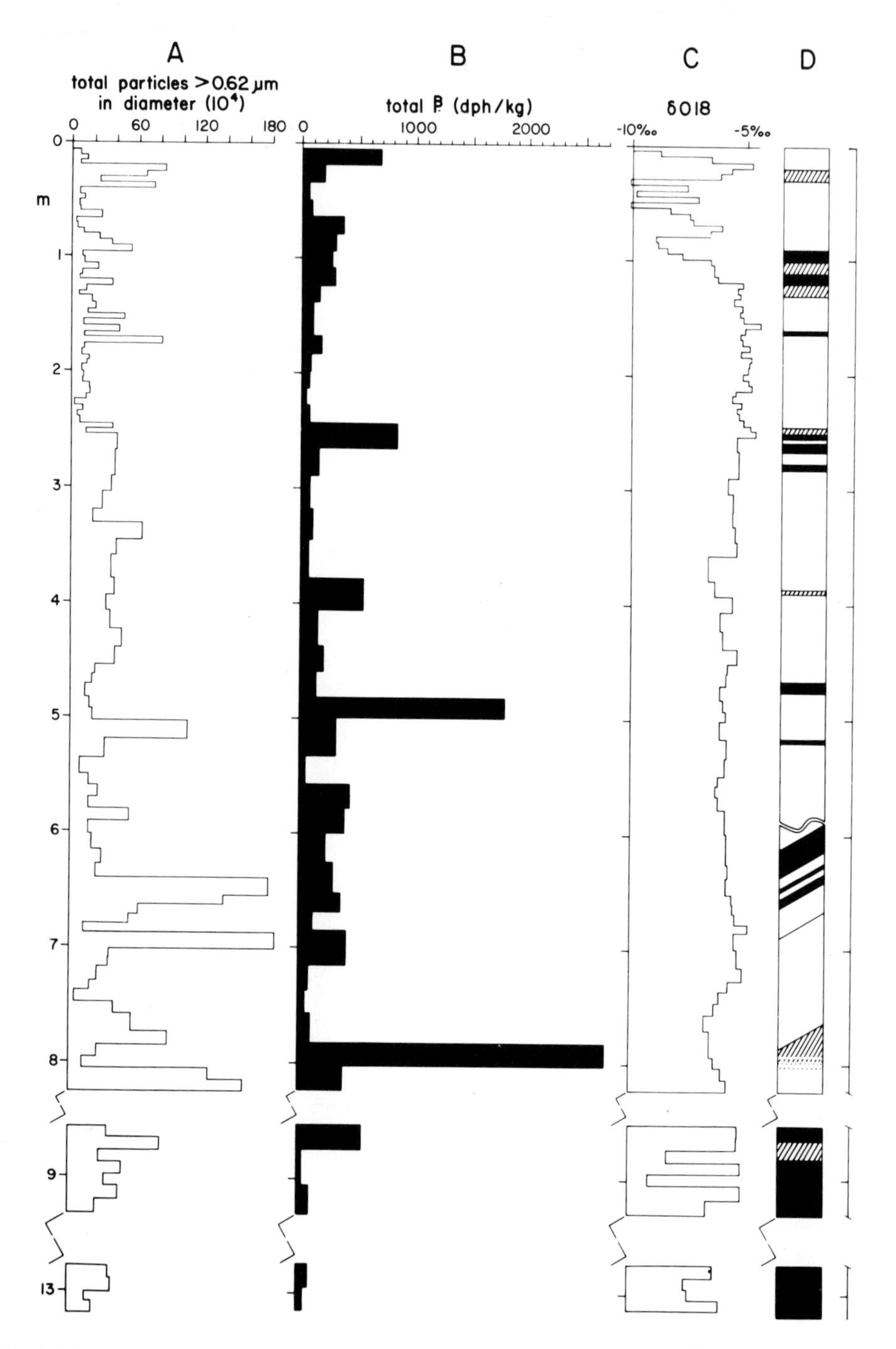

Fig. 5.5:7. Core profile at site I. A. concentration of microparticles > 0.62 μm in diameter per 500 μℓ of sample. B. total β radioactivity in dph/kg, shaded. C. oxygen isotope ratios, $\delta^{18}O$, unshaded. D. ice core stratigraphy: ice horizons and visually dirtier layers are indicated by solid lines and hatching, respectively.

Fig. 5.5:8. Core profile at site II. Symbols as for Fig. 5.5:7. Gap at 195–407 cm is due to lack of sampling bottles.

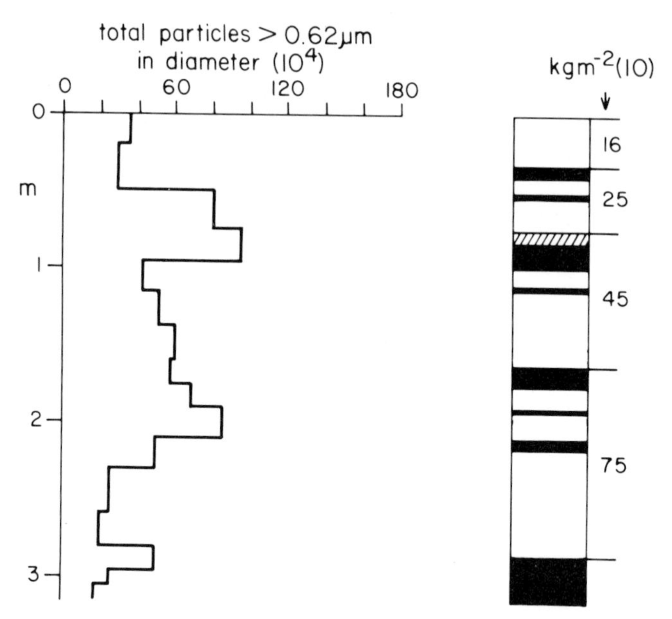

Fig. 5.5:9. Pit dug in general area of cores I and II in December 1979–January 1980, corresponding to rightmost profile in Fig. 5.5:1. Symbols as for Figs. 5.5:7 and 5.5:8.

negative $\delta^{18}O$ values in temperate glaciers (Sharp *et al.*, 1960). It is thought that rain and snow precipitated in the warmer season possesses less negative $\delta^{18}O$ values. Temperature records from the Kenya highlands exhibit little variation over the recent past, and precipitation on the col between the Lewis and Gregory Glaciers is as a rule in solid form. Thus an explanation in other terms must be sought for the $\delta^{18}O$ profile of core II, Fig. 5.5:8.

The Meren Glacier in New Guinea (Hope *et al.*, 1976, pp. 39–49) also shows a change towards more negative $\delta^{18}O$ values with depth. Hope *et al.* consider that most of the snowfall is associated with cyclonic activity, whereas daily convective clouds bring mainly rain. Assuming the air in convective systems is depleted in heavy isotopes, the percolation of rain water would yield more negative $\delta^{18}O$ values at depth. However, neither the snow nor rain was analyzed isotopically, leaving this explanation open to speculation.

Table 5.5:1 summarizes the oxygen isotope and microparticle data from four glacier sites in the tropics. No simple relation is apparent between $\delta^{18}O$ ratios and mean surface air temperature. Values from Mount Kenya and Kilimanjaro are of similar magnitude, reflecting their vicinity and similar climatic environments. The largest range is found on Quelccaya in the Peruvian Andes. The high particle concentrations at Lewis Glacier reflect the proximity of exposed rock massifs to the core sites. By contrast, particle concentrations are low at the Quelccaya summit, which is the highest point in the area, and distant from surfaces not covered by ice.

Stratigraphic control by snow pit and net balance studies is fundamental to the chronology and climatic interpretation of ice cores. These are provided for the last few years by our field program at Mount Kenya.

The annual march of precipitation in Central Kenya is characterized by two rainy seasons occurring about March–June and September–December (Sections 2.2 and

TABLE 5.5:1

Summary of oxygen isotope and microparticle data from tropical glaciers. A. Summit of Quelccaya Ice Cap, Peru, 13°56′S, 70°50′W, 5650 m; (source: Thompson *et al.*, 1979): B. Meren Glacier, Mount Carstenz, Indonesia, 0°9′S, 37°19′E, 4870 m (source: Hope *et al.*, 1976, pp. 39–49); C. Lewis Glacier, 0°9′S, 37°19′E, 4870 m; D. Kilimanjaro, 3°00′S, 37°20′E, 5895 m (source: Gonfiantini, 1970)

	Quelccaya	Meren	Lewis	Kilimanjaro
mean air temp (°C)	– 3	+ 0.5	– 0.5	
mean $\delta^{18}O$ near sfc (°/oo)	–21.0	–15.3	– 7.8	–3.7 at 4600 m to –6.8 at 5700 m
mean $\delta^{18}O$ all samples (°/oo)	–19.4	–15.7	– 6.0	
$\delta^{18}O$ max	–11.0	–12.7	– 4.8	
$\delta^{18}O$ min	–33.0	–17.0	–10.2	
$\delta^{18}O$ range	22.0	4.3	5.4	
	most negative in warm season			
concentration of particles > 0.62 μm (10^8 per liter)	2		8	

5.4.2). These are accumulation seasons at Mount Kenya. Precipitation activity and cloudiness is at a minimum during January–February and July–August, the ablation seasons on the mountain. Generally, the two ablation seasons are reflected in the pit profiles by pronounced ice horizons, and the accumulation seasons by more extended layers consisting predominantly of firn.

Figs. 5.5:7–5.5:9 indicate that the ice horizons, which may represent the two ablation seasons of the year, tend to be characterized by both high concentration of microparticles, and large total β radioactivity. By contrast, the relation of $\delta^{18}O$ to the seasonality of net balance is not apparent. It seems desirable to extend the comparison of ice cores versus net balance and precipitation conditions beyond the most recent years of our monitoring program of Lewis Glacier. To that end, precipitation index series were constructed for a group of long-term stations in Central Kenya, as listed in Table 5.4.2:3, namely (a) for the March–June and September–December, (b) for July–August, and (c) for January–February. The index (a) for the rainy seasons is of interest in relation to the mass content of identifiable layers in the core and pit profiles, Figs. 5.5:1 to 5.5:6, and 5.5:7 to 5.5:8. The latter two indices (b) and (c) are indicative of the interannually varying intensity of the 'dry' or ablation seasons, and are thus pertinent to the development of ice horizons.

Experience suggests no close correlation between precipitation activity on Mount Kenya as compared with the surrounding highlands. This may be due in part to the pronounced diurnal circulations. In the development of the January–February and July–August 'lulls' of precipitation, however, the mountain does seem to share with most of Central Kenya a large-scale general circulation control. Accordingly, the index series in Table 5.4.2:3 can only serve to a limited extent as proxies for the precipitation and net balance conditions at Lewis Glacier itself.

Ascribing successive ice horizons in Figs. 5.5:8, 5.5:9, and 5.5:1 to the sequence of ablation seasons, a reasonable correspondence is apparent for the more extreme conditions of ice horizon and precipitation indices (b) and (c). Conversely, the mass content of identifiable layers shows little relation to the index (a) of rainy season precipitation in Central Kenya, Table 5.4.2:3, except for the budget years March 1978–February 1979 and March 1979–February 1980, which are documented by our on site monitoring.

At this stage it is suggested that the major ice horizons in Figs. 5.5:1–5.5:5 and 5.5:7–5.5:9 mark primarily the twice-annual ablation seasons. In view of temperatures near 0°C and the concomitant melting and percolation effects, however, it is considered that ice horizons may originate through mechanisms other than surface ablation. Conditions are conceivably complicated further by more complex topography in other portions of the glacier. Furthermore, annual net balance can become negative for nearly all portions of the glacier as we measured during March 1978–March 1980, and March–July 1980, so that stratigraphic chronology is destroyed. On site concurrent monitoring of precipitation, net balance, pit profiles, and sampling for various laboratory analyses, over a series of years may elucidate this problem complex. If in fact the concomitant maxima of microparticle concentration and total β radioactivity mark the horizons of the twice-annual ablation seasons, then this would open the prospect for a net balance chronology from ice cores, similar to our exploits on Quelccaya.

The present findings in conjunction with our previous results from the Peruvian Andes (Thompson *et al.*, 1979) exemplify the potential of climate reconstruction from ice cores in the tropics. The Quelccaya Ice Cap meets simultaneously the following important conditions: (a) high elevation (low temperature, no percolation); (b) gentle topography (minimum effect of flow dynamics on stratigraphy); (c) location in outer tropics (some seasonality). Elsewhere in the tropics conditions are less favorable for ice coring. The concurrent study of modern net balance and stratigraphy is always imperative. The summit ice fields of Kilimanjaro are the obvious choice for further climatic ice core studies in equatorial East Africa.

5.6. Long-term Climatic Forcing and Glacier Response

The climate governs the heat and mass economy of the glacier, as expressed in Eqs. 5.4.1:(1) and 5.4.1:(3). In turn, the distribution of net balance with elevation (Section 5.4.3) along with the bedrock configuration (Section 5.2.6) control the dynamics of the ice flow, and then the ice extent. It is through this chain of causalities (Fig. 5.6:1) that climate variations are ultimately reflected in glacier advance or retreat. Understanding of this causality chain is needed for quantitative translation of the well documented and conspicuous glacier response into the unknown, and small climatic forcing.

Considerable insight into the characteristic magnitudes of forcing and response can be gained from simple sensitivity analysis for the glacier as a whole (Hastenrath, 1975). The pertinent terms in Eq. 5.4.1:(3) of the continuity of ice mass are solid precipitation P_s, sublimation Q_{ei}/L_s, melting Q_{mm}/L_m, and net balance Δ_s. The pertinent terms in the heat budget Equation 5.4.1:(1) are net allwave radiation $SWLW\uparrow\downarrow = SW\downarrow - SW\uparrow - LW\uparrow\downarrow$,

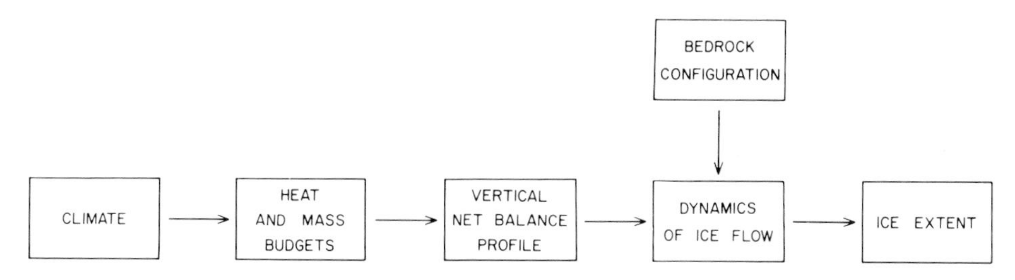

Fig. 5.6:1. Schematic causality chain of climate controlling ice extent.

sensible and latent heat flux at the glacier–air interface Q_s and Q_e, and heat used in melting Q_m; the latter two terms representing the linkage with the mass budget Equation 5.4.1:(3).

Reference is made to Section 5.4.11. Climatic elements and surface properties of immediate interest for long-term variations in the context of the heat budget Equation 5.4.1:(1) are as follows. Net allwave radiation $SWLW\uparrow\downarrow$ is controlled by cloudiness and surface albedo, as illustrated in Fig. 5.4.11:1. Changes in the sensible heat flux Q_s result primarily from variations in air temperature T_a, and changes in the latent heat flux Q_e from variations in T_a and relative humidity f. The resultant of changes in $SWLW\uparrow\downarrow$, Q_s, and Q_e, determines the changes in heat available for melting Q_m. The latter along with changes in solid precipitation P_s are of primary interest with regard to long-term variations in ice volume.

Based on these considerations and the analysis in Section 5.4.11, the long-term rate of decrease in ice volume as documented in Fig. 5.2.7:1 can be converted into changes in solid precipitation ΔP_s, cloudiness ΔC, albedo $\Delta\alpha$, air temperature ΔT_a, and relative humidity Δf, that could bring about the observed glacier-average imbalance Δ_s in the ice mass economy. Evaluations of data in Fig. 5.2.7:1 for the intervals 1899–1934, 1934–63, 1963–74, 1974–8, and 1899–1978, are summarized in Table 5.6:1. In these calculations the glacier surface is approximated by the arithmetic mean of values at the beginning and end of a given time interval. Entries refer always to changes in a single parameter.

Table 5.6:1 illustrates that the strongly nonlinear decrease in ice volume depicted in Fig. 5.2.7:1 is in part related to the secular shrinkage in glacier surface, in that $-\Delta V$ decreases strongly from the earlier to the later time intervals, but $-\Delta_s$ does not. Largest values of $-\Delta_s$ of the order of 100 cm a^{-1} are found for the intervals 1899–1934 and 1899–1978.

Entries for ΔP_s are identical to Δ_s and denote the change in solid precipitation that by itself could explain the observed mass change of the glacier. Comparison with Fig. 2.5:5 shows that precipitation variations of this magnitude are not borne out by rainfall gauges in the Kenya highlands in the course of the 20th century. For the era prior to systematic rainfall measurements, however, the water level variations of East African lakes discussed in Section 2.5 and depicted in Fig. 2.5:4 attest to drastic change in the hydrometeorological conditions. In particular, a precipitation decrease of the order of 150–200 mm a^{-1} can be inferred from the 1880s to the turn of the century (ref. Section 2.5). It is plausible that Mount Kenya shared this abrupt change of hydrometeorological conditions with East Africa at large. In accordance with Table 5.6:1, this event is

TABLE 5.6:1

Sensitivity analysis: changes in climatic elements and surface properties corresponding to observed glacier-average imbalance in ice mass economy for intervals 1899–1934, 1934–63, 1963–74, 1974–78, 1978–82, and 1899–1978. Volume change $\Sigma\Delta V$ in 10^4 m^3 and ΔV in 10^4 m^3 a^{-1}; area A in 10^4 m^2; mass imbalance $\Sigma\Delta_s$ in 10^3 kg (H_2O) m^{-2} (= 10^2 g cm^{-2} = cm of liquid water equivalent); and Δ_s in 10^1 kg (H_2O) m^{-2} a^{-1} (= g cm^{-2} a^{-1} = cm of liquid water equivalent per year); air temperature change commensurate with change $\Sigma\Delta_s$ in elevation of glacier surface, ΔT_a ($\Sigma\Delta_s$) in °C; Δ_s equivalent of heat of melting Q_m (Δ_s) in W m^{-2}; and corresponding changes in liquid water equivalent solid precipitation ΔP_s in cm a^{-1}, of cloudiness ΔC in tenths, of albedo $\Delta\alpha$ in percent, of air temperature ΔT_a in °C, and of relative humidity Δf in percent

		1899–1934	1934–63	1963–74	1974–78	1978–82	1899–1978
$\Sigma\Delta V$	[10^6 m^3]	− 23	− 8	− 1	− 1	− 1	− 1
ΔV	[10^4 m^3 a^{-1}]	− 65	−28	− 9	−25	−25	−41
A	[10^4 m^2]	54	41	33	30	28	46
$\Sigma\Delta_s$	[m]	− 38	−18	− 3	− 3	− 3	−63
ΔT_a ($\Sigma\Delta_s$)	[°C]	+ 0.2	+ 0.1	+ 0.0	+ 0.0	+ 0.0	+ 0.4
Δ_s	[cm a^{-1}]	−108	−61	−25	−75	−75	−80
Q_m (Δ_s)	[W m^{-2}]	+ 12	+ 7	+ 3	+ 8	+ 8	+ 9
ΔP_s	[cm a^{-1}]	−108	−61	−25	−75	−75	−80
ΔC	[tenths]	− 1.5	− 0.9	− 0.4	− 1.0	− 1.0	− 1.1
$\Delta\alpha$	[%]	− 4	− 2	− 1	− 3	− 3	− 3
ΔT_a	[°C]	+ 1.6	+ 1.0	+ 0.4	+ 1.1	+ 1.1	+ 1.2
Δf	[%]	+ 10	+ 6	+ 3	+ 7	+ 7	+ 8

considered a possible important factor for the onset of glacier recession in the latter part of the last century, and for its continuation to the present. As discussed in Section 5.4.11 decrements in cloudiness $-\Delta C$ and albedo $-\Delta\alpha$ concomitant with a precipitation decrease $-\Delta P_s$ combine with the latter factor in decreasing the glacier net balance Δ_s.

Table 5.6:1 furthermore shows that explaining the observed ice recession by changes in heat budget terms would require decrements in cloudiness $-\Delta C$ or albedo $-\Delta\alpha$, or increments in air temperature ΔT_a or relative humidity Δf, of the order of one tenth, 3%, 1°C, and 10%, respectively. Smaller values may be required with two or more elements changing in unison. Climatic forcings exemplified by ΔC and $\Delta\alpha$ are too small to be monitored by conventional sensing techniques, and prospects are little better concerning ΔT_a. For tropical Africa, the evaluations by Callendar (1961), Mitchell (1963), and Dronia (1967) suggest little temperature change prior to the turn of the century, followed by a gradual warming of the order of a few °C until around the 1950s. The latter could contribute in the proper sense to the observed glacier recession in the course of the 20th century – along with the sustained effect of an abrupt precipitation decrease in the latter part of the 19th century. Changes in cloudiness and albedo concomitant with those in precipitation may also be contributing factors. The gradual lowering of the glacier surface further complicates matters in that air temperature in 1–2 m above the ice–air interface becomes effectively higher, even without a positive local rate of change of temperature, or overall warming of the environment. Table 5.6:1 shows a lowering of the ice surface of the order of 60 m over the interval 1899–1978. With a lapse rate of the order of 0.65°C/100 m this would correspond to an effective temperature rise for the air at 1–2 m above the glacier surface of about 0.4°C.

This simple sensitivity analysis thus indicates that the conspicuous glacier response could be produced by climatic forcings exemplified by approximate changes in solid precipitation ΔP_s of 100 cm a^{-1}, in cloudiness ΔC of -1 tenth, in albedo of -3%, in air temperature ΔT_a of $+1$°C, and in relative humidity of +10%, as shown in Table 5.6:1. Comparison with other evidence of long-term precipitation and temperature variations then suggests that abrupt precipitation decrease with accompanying cloudiness change in the latter part of the 19th century are major causes for the onset of ice recession and its continuation into the 20th century; temperature increase and possibly changes in cloudiness and albedo in the course of the 20th century being further contributing factors.

While providing a general idea of the magnitude of climatic forcing, this straightforward sensitivity analysis for the glacier as a whole does not consider redistribution of ice mass related to internal flow characteristics, and is unable to ascertain the time sequence of climatic events. Kruss (1981, 1983 a, b) has comprehensively studied the response of Lewis Glacier to climatic forcing by computer simulation of the ice dynamics using numerical modelling as referred to in Sections 5.2.4 and 5.3.2, and 5.3.4. The modern vertical net balance profile along with measured surface velocity or bedrock topography are basic input. An equilibrium glacier of the documented maximum extent is generated and the corresponding net balance profile is ascertained. Then glacier retreat is simulated with the observed terminus positions at various epochs (Map 4.3:8*, Fig. 5.2.7:1) serving as constraints. The earlier modeling of the Carstensz Glacier of New Guinea by Allison and Kruss (1977) produced inference on the timing of climatic change and onset of glacier retreat as well as the time sequence of changes in the vertical net balance profile. Based on the earlier work, Kruss (1981, 1983b) pursued a novel approach for Lewis Glacier, by modeling in unison the entire causality chain sketched in Fig. 5.6:1, from climatic forcing to terminus response.

Kruss' (1981, 1983a) analysis of frequency response yields for net balance variations on the scale of decades to centuries a lag in terminus response of the order of 10–20 years; the amplitude response of the terminus increases rapidly with period on the scale of decades, further amplitude increase with period being small on the scale of centuries.

Climatic forcings in terms of precipitation, albedo, and cloudiness decrease, and temperature increase are considered. The onset of ice reterat from the innermost large moraine of Lewis Glacier is reconstructed at 1890, in response to climate change beginning around 1883 and continuing to the turn of the century. This is followed by two decades with little variation in climatic conditions, pronounced change extending from the 1920s to the early 1930s, a reversal in the 1940s, and a recommencement of the earlier climatic tendency in the 1950s. In principle, any of the four aforementioned climatic forcings alone or in various combination could have produced the observed ice retreat. Useful constraints are, however, available from independent observations of lake level and temperature variations as reviewed in Section 2.5.

Thus, for the latter part of the 19th century, a warming of the required magnitude can be excluded. It is noted, however, from Fig. 2.5:4 that the early 1880s onset of climatic change as inferred from the Lewis terminus record is bracketed by the water level maxima of Lake Tanganyika to the South in the 1870s and of Lake Turkana to the North in the 1890s. Likewise, the water level of Lake Victoria dropped from the early 1880s to the turn of the century. A corresponding precipitation decrement of about 150–200 mm a^{-1} can be inferred (ref. Section 2.5). This is expected to occur in concert with decrease in cloudiness and glacier albedo, which would accentuate the effect on the glacier mass economy.

Fig. 2.5:5 shows no pronounced long-term precipitation trends in the 20th century, while there are indications of a gradual warming until around the 1950s. Associated albedo decrease would enhance the effect on the ice budget. Independently of precipitation, cloudiness, and temperature change, albedo could vary in consequence of differing deposition of particulate matter on the glacier, but frequent snowfall would minimize this effect. Albedo changes may have played an important role in, but they do not seem plausible as sole cause of the ice recession.

In conclusion from Kruss' (1981, 1983b) analysis, the following most plausible climate change scenario emerges as cause of the Lewis terminus retreat during the last 100 years: precipitation decrease of the order of 150 mm a^{-1} accompanied by cloudiness and albedo reduction of the order of one tenth and 2%, respectively, in the latter part of the 19th century; and warming of the order of a few tenths of a centigrade degree with concomitant albedo decrease of around 1%, from the turn of the century to the 1950s, this change being concentrated in the 1920s to early 1930s. The latter detail is interesting in relation to the small 20th century moraine ridges mentioned in Chapter 4.

Kruss' (1981, 1983b) inference of the onset of Lewis glacier retreat in the early 1890s and of the beginning of the climate change causing it in the early 1880s, should be viewed in context with the chronology for other glaciers in the tropics. For the Carstensz Glacier in New Guinea, Allison and Kruss (1977) inferred from numerical modeling of the ice retreat an onset of climate change around 1850. For the Ecuadorian Andes, Hastenrath (1981) evaluated historical sources since the era of Spanish colonization. Perennial ice and snow cover during the 16th and the first half of the 18th centuries was much more extensive than at present. There is evidence for a gradual rise of the ice equilibrium line since at least the middle of the 19th century. Furthermore, around 1870 glaciers were recorded already significantly inside the terminal moraines marking their maximum extent. Considering the time lag between climatic forcing and glacier response, the corresponding onset of climate change is to be placed no later than around the middle of the 19th century. Accordingly, the onset of climate change as prelude to the drastic and presently continuing glacier recession throughout the tropics, occurred distinctly later in East Africa than in the two other glaciated high mountain regions under the Equator.

CHAPTER 6

EAST AFRICAN GLACIERS AND THE GLOBAL TROPICS

"Bescheiden wir uns . . . damit, dass wir . . . die Lösung der . . . Eiszeitfrage um einen . . . Schritt gefördert haben."

(Hans Meyer, *Kilimanjaro*, 1900)

Exploration in the course of a century has drawn attention to the existence of glaciers near the Equator and has brought to light evidence of variations in ice extent in the geological past. A wealth of field evidence of former glaciations is accumulating from the various high mountain regions of the tropics, but absolute dates to confirm spatial correlations are few. Concerning the modern glaciation, a preliminary global inventory of ice extent is being created and the need for the monitoring of glacier fluctuations is recognized. During the very interval with repeated visits to the peak regions of the East African mountains since the late 1800s a drastic ice recession is borne out by historical documentation, in good spatial consistency. A similar secular glacier behavior is now known from the other two glaciated high mountain regions under the Equator, New Guinea (Hope *et al.*, 1976) and the Ecuadorian Andes (Hastenrath, 1981). The study of East African glaciers shall here be placed in perspective with the ice and climate variations in the global tropics.

6.1. Pleistocene and Early Holocene Variations

An abundance of geomorphic evidence attests to the formerly much larger ice extent in the mountains of East Africa. Absolute datings are few. At Kilimanjaro, 'older' glaciations have been dated at 100 000 years B.P. and older; possible corollaries suggested in the Ruwenzori and on Mount Kenya being without age determination. Regarding later glacial events, the retreat of the ice from the lowest-reaching large moraines in the Ruwenzori is dated at >14 700 years B.P., broadly concordant with the disappearance of ice from the summit region of the Aberdares and from cirques at Mounts Elgon and Badda at >12 200, >11 000, and >11 500 years B.P., respectively. This is the scanty time frame within which the glacial history of East Africa is to be accommodated.

A relative chronology offers itself from elevation, appearance, and spatial arrangement of moraines at individual mountains, in the first place (Tables 3:1 and 3.2:1). Further, a spatial correlation is suggested from comparison between mountains (Table 3:1). Thus three moraine stages are distinguished in order of decreasing age and ice extent. Stage I comprises the lowest-reaching large moraines in the Ruwenzori, from which the ice began to retreat around 14 700 years B.P. and presumed corollaries on Kilimanjaro and Mount Kenya, as well as on Mount Elgon and in the Aberdares. For this stage, the total ice extent in East Africa is estimated at about 800 km^2, of which Kilimanjaro, Ruwenzori, and Mount Kenya contribute about one fourth each (Table 3:2). From stage II onward snow and ice cover is limited to these three presently glaciated mountains, totalling less than 200 km^2. At stage III, a few centuries old, the total ice cover is reduced to about 30 km^2, of which Mount Kenya contributes only a small fraction.

If moraines of stage I at the other mountains are indeed corollaries of the lowest-reaching large moraines in the Ruwenzori, then they would mark a largest ice extent in East Africa around 15 000 years B.P. No absolute dating is available for moraine complex II. Moraine complex III is thought to be a few centuries old, since the ice was not far from these moraines at the end of the 19th century. A plausible spatial consistency of moraine complexes is suggested between the mountains of East Africa and the High Semyen of Northern Ethiopia.

In terms of their climatic significance, the lake level variations and vegetation changes reviewed in Sections 2.5 and 2.3 are interesting in relation to the evidence on East African glaciers. Paleolimnological findings suggest humid conditions from $\geqslant$ 27 000 to 20 000 years B.P., and a lowering of vegetation limits is indicated from around 30 000 to 20 000–15 000 years B.P. This should be compared with the evidence of ice retreat from the lowest-reaching large moraines in the Ruwenzori at > 14 700 years B.P. Proceeding in time, the latter glacial event and the disappearance of ice from the summit region of the Aberdares and from cirques at Mounts Elgon and Badda at 12 200, > 11 000, and 11 500 years B.P. are to be viewed in context with the marked rise of lake levels around 13 000–11 000 years B.P., the upward shift of vegetation limits from around 15 000 to after 10 000 years B.P., and the pollen evidence of drier and cooler conditions than presently from 14 700 to sometime between 12 500 and 9500 years B.P. Whether the complex II moraines reflect a glacial event related to the 10 000–8000 years B.P. episode of enlarged lakes remains to be ascertained by absolute dating. On the scale of the last few centuries, which is interesting with regard to the complex III moraines, pollen and lake records lack detail. As a prerequisite for evaluating the palynological, paleolimnological, and glacial morphology information together in terms of climate, an absolute chronology of glacial events is needed in particular.

This very incomplete picture of the glacial history of East Africa shall in the following be compared with field evidence from other high mountain regions of the tropics. The synopsis (Table 6.1:1) is based on the review in Hastenrath (1981). Table 6.1:1 indicates a relative abundance of moraine observations for various mountain regions in low latitudes, but a scarcity of absolute dating. Spatial correspondence is suggested from the elevation, appearance, and arrangement of moraine systems. In Table 6.1:1, the organization into columns is intended to suggest possible spatial correlations, but more importantly to underline the need for the establishment of an absolute chronology through direct dating of glacial events throughout the global tropics.

Table 6.1:1 contains for Eastern Africa a resume of the information presented in Tables 3.1 and 3.2:1, and in addition data for the Americas and Australasia. Within each of these three domains, there is considerable uncertainty concerning the spatial

TABLE 6.1:1

Synopsis of moraine stages in the high mountains of the tropics. H, SL, lat, refer to elevation (in mm) of highest peak and modern snowline, and latitude, respectively. Ages in parentheses are assumed, not measured. For Eastern Africa, the content of Tables 3:1 and 3.2:1 is summarized

	H lat	SL						
a. *Africa*			IV	III		II	I	
b. *Americas*								
Mexico Heine, 1975	5670 19N	(5000)		M V 5100– 4300 m 1800+ A.D.	M IV 4200– 3800 m 2000 B.P. weak	M III 1. 4000– 2900 m 9000 B.P. 2. 3500– 3000 m 10 000 B.P.	M II 3200– 2600 m 12 100 B.P.	M I 2800– 2600 m 34 000– 32 000 B.P.
Guatemala Cuchumatanes; Hastenrath 1974a	3837 15 N	–				3600– 3500 m 4 separate ridges		
Costa Rica Talamanca; Hastenrath 1973b	3819 10 N	–				1. 3500– 3380 m up to three ridges, small 2. 3350– 3300 m 3 separate large ridges		
Venezuela; Schubert, 1972a, b 1974, 1975	5002 9 N	(4750)			4300– 4000 m weak	3700– 3000 m > 10 000 B.P. or > 13 000 B.P.	3000– 2600 m	
Salgado- Labouriau *et al.*, 1977								deglaciation 3600 m > 12 650 B.P.

Table 6.1:1, continued

	H lat	SL	IV	III		II	I	
Colombia Santa Marta; Gansser, 1955 Raasveldt, 1957	5775 11 N	(4800)		Bolivariano		Mamancanaca	Aduriameina	
Colombia Cocuy; Gonzalez *et al.*, 1965	5490 7 N	4600		4. 4200 m (1700– 1800 A.D.?)	3. 4200 m 2900– 2300 B.P.	2. 4050 m > 7500 B.P.	1. 4000 m	deglaciation 3880 m 12 320 B.P.
Colombia Ruiz-Tolima; Herd/Naeser, 1974	5420 5 N	(4600)				3450 m three phases < 13 760 B.P.		3450 m → 100 000 B.P. →
Ecuador; Hastenrath 1981	6310 0–5 S	(4900– 4300)	I. 4900–4600 m 1900+ A.D.	II. 4700–4200 m ~ 1500– 1800 A.D.		III. 4500–3400 m 1. upper complex, smaller 2. lower complex, large, up to 3 separate ridges	glacier striations down to ≲ 3000 m	
Peru C. Blanca; Clapperton, 1972	6768 8 S	5000	4. 4750–4250 m –1900 A.D.	3. 4650–4200 m (1750–1800 A.D.?)		2. 4200–2000 m (6000–4000 B.P.?)	1. 4000–3000 m	
Lliboutry *et al.*, 1977			Safuna VI 4000–4380 m 1900+ A.D.	Safuna V/ Huaraz m8 4600–4400 m		Safuna IV–III/ Huaraz m7 4500–4240 m	Safuna II–I/ Huaraz m6–m5 4400–3900 m (7000 B.P.)	Safuna Huaraz m3–m2 3900–3500 Safuna Upper Terrace Huaraz m4–m1 3900–3200 m

Table 6.1:1, continued

	H lat	SL	IV	III		II	I	
Peru Quelccaya/ Vilcanota Mercer *et al.*, 1975 Mercer and Palacios, 1977	5645 14S	5400		5100 m (Q) < 270 B.P. 4550 m (V) < 630 B.P.	4500 m (V) > 2830 B.P.	5050 m (Q) < 10 900 B.P.	5050 m (Q) > 12 240 B.P.	4450 m (V) 28 000– 14 000 B.P.
Chile Elqui Valley; Caviedes/ Paskoff, 1975	6252 30 S	5500– 5000		La Laguna 3100 m		Tapado 2500 m		
c. *Australasia*								
Borneo Kinabalu; Koopmans/ Stauffer 1968	4100 6 N	–					2800 m	
Irian Jaya Carstensz; Peterson/Hope, 1972; Galloway *et al.*, 1973; Hope *et al.*, 1976;	5029 4 S	4400			4200 m 3500 B.P. + three more advances > 1600 B.P.		3500 m < 11 300 B.P.	deglaciation 3680 m > 14 000 B.P.
Papua New Guinea; Löffler, 1972, 1976, 1982;	4509 6 S	–					3700 m > 10 700 B.P.	deglaciation 3600 m Mt. Wilhelm > 12 600 B.P. glaciation 290 000 B.P. glaciation > 380 000 B.P.

correspondence, that can be resolved conclusively only by absolute dating. Attempts at correlations between continents are all the more open to question.

Of primary interest in the inter-continental comparison is the very limited information on the absolute age of glacial events in equatorial East Africa. Thus, the evidence of glaciations at > 100 000, 300 000 and 500 000 years B.P. at Kilimanjaro should be compared with the findings of a glaciation at >100 000 years B.P. in the Andes of Colombia, and of glaciations around 290 000 and 380 000 years B.P. in New Guinea. Furthermore, the onset of deglacierization of the Ruwenzori at >14 700 years B.P. agrees with the beginning of ice recession at >14 000 years B.P. in New Guinea, and is broadly consistent with evidence of glaciation in the South American Andes, namely at 28 000–14 000 years B.P. in Peru, at <13 760 years B.P. in Colombia, and at >13 000 years B.P. in Venezuela.

Beyond these absolute dates, spatial correlation of moraine complexes between Africa, the Americas, and Australasia is at best speculative. Thus the lowest-reaching East African moraine complex I (Tables 3:1 and 6.1:1) may have its corollaries in the most extensive and continuous moraine system in the Americas and in Australasia, as indicated in the eighth (second-to-last) column of Table 6.1:1. The moraine stage II of East Africa may correspond to the next higher moraine complex in the Americas given in the seventh column of Table 6.1:1. We are on firmer ground concerning the East African moraine complex III inasmuch as the ice rim is known to have been close to it in the latter part of the 19th century. A similar state of affairs in the 19th century is documented for moraine complex II in the Ecuadorian Andes and complex M V on the Mexican volcanoes (fifth column of Table 6.1:1). The correlation between Ecuador and Mexico and other regions of the Americas is inferential. Finally, the moraine arcs (IV) which formed on the East African mountains during the 20th century have a well-confirmed corollary (I) in the Ecuadorian Andes (fourth column in Table 6.1:1). Thus a correspondence between East Africa and the Americas is firmly established for the East African moraine stages III and IV, while correlation is dubious for stage I, and especially stage II, pending the development of an absolute chronology.

In a comparison throughout the global tropics, the age of deglaciation is of particular interest. For 9 of the 13 tropical locations listed in Hope *et al.* (1976, p. 195), Flenley and Morley (1978) plotted elevation in dependence of deglaciation age and found an approximately linear negative relationship. In Fig. 6.1:1 all 13 sites given in Hope *et al.* (1976, p. 195) are entered, in addition to three newly published locations in Africa (Gasse and Descourtieux, 1979; Hamilton and Perrot, 1978, 1979; Perrot, 1982a) and two in the Colombian (Herd and Naeser, 1974) and Venezuelan Andes (Salgado-Labouriau *et al.* (1977), respectively. Fig. 6.1:1 shows an overall decrease of deglaciation age with elevation, and suggests a tendency for a somewhat earlier timing in Africa than in the Australasian sector.

The field evidence of glacial events in the global tropics should be considered together with the vegetation history (Flenley, 1979a, b) and the variations of lake levels (Street and Grove, 1979). For the Colombian Andes a lowering of vegetation limits is indicated from a maximum at about 30 000 to around 15 000 years B.P., followed by an upward shift until about 6000 years B.P. In the New Guinea highlands the forest limit dropped from a maximum around 27 000 to a broad minimum between 24 000 and 14 000 years B.P. A pronounced upward shift of vegetation limits took place from around 10 000 to

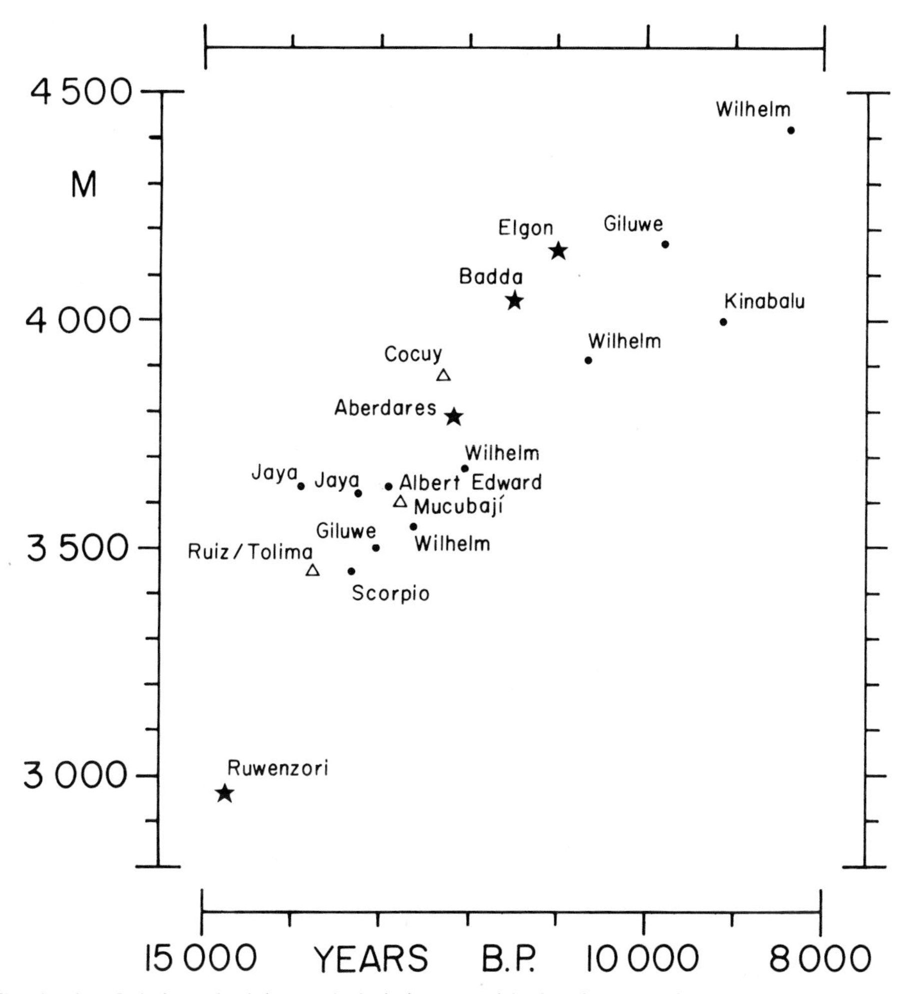

Fig. 6.1:1. Variation of minimum deglaciation age with elevation. (a) Africa (sources: Livingstone, 1962; Gasse and Descourtieux, 1979; Hamilton and Perrott, 1978, 1979; Perrott, 1982a), stars; (b) Americas (sources: Gonzalez *et al.*, 1965; Herd and Naeser, 1974; Salgado-Labouriau *et al.*, 1977), triangles (the entry for Ruiz/Tolima is maximum age of moraines rather than minimum age of deglaciation); (c) Australasia (source: Hope *et al.*, 1976, p. 195), dots.

the absolutely highest elevation around 7000 years B.P. In very broad terms, the synopsis by Flenley (1979a, b) thus shows for the South American Andes and Australasia vertical displacements of vegetation belts remarkably similar to those on the East African mountains as summarized above, although timing and amount of changes differ in detail. Consistent with the summary in Section 2.5, Street and Grove's (1979) review of lake variations indicates for tropical Africa humid conditions prior to about 21 000 years B.P., followed by a tendency for low lake stands until about 13 000 years B.P., and an episode of greatly enlarged lakes around 10 000–7000 years B.P. A similar pattern is found for Australia.

The lowering of vegetation limits in Colombia from about 30 000 to 15 000 years

B.P. is interesting in relation to the evidence of glaciation in the South American Andes between about 28 000 and 14 000 years B.P. Furthermore, the low elevation of the forest limit in the New Guinea highlands during 24 000 to 14 000 years B.P. and the lowering of lake levels in Australia from about 20 000 to 12 500 years B.P. should be viewed in context with the onset of ice recession in New Guinea at >14 700 years B.P.

In summary, and considering Africa, the Americas, and Australasia in context, major parallels are apparent in the late Pleistocene to early Holocene environmental changes manifested in glaciers, vegetation, and lakes. For the equatorial belt as a whole, deglaciation progressed from around 3000 m after about 15 000 years B.P. towards the 4500 m level at 8000 years B.P. (Fig. 6.1:1); the partial evidence for Africa and the Americas is at any rate consistent with the more continuously documented sequence for Australasia. Over about the same time span vegetation belts (Fig. 2.3:3 and Flenley, 1979a, b) shifted upward by more than 1000 m not only in East Africa, but also in the South American Andes and in New Guinea. Lake levels rose nearly concurrently, namely from around 13 000 to 9000 years B.P., both in tropical Africa and in Australia (Street and Grove, 1979). For the upward shift of the ice rim and of the vegetation belts, a warming at the respective altitudes throughout the tropics appears a plausible cause. For the glacier variations the elevation of the 0°C isothermal surface is of particular interest, inasmuch as at higher temperatures ablation can take place predominantly through melting, which is energetically almost by an order of magnitude more economical than sublimation. It should be noted, however, that temperature changes at the upper vegetation limits or at the ice rim – that is in the altitudinal domain from 2000 to more than 4000 m – are not as a rule equal to those at sea level. In fact, altered precipitation and humidity conditions such as indicated by the lake level variations are typically associated with changes of lapse rate. Starting with the same temperature at sea level, a change in lapse rate of 1°C per 1000 m results in a temperature change of 5°C at the 5000 m level. In interpreting changes in the altitudinal zonation of high mountains, changes of lapse rate may thus be no less important than temperature trends at sea level. The direct effect of precipitation changes on vegetation and glaciers further complicates the problem.

6.2. The Recent Glaciation

An inventory of the present glaciation is presented in Chapter 4 (Maps 4.1:1, 4.2:1*, and 4.3:1; Tables 4.1, 4.2:1, and 4.3:1; Appendix 3). At Kilimanjaro only the Kibo cone is glaciated at present. There are 20 separate ice entities with a total area of about 5 km^2. In the Ruwenzori, 43 glaciers are recognized at present covering an area of about 4 km^2, while another 5 glaciers have disappeared since the beginning of the century. On Mount Kenya, of the 18 glaciers at the turn of the century, 11 have survived to the present, with an overall area of less than 1 km^2. Early expedition reports, drawings, and photographs are evaluated for the reconstruction of glacier variations since the end of the 19th century. The chronology of varying ice extent can be established in considerable detail for Mount Kenya, partly due to the relative abundance of historical photographs, but especially because an excellent topographic map at scale 1:5000 is available. Creation of a comparable cartographic background for Kilimanjaro and the Ruwenzori would open the prospect of mapping the long-term glacier variations in these areas in considerable detail. For all three presently glaciated mountain regions of East Africa a drastic and essentially continuous ice recession is borne out since the earliest observations in the latter part of the past century.

Regarding the other high mountain regions of the tropics, glaciers still exist in the Irian Jaya part of New Guinea (Hope *et al.*, 1976, pp. 29–35). On Mount Jaya (Carstensz) the total area covered by the 8 identifiable ice entities extending between 4800 and 4400 m amounts to about 7 km^2. Smaller remnants of perennial ice are reported for two other mountains in Irian Jaya. Historical documents bear out a monotonic recession since the earliest visit to the peak region of Mount Jaya in 1913, but numerical modeling (Allison and Kruss, 1977) places the onset of retreat of the Mount Jaya glaciers at the middle of the 19th century.

In the Ecuadorian Andes (Hastenrath, 1981), the total present ice cover is estimated at about 220 km^2. Well over a 100 individual glacier tongues can be distinguished, which typically descend from ice caps crowning the various volcanic summits. At exceptional locations glacier snouts extend to about 4200 m, but more commonly they stay around 4800 m; glaciated summits range from around 5000 to more than 6000 m. Historical sources extend back to the era of Spanish colonization. Perennial ice and snow cover during the 16th and the first half of the 18th centuries was much more extensive than at present. There is evidence for a gradual rise of the ice equilibrium line since at least the middle of the 19th century. Furthermore, at that time glaciers were recorded already significantly inside the terminal moraines marking their maximum extents.

Glacier inventories have been compiled or are nearing completion for the other countries of the tropical Andes. The largest ice masses are located in the mountains of Peru and Bolivia. In general, monotonic ice recession is indicated since the 19th century. Historical documentation for earlier epochs may be less continuous than for the Ecuadorian Andes, but a systematic search has yet to be undertaken.

A remarkable result of Kruss' (1981, 1983a) numerical modeling of Lewis Glacier, Mount Kenya, is that the climatic forcing causing the onset of glacier retreat in East Africa is timed distinctly later (1880s) than in the other two high mountain regions under the Equator.

6.3. Spatial Patterns of Ice Equilibrium Elevation

On a planetary scale, the elevation of the ice equilibrium level varies most conspicuously with latitude. On the regional scale, there can be considerable East-West asymmetries in the altitudinal zonation of perennial ice (Chapters 3 and 4), soil frost phenomena (Section 2.4), and vegetation (Section 2.3), as is the case in East Africa. Because of the paucity of quantitative glaciological field studies and a variety of topographic effects, estimates of the modern ice equilibrium level (MEL) are of limited accuracy. The state of affairs is even less satisfactory regarding the Pleistocene ice equilibrium level (PEL), as a consequence of the scarcity of absolute dating and the aforementioned factors. The PEL is then estimated from cirque and moraine morphology with the assumption that the fossil geomorphic evidence on the various mountains belongs to a contemporaneous glacial episode. The large-scale meridional and zonal variations of MEL and PEL traced in the present section are to be understood with these qualifications. The synopsis is supported by the schematic meridional profiles, Figs. 6.3:1, 6.3:2, and 6.3:3, for Africa, the Americas, and the Australasian sector, respectively.

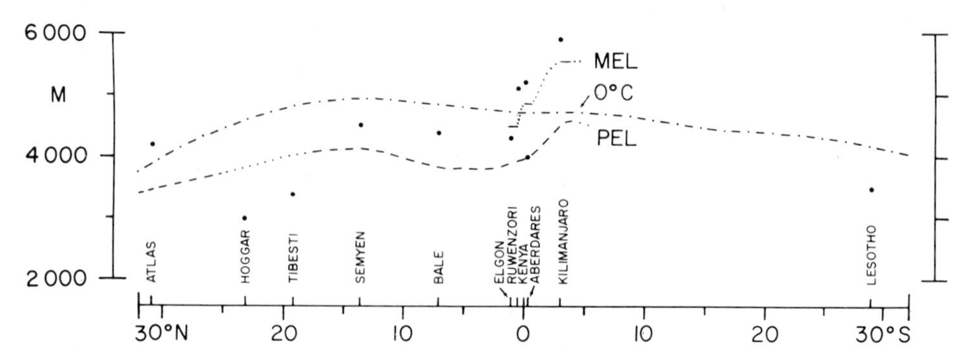

Fig. 6.3:1. Schematic meridional profile for Africa along approximately 30°E, except for about 5°W–5°E in Northern portion of transect. Mean annual 0°C isothermal surface, 0°C dash-dotted; modern ice equilibrium level, MEL, solid; Pleistocene ice equilibrium level, PEL, broken line.

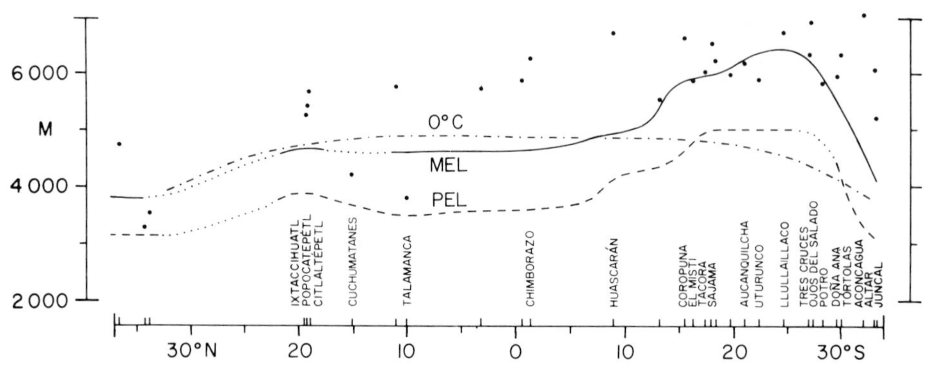

Fig. 6.3:2. Schematic meridional profile for the Western Cordillera of the Americas along approximately 75°W. Symbols as for Figs. 6.3:1 and 6.3:2 (source: Hastenrath, 1971).

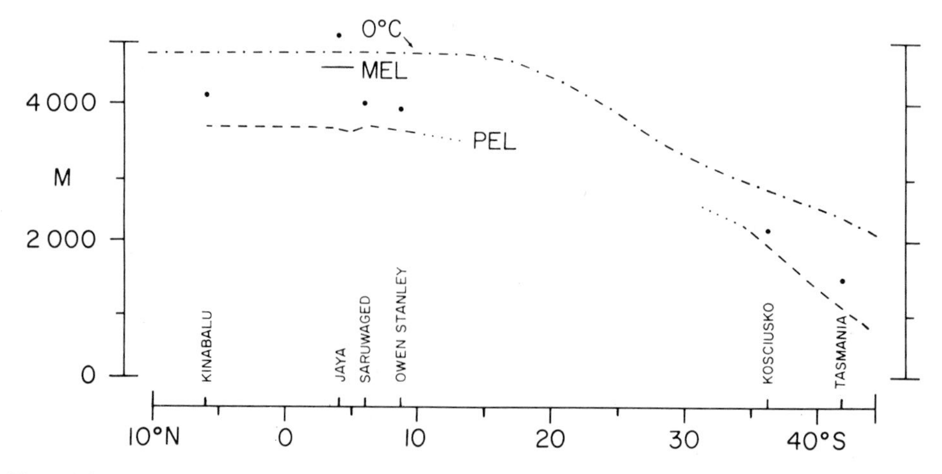

Fig. 6.3:3. Schematic meridional profile for the Australasian sector along approximately 145–120°E. Symbols as for Figs. 6.3:1 and 6.3:2 (sources: Hastenrath, 1972; Koopmans and Stauffer, 1968; Galloway *et al.*, 1972; Löffler, 1972, 1977, pp. 168–171).

Within East Africa the ice equilibrium level depends strongly on aspect and shows marked zonal asymmetries, whereas the latitude variation is of subordinate importance.

On the Kibo cone of Kilimanjaro (Section 4.1, Map 4.1:1, Appendix 3, Table 4), the MEL stands above the summit level, or at more than 5900 m, on the East side, while it is estimated at somewhat above 5000 m in the West. Overall, the MEL stands rather higher than the 0°C isothermal surface (Fig. 2.2:1). Apart from the aforementioned zonal asymmetry resulting from the diurnal cycle of winds and cloudiness, the radiation geometry at this Southern hemisphere location makes for the somewhat more extensive ice cover on the South as compared to the North side of Kibo. A similar azimuth dependence is indicated for the Pleistocene glaciations. Referring to the lowest-reaching moraine complex I, the PEL is estimated at about 1000 m lower than the MEL.

In the perennially cloudy Ruwenzori (Section 4.2, Map 4.3:1*; Appendix 3, Table 5), azimuth dependence of glaciation is not pronounced. The MEL is estimated at ranging between about 4700 and 4400 m on the various mountains, so that it stays distinctly below the annual mean 0°C isothermal surface in the free atmosphere (Fig. 2.2:1). With reference to the moraine complex I, the PEL is estimated to be about 1000 m lower than the MEL. Osmaston (1965) inferred a rise of the PEL from Southeast to West.

On Mount Kenya (Section 4.3, Maps 4.3:1 and 4.3:2; Appendix 3, Table 6) the modern and past glaciations are most vigorous on the Southwest side, as a result of diurnal circulations and radiation geometry. The MEL is estimated at around 4800 m on the Southwest side, being higher to the Northeast of the crest. This is near the annual mean elevation of the 0°C isothermal surface (Fig. 2.2:1). With reference to the moraine complex I, the PEL is estimated at about 700 m lower than the MEL.

Referring to moraine complex I, the PEL is estimated at about 3900 m for the Aberdares (Section 4.4, Map 3.4:1*) and at about 3800 m for Mount Elgon (Section 4.5, Map 3.5:1*).

This large diversity of altitudinal zonation within East Africa cannot be represented in the meridional profile, Fig. 6.3:1, intended to facilitate a large-scale synopsis for the African continent.

Proceeding from East Africa to Ethiopia (Section 4.6), the PEL in High Semyen (13°N) ie estimated at around 4100 m, with lowest values in the Northwestern and Western quadrants. In Southern Ethiopia (7°N) the PEL may have been as low as about 3800 m, in good spatial consistency with Northern Kenya.

Progressing further northward and disregarding the very marginal field evidence for the Hoggar (23°N), the PEL is reported at around 3500 m in the High Atlas of Morocco (31°N; Messerli, 1967, pp. 181–4).

Regarding Southern Africa (Section 4.8), the mountains of Lesotho (30°S) seem to be too low to have reached even the PEL.

The meridional variation of mean annual 0°C isothermal surface, MEL, and PEL, over the Americas is summarized in Fig. 6.3:2. The profile refers to the Western Cordillera and is simplified from Hastenrath (1971). The 0°C isothermal surface drops from the equatorial belt towards the subtropics of either hemisphere. The MEL stays below the annual mean 0°C isothermal surface in the relatively humid zone between about 10°S and 10°N, and close to it in the outer tropics of the Northern hemisphere. By contrast, in the hyper-arid Andes from Southern Peru through Bolivia to Northern Chile the MEL stands more than 1000 m above the 0°C isothermal surface. In fact, in this desertic zone, the MEL reaches elevations in excess of 6000 m. The PEL is also lowest in the equatorial belt, and highest in the outer tropics to subtropics of either hemisphere, similar to the MEL. The height differential between the MEL and PEL is largest in the desertic belt of the Southern hemispheric Andes, where it amounts to about 1500 m. In the regions where the MEL is lower than the 0°C isothermal surface glaciers may respond strongly to changes in temperature, but in situations where the MEL stands well above the 0°C isothermal surface, the ice cover would be most sensitive to precipitation.

In addition to the broad meridional patterns illustrated in Fig. 6.3:2, there are marked variations of the MEL and PEL in the zonal direction (Hastenrath, 1971). In the less humid regions, such as on the Mexican volcanoes and the mountains of Southern Peru, a zonal asymmetry of the MEL is apparent, related to the decreased insolation at the Westerly azimuths as a result of diurnal circulation, and cloudiness. However, outside these drier regions and on a larger scale, the general circulation exerts a dominant control on the ice distribution. Thus, in the belt of the tropical Easterlies both MEL and PEL tend to rise from East to West, while the reverse holds for the zone of the temperate latitude Westerlies. A transitional pattern is found around 30°S, where the MEL rises from East to West while the PEL drops in the same direction. The resulting height difference between MEL and PEL is more than 1500 m on the Pacific side of the Andes, but less than 500 m to the East. This may reflect the equatorward expansion of the temperate latitude Westerlies during the Pleistocene.

The meridional pattern in the Australasian sector is illustrated in Fig. 6.3:3, in part based on Hastenrath (1972). Additional sources include Koopmans and Stauffer (1968), Galloway *et al.* (1972), Loeffler (1972; 1977, pp. 168–171) and Hope *et al.* (1976, pp. 39–53, 173–5). The mean annual 0°C isothermal surface drops from the equatorial belt towards the extratropics of the Southern hemisphere. The MEL is reached only in the Irian Jaya part of New Guinea, where it stays with about 4600 m distinctly below the 0°C isothermal surface. The PEL ranges from 3600–3700 m at Mount Kinabalu on Borneo and at Mount Jaya on New Guinea to around 3550 m in Central New Guinea and to 3600–3700 m in the Southeastern part of Papua New Guinea. The height differential between the MEL and PEL in Irian Jaya amounts to about 1000 m. After a gap of nearly 30 degrees of latitude, the PEL profile resumes with values around 2000 m over Southeastern Australia, from where a further dropoff is indicated to 600–1300 m over Tasmania.

Figs. 6.3:1, 6.3:2, and 6.3:3 show similar large-scale meridional patterns, but considerable diversity in the absolute elevations of MEL and PEL. In particular, the elevation of the MEL relative to the 0°C isothermal surface is characteristic of the varied climatic conditions at the glaciers. Accordingly, a certain change in precipitation or in temperature affects the glacier in different ways depending on the background climate at the ice equilibrium level. As discussed in Section 6.1, a further complication arises from altered lapse rates that may be concomitant with changes in paleo-precipitation. Thus the popular exercise of converting changes of the ice equilibrium line into paleotemperatures at sea level may lead to more confusion than insight.

6.4. Lewis Glacier

In view of the drastic ice recession since the latter part of the last century borne out for all three high mountain regions under the Equator, a detailed study for selected glaciers is called for, aimed at the quantitative inference of the unknown climatic forcing from the observed cryospheric response. This requires (a) detailed historical documentation of long-term glacier behavior; (b) the knowledge of the present morphology, kinematics, and heat and mass budget characteristics; and involves (c) the numerical modelling of ice dynamics and climate-glacier relationships. Evaluation of ice cores in terms of oxygen isotope, total β radioactivity, and microparticle content, provides an alternative approach to climate reconstruction.

Lewis Glacier on Mount Kenya possesses a historical documentation of long-term glacier bahavior since the end of the last century which is unique for all of the tropics. The catchment area is well defined, and it is of reasonable access. Therefore Lewis Glacier was chosen for the field study of morphology, kinematics, mass and heat budget, as described in Chapter 5. Field observations serve as input to the computer modeling of ice dynamics, long-term glacier variations, and the quantitative inference of climatic change.

As the compilation of the first World Glacier Inventory draws to a close, international efforts at global environment and climate monitoring would do well to consider a long-term commitment to monitor in detail a few well selected glaciers in the tropics. In addition to the study of bedrock configuration and repeated mappings of the surface topography, the field programs would include the continuous assessment of the mass and heat budget and kinematics. In conjunction with adequate historical documentation of past variations in ice extent, this would provide the observational basis for the quantitative reconstruction of the climatic forcing. The detailed account of the Lewis Glacier field project in Chapter 5 may serve as framework for the design of monitoring programs at other selected glaciers in the tropics.

6.5. Glacier-Climate Relationships

Climate variations control the long-term glacier behavior through a causality chain involving the heat and mass budget, the bedrock configuration, and the ice dynamics. Conversely, the unknown and small climatic forcing must be inferred from the well documented and conspicuous glacier response.

Sensitivity analysis for the glacier as a whole indicates the typical magnitude of changes in climatic elements and surface properties that could account for the observed glacier variations. For Lewis Glacier it is found that the documented glacier response could be produced by approximate changes in solid precipitation of 100 cm a^{-1}, in cloudiness of 1/10 or in albedo of 3%, in air temperature of 1°C, or in relative humidity of 10%. Comparison with various meteorological and hydrological evidence then suggests abrupt precipitation and cloudiness decrease in the latter part of the 19th century as major factors for the onset of ice recession and its continuation into the 20th century; of further importance being a slight warming and possibly changes in cloudiness and albedo in the course of the 20th century.

Furthering the results from sensitivity analysis for the glacier as a whole, computer simulation of the ice dynamics allows to consider the redistribution of ice mass related to internal flow characteristics and the time sequence of climatic events. Results of the numerical modelling are interpreted in context with independent hydrometeorological information. On this basis, the most plausible sequence of climate causes for the observed ice recession appears to be as follows: precipitation reduction of the order of 150 mm a^{-1} associated with small cloudiness increase and albedo decrease in the last two decades of the 19th century; and a warming of a few tenths of a centigrade degree along with a slight albedo decrease since the turn of the century, effects being concentrated in the 1920s to early 1930s.

This exercise in climate reconstruction emphasizes that the conspicuous and well observable glacier response is caused by only small climatic forcing. In fact, the pertinent changes in temperature, cloudiness and albedo may be too small to be monitored directly by conventional modern sensing techniques. Kruss (1981) has pioneered the translation of glacier response into climatic forcing. Accordingly, strategies for the large-scale monitoring of long-term climatic variations may judiciously circumvent direct measurement of the very small changes in meteorological elements, and instead rely on systematic observation of environmental effects, such as the glacier behavior.

The detailed quantitative translation of observed glacier response into the climatic forcing should caution against the popular belief that glacier variations can be understood simply in terms of temperature trends. An instructive example is the drastic precipitation decrease over the last two decades of the 19th century, which must be considered as the main cause for the onset of glacier recession in East Africa. Kraus (1955a, b) found a rainfall decrease over this interval affecting much of the tropics, especially the East coasts of the continents, but excluding the region influenced by the Australian-Asiatic monsoon system. Note that there are regional differences both in the change of rainfall regime and in the onset of glacier recession. The general circulation causes of the marked dislocation of hydrometeorological regimes in the latter part of the past century remain a topic of further investigations.

6.6. Recommendations

As one of the three regions with perennial ice near the Equator, East Africa merits particular attention in the study of glaciers and climate. Building on the pioneering explorations at the end of the last and during the early decades of this century, and the initiation of systematic aerial photography and topographic mapping, studies in the course of recent decades have greatly enhanced our knowledge of East African glaciers. However, from the synopsis presented here, important gaps are apparent in the understanding of glacier behavior. Research is needed on (a) Pleistocene and early Holocene glacier and climate variations; and (b) the recent glaciation, fluctuations in ice extent, and climate-glacier relationships.

At the time scale of the Pleistocene to early Holocene, Table 3:1 attempts a synopsis of glacial events. The proposed spatial correlations must await further confirmation by absolute dating. The dates of ice retreat from the lowest-reaching large moraines in the Ruwenzori, and of the disappearance of ice from the summit region of the Aberdares and from cirques at Mounts Elgon and Badda, respectively, are of interest in relation to moraine complex I. The approximate age of complex III is inferred from the proximity to a historically documented ice extent. Finally, the timing of the complex II moraines is altogether open. Only with these reservations does Table 3:2 provide a partial history of varying ice cover. Since climatic variations affect vegetation and lakes as well as glaciers, it is desirable to evaluate palynological, paleolimnological, and geomorphic evidence in context. The different kinds of information are expected to be complementary rather than redundant. In order to link the glacial morphology in the mountains with the other kinds of potential climate indicators at lower elevations, absolute dating of moraines by carbon-14 or other methods is called for; prospects of an intertwining through stratigraphy appearing remote. The establishment of an absolute glacial chronology is considered a task of high priority in that this would open the prospect for interpreting various climatic indicators in context.

For the recent glaciation, an inventory has been established here, commensurate with the scale of available topographic maps and field observations. Topographic maps of a quality comparable to that available for the peak region of Mount Kenya would allow considerable improvement of the glacier inventory and an actual mapping of varying terminus positions for Kilimanjaro and the Ruwenzori. Particularly important is the prospect of systematic glacier monitoring. Variations in terminus positions with respect to an appropriately established reference base could be measured for numerous glaciers at intervals of a year or more. This effort could include the easier accessible large glaciers in all three presently glaciated high mountain regions of East Africa. More detailed field studies limited to very few selected glaciers should include the aero-photogrammetric mapping of ice topography at a scale of 1:2500 or 1:5000, repeated at intervals of about 5 years; measurements of surface ice flow velocity by surveying of an array of stakes, carried out about once a year; recordings of net balance at this stake network; regular precipitation measurements; and various more detailed investigations including ice core studies. The observation program developed on Lewis Glacier, Mount Kenya, is presented as an example.

The goals proposed here could be achieved through cooperation between various research groups. The systematic monitoring of the recent glaciation, in particular, calls for a definitive commitment at the national or international levels. In view of the unique location near the Equator, systematic glacier observations in East Africa would constitute an important contribution to global environment monitoring. The present account is intended as a foundation for the tasks that lie ahead.

SUMMARY

Abstract

Major objectives of the study are a synopsis of geomorphic evidence of former glaciations; an inventory of the modern ice extent; documentation of glacier variations since the earliest observations at the end of the 19th century; assessment of the present morphology, kinematics, and heat and mass budget of Lewis Glacier, Mount Kenya; quantitative inference of the long-term climatic forcing from the observed glacier response; and an appraisal of East African glaciers in context with environmental change throughout the global tropics.

Maps of moraine stages at scale 1 : 100 000 are presented for Kilimanjaro, Ruwenzori, Mount Kenya, Aberdares, and Mount Elgon. Given the paucity of absolute datings, a correlation of moraine complexes at the various mountains is proposed from elevation, appearance, and spatial arrangement. Thus, three moraine stages are distinguished in order of decreasing age and ice extent. At stage I the total ice cover in East Africa amounts to about 800 km^2, including Mount Elgon and the Aberdares. From stage II onward ice cover is limited to the three presently glaciated mountains, total ice cover at stage II being less than 200 km^2, and Mount Kenya and Kilimanjaro still possessing the largest shares. At stage III, a few centuries old, the total ice cover is reduced to about 30 km^2, of which Mount Kenya contributes only a small fraction.

An inventory is presented of the recent glaciation. At Kilimanjaro 20 ice entities are recognized, with a total area of about 5 km^2. In the Ruwenzori, 43 glaciers exist at present, covering an area of about 4 km^2, another 5 glaciers having disappeared since the beginning of the century. On Mount Kenya, of the 18 glaciers at the turn of the century, 11 have survived to the present, with an overall area of less than 1 km^2. Historical documentation since the earliest observations at the end of the 19th century bear out a drastic and essentially continuous ice recession in all three presently glaciated mountain regions. A detailed mapping of varying terminus positions at scale 1 : 5000 is presented for Mount Kenya.

A multi-annual field study of morphology, kinematics, mass and heat budget is being conducted on Lewis Glacier, Mount Kenya. This includes repeated aero-photogrammetric mappings of the glacier surface at scale 1 : 2500; determinations of ice thickness and bedrock topography by the seismic and gravimetric methods and through numerical modelling; measurements of ice surface flow velocities by repeated surveying of a network of stakes; measurements related to the heat budget; establishment of the mass budget characteristics through monitoring of net balance at an array of stakes, measurements of (water) runoff, and monthly gauging of precipitation. This field investigation of the present glacier conditions serves as the empirical basis for the study of the ice dynamics and long-term glacier behavior, and quantitative inference of earlier climate variations. The most plausible sequence of climate change covering the well-documented recession of Lewis Glacier is as follows: precipitation reduction of the order of 150 mm a^{-1} associated with cloudiness decrease and albedo increase of the order of one tenth and 2%, respectively, in the latter part of the 19th century; and warming of the order of a few tenths of a degree along with an albedo decrease of about 1% from the turn of the century to the 1950s with change concentrated in the 1920s to early 1930s.

The study of East African glaciers is important for the understanding of climate and ice variations in the global tropics. A concordance of major glacial events, vegetation changes, and lake-level variations from the late Pleistocene to the early Holocene is suggested from a comparison with the Americas and Australasia, although spatial correlations are hampered by the paucity of absolute datings. Concerning the recent glaciation, continuous measurements at selected glaciers in various high mountain regions of the tropics are regarded an essential component of global environment monitoring.

Contents of Chapters

1. The missionaries Rebmann and Krapf reported the existence of snow and ice in East Africa as early as 1848–9, but the first visits to the peak regions of Kilimanjaro, Mount Kenya, and the Ruwenzori, materialized only towards the end of the 19th century. The present monograph is based on the evaluation of historical accounts, air photographs and topographic maps, and own field research in the course of the 1970s. Objectives include a synopsis of the geomorphic evidence of former glaciations; an inventory of the present ice extent; assessment of recent glacier variations; a field study of morphology, kinematics, mass and heat budget of Lewis Glacier, Mount Kenya, being the basis for the reconstruction of the unknown climatic forcing from the observed glacier response; and an appraisal of East African glaciers in context with environmental change throughout the global tropics.

2. The gross physiography of the East African highlands is dominated by two broadly meridionally oriented rift systems. Mountains high enough to allow pleistocene or recent glaciations are of volcanic origin, except for the crystalline block of the Ruwenzori.

The large-scale circulation over East Africa is dominated by the alternation between winter and summer monsoons in the realm of the Indian Ocean. A simple rainfall maximum mostly between March and May is found in much of the Northern and coastal regions, broadly coincident in timing with the passage of the equatorial trough of low pressure and winds of Southerly component. Large parts of central Kenya and Uganda show a second seasonal peak around September to November, timed similar to the southward shift of low pressure trough and confluence zone. Rainfall in the mountains is strongly topographically controlled.

Pollen profiles point to conditions drier and colder than at present from 14 700 to between 12 500 and 9500 years B.P. A change to drier, or more seasonal conditions, or a combination of both is indicated for around 6500 years B.P. A marked change in pollen types since about 3000 years B.P. may be related to the advent of agriculture. Proceeding from the savanna or steppe upward, Hedberg (1951) distinguishes three altitudinal zones of vegetation: (a) Montane Forest Belt, (b) Ericaceous Belt, and (c) Alpine Belt.

Needle ice, turf exfoliation, and vegetation terracettes occur commonly above 3500 m, and stone polygons and stripes, fine earth polygons and bands, and fine earth buds and ribbons upward of 4200 m. East African lakes underwent a rise in water levels around 13 000–11 000 years B.P., and were greatly enlarged between 10 000 and 8000 years B.P. Regarding the recent history, a marked drop of lake levels and of the water discharge of the Nile occurred in the latter part of the 19th century.

3. Geomorphic evidence of formerly much larger ice extent is found on the presently glaciated mountains, Kilimanjaro, Mount Kenya, Ruwenzori, but also in the Aberdares and at Mount Elgon. 'Older' glaciations are suggested for the former three mountains, but absolute dates have been obtained only for Kilimanjaro, indicating ages of 100 000 years B.P. and older. Of later events, the retreat from the lowest-reaching large moraines in the Ruwenzori is dated at $>14\,700$ years B.P., and the disappearance of ice from the summit region of the Aberdares and from cirques at Mounts Elgon and Badda at $>12\,200$, $>11\,000$ and $>11\,500$ years B.P. In the absence of a spatially comprehensive absolute chronology, a correlation of moraine complexes on the various mountains is proposed from elevation, appearance, and spatial arrangement. Thus three moraine stages are distinguished in order of decreasing age and ice extent. At stage I the total ice cover in East Africa amounts to about 800 km^2, including Mount Elgon and the Aberdares. From stage II onward snow and ice cover is limited to the three presently glaciated mountains, total ice cover at stage II being less than 200 km^2, and Mount Kenya and Kilimanjaro still possessing the largest shares. At stage III, a few centuries old, the total ice cover is reduced to about 30 km^2 of which Mount Kenya contributes only a small fraction. This compares with a total ice cover in the high mountains of East Africa at present of less than 10 km^2.

4. An inventory is presented of the recent glaciation. At the Kibo cone of Kilimanjaro 20 ice entities are recognized, with a total area of about 5 km^2. In the Ruwenzori, 43 glaciers exist at present covering an area of about 4 km^2; another 5 glaciers having disappeared since the beginning of the century. On Mount Kenya, of the 18 glaciers at the turn of the century 11 have survived to the present, with an overall area of less than 1 km^2. Historical documentation is presented of glacier variations since the earliest observations at the end of the 19th century. A drastic and essentially continuous ice recession is borne out for all three presently glaciated mountain regions. A detailed mapping of varying terminus positions at scale 1:5000 is presented for Mount Kenya. Glaciological observations of earlier expeditions are reviewed.

5. A multi-annual field study of morphology, kinematics, mass and heat budget is being conducted on Lewis Glacier, Mount Kenya. Results to date include aero-photogrammetric mappings of the glacier surface at scale 1:2500 in February 1974, February 1978, and March 1982; determinations of ice thickness and bedrock topography based on seismological and gravimetric measurements and numerical modelling; assessment of the spatial pattern of surface ice flow velocity by repeated surveying of a network of stakes; estimate of major heat budget terms during intensive short-term observation programs; establishment of the mass budget characteristics through monitoring of net balance at an array of stakes, measurement of (water) runoff, and monthly gauging of precipitation. This field investigation of the present glacier conditions serves as the empirical basis for

the study of the ice dynamics and long-term glacier behavior, and quantitative inference of earlier climate variation. The most plausible sequence of climate change causing the well-documented recession of Lewis Glacier is as follows: precipitation reduction of the order of 150 mm a^{-1} associated with cloudiness decrease and albedo increase of the order of one tenth and 2%, respectively, in the latter part of the 19th century; and warming of the order of a few tenths of a degree along with an albedo decrease of about 1% from the turn of the century to the 1950s with change concentrated in the 1920s to early 1930s.

6. The abundant fossil glacial morphology attests to a formerly much larger ice extent in the high mountains of East Africa. 'Older' glaciations are suggested for Kilimanjaro, Ruwenzori, and Mount Kenya, but only for the former have absolute dates been obtained, indicating ages of 100 000 years B.P. and older. Regarding later glacial events, the retreat from the lowest-reaching large moraines in the Ruwenzori is dated at 14 700 years B.P., broadly concordant with the disappearance of ice from the summit region of the Aberdares and from cirques at Mounts Elgon and Badda at >12 200, >11 000, and >11 500 years B.P., respectively. In the absence of an absolute chronology, a correlation of moraine complexes on the various mountains is proposed from elevation, appearance, and spatial arrangement. Thus ice moraine stages are distinguished in order of decreasing age and ice extent; total ice cover in all of East Africa being estimated at 800, less than 200, and about 30 km² for stages I, II, and III, respectively. The latter moraine complex is a few hundred years old.

A few absolute dates are available for a comparison with the Americas and Australasia. The evidence of glaciations on Kilimanjaro at >100 000 years B.P. and earlier and of the onset of deglaciation in the Ruwenzori at >14 700 years B.P. and somewhat later at other mountains of Eastern Africa, has corollaries in the South American Andes and in New Guinea. Beyond these absolute dates, spatial correlation of glacial events between these three regions of the tropics is speculative. Vertical shifts of vegetation belt and lake level variations from the late Pleistocene to the early Holocene are in part concordant in the three regions.

Only the three highest mountains of East Africa are presently glaciated: Kilimanjaro carries 20 separate ice entities with a total area of about 5 km²; the Ruwenzori have 43 glaciers totalling some 4 km²; and at Mount Kenya 11 glaciers are still in existence, with an overall sea of less than 1 km². In the Irian Jaya part of New Guinea, the 8 identifiable ice entities cover about 7 km². In the Ecuadorian Andes the total ice cover is about 220 km² and more than 100 glacier tongues can be identified, that mostly descend from large ice caps. Glacier inventories are nearing completion for all of the Andes. Historical documentation since the end of the 19th century bears out a drastic and essentially continuous ice recession in all three presently glaciated mountain regions of East Africa. The mapping of varying glacier termini at Mount Kenya is unprecedented in its detail and scale. Lewis Glacier, in particular, is now the ice body with the most complete historical documentation of long-term glacier behavior in all of the tropics. The recent glacier recession began in East Africa in the 1880s, but around the middle of the 19th century in the Ecuadorian Andes and in New Guinea. The onset of ice recession in East Africa is due to a drastic precipitation decrease during the last two decades of the 19th century. This affected other tropical regions, but its general circulation causes remain to be explored.

On a planetary scale, the elevation of the ice equilibrium level varies most conspicuously with latitude. The modern ice equilibrium level is near or below the 0°C isothermal surface in the humid equatorial belt, but well above it in the arid region of the Southern hemispheric Andes. The background climate of the ice equilibrium level implies a differing sensitivity to temperature and precipitation changes. In particular, temperatures above 0°C allow ablation through melting, which is energetically much more economical than sublimation. The height differential between the modern and Pleistocene ice equilibrium levels ranges between about 700 and more than 1500 m in the various high mountain regions of the tropics. The field evidence of modern and Pleistocene equilibrium line elevations shows a spatially consistent drop from the equatorial belt to the higher latitudes of either hemisphere.

Because of the excellent documentation of secular ice variations, the well-defined catchment area, and the reasonable accessibility, Lewis Glacier was chosen for the quantitative study of climate-glacier relationships. Knowledge of the present glacier characteristics is essential to the study of ice dynamics and long-term climate-glacier interrelation. On this basis it has been possible to infer quantitatively the climatic forcing for the observed long-term glacier response. In view of the broadly concurrent secular glacier recession in distant high mountain regions of the tropics, results are believed to be of general importance. The multi-annual field program underway on Lewis Glacier, Mount Kenya, is focused on the modern morphology, kinematics, mass and heat budget, and may serve as example for the design of monitoring programs at selected tropical glaciers.

APPENDIX 1

TOPOGRAPHIC MAPS, AIR PHOTOGRAPHS, AND SATELLITE IMAGERY

Maps		**Air Photographs**
MOUNT MERU	4570 m	
1:50 000		
Ta 55/1	Oldonyo Sambu	60/TN/5, 30 Jan 62: #018–021, 055–057.
55/2	Ngare Nanyuki	
55/3	Arusha	
55/4	Usa River	
KILIMANJARO * 5899 m		
1:50 000		
Ta 56/1	West Hai	V 13A/595, 13 Feb 57: #0024–0030.
56/2	Kilimanjaro	V 13A/RAF/686, 25 Feb 58, #0080–0085;
57/1	Rombo	13A/RAF/688, 2 March 58: #0009–0019, 0055–0063, 0099–0108.
Ta 42/3 (Ke 181/3)	Olmolog	
Ta 42E/3 (Ke 182/3)	Oloitokitok	
Ta 42/4 (Ke 181/4)	Rongai	
1:100 000		
D.O.S. 522, Ed. 1–D.O.S., 1965. 'Kilimanjaro'		
RUWENZORI	* 5113 m	
1:50 000		
Ug 56/3	Bundibugyo	15 UG 13, June 55: #010–027, 040–051;
65/2	Margherita	15 UG 14, June 55: #008–020;
65/4	Nyabirongo	15 UG 31, 20 Oct 55: 05–13, 21–29;
66/1	Mubuku	15 UG 33, 24 Sept. 55: #015–026.
1:25 000		
U.S.D. 15, Ed. 2–U.S.D., 200–5/20, 1970 'Central Ruwenzori'		

Maps		Air Photographs
MOUNT KENYA	* 5199 m	
1:50 000		
Ke 107/3	Nanyuki	V 13A/RAF/14, 14 Feb 47: #5063–5098, 5114–5135, 5140–5153, 5170–5171.
107/4	Maranya	V 13A/RAF/20, 21 Feb 47: 5104–5119, 5129–5138, 5154–5163.
108/3	Meru	13B/RAF/341, 29 Jan 63: #074–076.
121/1	Naro Moru	VI 3B/RAF/627, 10 Feb 67: #0060–0062.
121/2	Mount Kenya	
121/3	Karatina	
121/4	North Embu	
122/1	Nkubu	
122/3	Chuka	

1:25 000
Survey of Kenya, SK 75 (D.O.S. 302), Ed. 4–SK, 1971

1:5000
Forschungsunternehmen
Nepal-Himalaya

Maps		Air Photographs
ABERDARES	4002 m	
1:50 000		
Ke 120/1	Ndaragwa	104/KE/2, 6–7 Jan 69: #023–027, 117–121.
102/3	Kipipiri	

Maps		Air Photographs
MOUNT ELGON	4322 m	
1:50 000		
Ke 74/3 (Ug 55/3)	Elgony	V 13A/RAF/814, 29 Jan 59: #0064–0066, 0078–0082.
Ke 74W/4 (Ug 54/4)	Budadiri	13B/RAF/603, 26 Jan 67: #0016–0020, 0037–0041.
Ke 87/2 (Ug 64/2)	Bubulo	
Ke 88/1 (Ug 55/1)	Kimilili Kaproron	

1:125 000
Dept. of Lands and Survey, Uganda,
U 1 U.S.D., 3000–5/671, Ed. U.S.D., 1967,
'Mount Elgon'.

Satellite Imagery

Image no.	*Date*	*Hour GMT*	*MSS bands*
Kilimanjaro			
180/62:2205–0700	15 Aug 1975	07:00	7
			45 7 color
180/62:2367–06582	24 Jan 1976	06:58	7
			45 7 color
Ruwenzori			
186/60:2049–07344	12 March 1975	07:34	7
			45 color
Mount Kenya			
180/60:2367–06573	24 Jan 1976	06:57	7
			45 color

APPENDIX 2

LIST OF HISTORICAL PHOTOGRAPHS AND DRAWINGS

KILIMANJARO

Date	Source			Remarks
Overall				
1972 March 18	courtesy of Peter Gollmer, Nairobi			vertical air photograph, by Geosurvey
1974 March 10	courtesy of Alan Root, Nairobi			aerial, by Alan Root present frontispiece
1975 Aug 15	NASA			LANDSAT, present Photo 4.1:1.
Eastern Sector				
1887 Sept approx.	Meyer	1888a	photo 8	
1887 Sept approx.	Meyer	1888a	photo 9	from NE
1889 Oct approx.	Meyer	1890d	photo facing pp. 123, 158	
1889 Oct approx.	Meyer	1891b	photos facing pp. 141, 183	same as Meyer 1890d, photos facing p. 123, 158
1898 Aug approx.	Meyer	1900a	photos facing pp. 134, 140, 353, 373	
1898 Aug approx.	Meyer	1928	photo facing p. 33	
1901 Oct approx.	Uhlig	1904	fig. 44	
1912 Dec	Klute	1920	plate 4	
1926 July	Latham	1926	photo facing p. 492	
1928 Jan	Nilsson	1931	fig. 26	
1930s	Wyss-Dunant	1938	photo facing p. 48, bottom	
1937 Dec/1938 Jan	Light	1941	photos 170, 171, 175	aerial
1943 July	Spink	1944	fig. facing p. 226, bottom	aerial, 803 Sqdn. RNAS, Tanga; present Photo 4.1:2.
1943 July	Spink	1945	fig. facing p. 212, bottom	aerial, 803 Sqdn. RNAS, Tanga
1943 July	Spink	1952	plate 35	aerial, from NE

Date		Source			Remarks
1953–7	approx.	Downie	1964	photo facing p. 7	aerial from SE, Tanganyika Air Survey
1953–7		Humphries	1972	plate 4a	aerial from SE; see Downie 1964, photo facing p. 7

Southern Sector

Date		Source			Remarks
1898 Aug	approx.	Meyer	1900a	fig. on p. 200	from SW, by Müller
1901 Oct		Uhlig	1904	figs. 50, 53–8	
1901		Jaeger	1909	fig. 19	by Uhlig; present Photo 4.1:4.
1912 Dec	approx.	Klute	1920	fig. 1	from SSW, by Oehler; present Photo 4.1:5.
1912		Reck	1921–2	fig. 1	by Oehler, see Klute
1912 Dec	approx.	Klute	1929	plate 4	from SSE, by Oehler
1921 Oct		Gillman	1923b	photo facing p. 3	
1930	approx.	MCEA	1932	photo facing p. 36, top	by Perkins
1930 Jan		Mittelholzer	1930	fig. 113	aerial, from SW, present Photo 4.1:6.
1934 Nov		Geilinger	1936	fig. 1	from SE
1960 March		Jaeger	1968	fig. 4	from SW, by Widmer
1960s		Nicol	1964	fig. 27	
1953–7	approx.	Downie	1964	photo facing p. 7	aerial from SE, Tanganyika Air Survey
1953–7	approx.	Humphries	1972	plate 4b	aerial from SE, see Downie, 1964, photo facing p. 7

Western Sector

Date		Source			Remarks
1898 Aug	approx.	Meyer	1900a	fig. facing p. 167 fig. p. 180 figs. facing pp. 171, 174, 177	painting, from WNW drawing, from WNW photographs, from WNW
1901 Oct		Jaeger	1909	fig. 20	by Uhlig; present Photo 4.1:8.
1906		Jaeger	1909	figs. 6, 7, 15	
1912		Jeannel	1950	fig. 13, p. 74	by Alluaud
1912 Dec	approx.	Klute	1920	fig. 6	by Oehler
1912		Reck	1921–2	fig. 4	by Oehler; see Klute, 1920, fig. 6
1912		Klute	1920	fig. 19	by Oehler
1912 Dec	approx.	Klute	1929	plates 1–3	by Oehler
1914 Apr		West	1915	fig. facing p. 6	
1930 Jan		Mittelholzer	1930	figs. 107, 111, 114	aerial, from NW

Date		Source			Remarks
1930		Kurz	1948	plate 93, bottom	see Mittelholzer, 1930, fig. 114
1930 Jan		Mittelholzer	1930	fig. 112	aerial, from W; present Photo 4.1:9.
1930		Kurz	1948	plate 93, top	see Mittelholzer 1930, fig. 112
1934 Jan		Geilinger	1936	fig. 3	
1943 July		Spink	1944	fig. facing p. 226, top	aerial, from SW, 803 Squadron, RNAS, Tanga;
1943 July		Spink	1945	fig. facing p. 212, top	aerial, from SW, 803 Squadron, RNAS, Tanga; present Photo 4.1:7.
1948 Nov		Salt	1951	fig. 4	by Swynnerton; present Photo 4.1:10.
1948 Nov		Salt	1951	figs. 5, 6	
1953–7		Downie	1964	fig. facing p. 6	aerial
1953–7		Humphries	1959	fig. 2	
1953–7		Downie and Wilkinson	1972	plate 1a	see Humphries, 1959, fig. 2
1960 March		Jaeger	1968	fig. 4	from SW, by Widmer

Northern Sector

Date		Source			Remarks
1898 Aug	approx.	Meyer	1900a	fig. facing p. 126	from NE
1912 Dec	approx.	Klute	1920	fig. 7	by Oehler
1930 Jan		Mittelholzer	1930	figs. 111, 114, 115	aerial; fig. 114 is present Photo 4.1:12.
1930		Kurz	1948	plate 23, bottom	aerial, see Mittelholzer, 1930, fig. 114
1931		Tuckett	1931	two photos	aerial

Kibo Crater

Date		Source			Remarks
1889 Oct		Meyer	1890a	plate 2	sketch
1898 Aug	approx.	Meyer	1900a	fig. facing p. 144	panorama from Hans Meyer Notch, drawing by Platz
1898 Oct	approx.	Johannes	1899	fig. on bottom of p. 270	
1901 Oct		Uhlig	1904	figs. 47–9	
1904 Aug		Uhlig	1908	figs. 16, 17	
1904 Aug		Jaeger	1909	plate 13	panorama from S of Johannes Notch, by Uhlig, present Photos 4.1:15a, b.
1904 Aug		Jaeger	1968	figs. 1, 2	by Uhlig, panorama from Gillman's Point
1927 March		Mostertz	1929–30	fig. 2	
1928 Jan		Nilsson	1931	fig. 3	panorama from SW

Date	**Source**			**Remarks**
1929 Jan	MCEA	1932	end of issue	panorama from Uhuru Peak (K.W.S.), by Rice, present Photos 4.1:16a–d.
1920s	Geilinger	1930	figs. 147, 149, 150	
1929 Jan	Geilinger	1936	fig. 6	
1930 Jan	Mittelholzer	1930	figs. 116–119	aerial
1930 Jan	Mittelholzer	1932	one photo	aerial
1930	Kurz	1948	plate 94, bottom	aerial, see Mittelholzer, 1930, fig. 119
1935 Feb	Geilinger	1936	fig. 8	
1935 Feb	Geilinger	1936	figs. 9, 10	
1937 Dec/1938 Jan	Light	1941	photos 172, 173	aerial
1940s	Spink	1947	fig. facing p. 328, top	
1943 July	Spink	1947	fig. facing p. 328, bottom	from SW, same as 1944, facing p. 226 top, partial
1943 July	Spink	1945	figs. facing p. 212 top 213 bottom, 215 top	fig. facing p. 212, top is present Photo 4.1:7.
1950	Spink	1952	plate 36	
1953–7 approx.	Humphries	1959	fig. 1	
1960 March	Jaeger	1968	fig. 3	by Widmer

RUWENZORI

Overall

1906	Roccati	1909		map, present Fig. 4.2:1.
1938	Eisenmann	1939a	p. 46	map sketch by Stumpp, present Map 4.2:3.
1938	Stumpp	1952		map, present Map 4.2:4.*

MOUNT EMIN (A)

1906 July	Filippi	1908b	panorama facing p. 268	from Gessi, by Sella
1906 July	Filippi	1908b	fig. on p. 241	from Gessi, by Sella
1945	Hicks	1946–7	fig. on p. 15, bottom	
1959 July	Whittow *et al.*	1963	fig. 4	

MOUNT GESSI (B)

1906 July	Filippi	1908b	fig. on p. 269	from Edward, by Sella
1906 July	Filippi	1908b	panorama facing p. 142	from Grauer, by Sella
1944	Hicks	1947	figs. on p. 14 16 top	by Haddow
1959 July	Whittow *et al.*	1963	fig. 3	

Date	Source			Remarks
MOUNT SPEKE (C)				
1906	Filippi	1908b	panoramas facing p. 142	from Baker
			p. 269	from Gessi
			fig. on p. 235	Speke Gl., from Scott-Elliot; present Photo 4.2:1.
1920s	Humphreys	1927	fig. facing p. 521, bottom	
1926	Osmaston	1961	fig. 1	Speke Gl., by Humphreys
1932	Whittow *et al.*	1963	ref. p. 589	from E, by Humphries
1934	Synge	1937	plate 38	
1934–5	British Museum of Nat. Hist.	1939	plate 13	Speke Gl.
1937 Dec	Light	1941	photo 197	aerial
1941	Hodgkin	1944	p. 304	Speke Gl.
1943	Firmin	1944–5	p. 33, top	Speke Gl.
1945	Hicks	1946–7	p. 20, top	
1951	Osmaston	1961	fig. 2	Speke Gl.
1950s	Whittow and Shephard	1959	fig. 1	Speke Gl., by Henderson
1959 July	Whittow *et al.*	1963	refs. p. 589	
1906–1960s	Whittow *et al.*	1963	fig. 2	from E, sketch
1960s	Temple	1968	fig. 3	aerial, Grant Gl.
1960s	Temple	1968	fig. 4	Speke Gl.
1967 Jan	Temple	1968	fig. 6	Speke snout
MOUNT STANLEY (D)				
1891	Stuhlmann	1894	facing p. 288	from W, present Photo 4.2:5.
1891	Meyer	1900a	p. 379	from W, by Stuhlmann
1891	Filippi	1908b	p. 206	from W, by Stuhlmann
1906	Filippi	1908b	panorama facing p. 142	
1906	Filippi	1908b	panorama facing p. 152	partially present Photo 4.2:7.
1906	Filippi	1908b	p. 200	Coronation Gl., present Photo 4.2:6.
1906	Filippi	1908b	fig. on p. 149, 151	
1906	Filippi	1908b	panoramas facing p. 152, 235	
1906	Filippi	1908b	p. 250	Savoia Gl.
1906	Roccati	1909	pl. 15, 28	Stanley from SE; pl. 28 is present Photo 4.2:8.

Date	Source			Remarks
1906	Roccati	1909	pl. 16	Helena and Coronation Gl.
1906	Roccati	1909	pl. 24	from Luigi di Savoia
1926	Chapin	1927	p. 615, 617	from W
1927	Humphreys	1927	fig. facing p. 517	Coronation Gl.
1931	Humphreys	1933	fig. facing p. 532	Savoia Gl.
1931	Humphreys	1933	fig. facing p. 482	Savoia Gl.
1932	Grunne *et al.*	1937	fig. 73	from W
1932	Grunne *et al.*	1937	fig. 99	from N
1932	De Heinzelin	1953	pl. 1	West Stanley Gl.
1932	Meersch, G. van der	1933	p. 108 top	Alexandra from W
1932	Meersch G. van der	1933	p. 135	aerial, from W, by Humphreys
1932	Meersch, G. van der	1933	p. 137	from W
1932	Grunne	1933	figs. facing p. 275, 276, 280, and on p. 285	from W
1934	Synge	1937	plates 37, 38	
1935 July	de Heinzelin	1953	plate 3	West Stanley Gl., by Bredo
1937 Dec	Light	1941	photos 197, 198, 199	aerial
1938	Eisenmann	1939a	plate 11, top	Stanley from NE
1939 July	de Heinzelin	1953	plate 3	West Stanley Gl., by Gilliard
1940 Sept	de Heinzelin	1953	plate 3	West Stanley Gl., by Philips
1943 Jan	Firmin	1944–5	p. 25, bottom	
1945	Hicks	1946	facing p. 215, top	from E. Bukurungu L., by Haddow
1945	Hicks	1946	facing p. 87	from E
1950 July	de Heinzelin	1951	facing p. 205	West Stanley and Alexandra, from W
1950 July	de Heinzelin	1953	plate 1	West Stanley Gl.
1952 July	de Heinzelin	1953	plate 1	West Stanley Gl.
1953	Busk	1954	facing p. 145	Coronation Gl.
1906–1960s	Whittow *et al.*	1963	fig. 18	Savoia Gl., sketch
1960s	Temple	1968	fig. 3	aerial, from N
1961	Fantin	1968	fig. 115	Coronation Gl.
1966 Dec	Temple	1968	fig. 7	Savoia Gl., from Freshfield

MOUNT BAKER (E)

Date	Source			Remarks
1906	Wollaston	1908	facing p. 68, 96	Moore Gl.
1900	Moore	1901	p. 15, 247	Baker Gl.

Date	Source			Remarks
1906	Roccati	1909	pl. 30, fig. 2	Moore snout; present Photo 4.2:14.
1906	Filippi	1908b	on p. 143, 147	Moore Gl.; fig. on p. 143 is present Photo 4.2:15.
1906	Filippi	1908b	panoramas facing p. 142, 152; fig. on p. 149	
1906	Filippi	1908b	fig. on p. 153	Semper and Y Gl.
1906	Roccati	1909	pl. 18, fig. 2	Semper and Y Gl.
1906	Whittow *et al.*	1963	ref. p. 608	Moore Gl., by Fisher
1908 Oct	Kassner	1911	facing p. 229	Moore Gl.
1920s	Whittow *et al.*	1963	refs. p. 589–590	Wollaston and Semper Gl., by Humphreys
1937 Dec	Light	1941	photo 197	aerial
1943	Firmin	1944–5	p. 33, bottom	Semper Gl.
1945	Bere and Hicks	1946	plate 4	N face of Baker
1950s	Whittow	1959	facing p. 373	Moore Gl.
1958 June	Whittow *et al.*	1963	fig. 19	Moore Gl.; present Photo 4.2:16.
1958 June	Temple	1968	fig. 8a	Moore snout
1966 Dec	Temple	1968	fig. 8b	Moore snout; present Photo 4.2:17.
1966 Dec	Temple	1968	fig. 9, 10	Moore snout

MOUNT LUIGI DI SAVOIA (F)

Date	Source			Remarks
1906	Filippi	1908b	panoramas facing p. 142, 246; on p. 190, 191, 251	

MOUNT KENYA

Overall

Date	Source			Remarks
1893	Gregory	1894	fig. 2	map, present Map 4.3:2.
1893	Gregory	1900	plate 10	map, present Map 4.3:3.
1899	Mackinder	1900		map, present Map 4.3:4.
1920 approx.	Royal Geographical Society, archive			map, present Map 4.3:5.
1926	Dutton	1929		map
1926	MCEA	1932	end of issue	reproduction of Dutton's map, present Map 4.3:6.
1947 Feb	Appendix 1			vertical air photograph
1960s	Mountain Club of Kenya	1971		map, present Map 4.3:7.

Date	Source			Remarks
1963 Jan	Forschungs-unternehmen	1967		map
1976 Jan 24	NASA			LANDSAT, present Photo 4.3:1.

Krapf Glacier (1)

Date	Source			Remarks
1899 Aug–Sept	Mountain Club of Kenya, archive			by Hausburg
1919–20	Arthur	1921	facing p. 20 bottom	
1920	Gregory	1921	plate 12a	by Melhuish
1926	Dutton	1929	fig. 39	
1930 Jan	Mittelholzer	1930	figs. 87, 88	aerial; fig. 87 is present Photo 4.3:2.
1938 Jan 13	Light	1941	photo 182	aerial
1944 July	Hicks	1945–6	facing p. 80	
1944 Aug	Spink	1945	facing p. 216, top	by Waddington; present Photo 4.3:3.
1945	Spink	1949	p. 281, top	aerial, by RAF
1947 Feb	Appendix I			vertical air photograph
1957 Dec	Charnley	1959	fig. 8	aerial; present Photo 4.3:4.

Gregory (2)

Date	Source			Remarks
1930	Mittelholzer	1931	figs. 87, 88	aerial; fig. 87 is present Photo 4.3:2.
1938 Jan 13	Light	1941	photo 182	aerial
1944 July	Hicks	1945–6	facing p. 80	
1944 Aug	Spink	1945	facing p. 216, top	by Waddington; present Photo 4.3:3.
			facing p. 217, top	by Waddington; present Photo 4.3:7.
1945	Spink	1949	p. 281, top	aerial, by RAF
1947 Feb	Appendix 1			vertical air photography
1957 Dec	Charnley	1959	fig. 8	aerial; present Photo 4.3:4.

Kolbe (3)

Date	Source			Remarks
1899 Aug–Sept	Mackinder	1900		map, present Map 4.3:4.
1920 approx.	Royal Geographical Society, archive			map, present Map 4.3:5.
1926	Dutton	1929		map, present Map 4.3:6.

Date	Source			Remarks
1927	Nilsson	1931	fig. 16	
1930	Mittelholzer	1930	figs. 87, 88	aerial; fig. 87 is present Photo 4.3:2.
1938 Jan 13	Light	1941	photo 182	aerial
1947	Appendix 1			vertical air photography

Lewis (4)

Date	Source			Remarks
1893 Jan	Gregory	1894	figs. 1, 2, 4	fig. 4 is present Photo 4.3:10.
1893 Jan	Gregory	1896	facing p. 176, and fig. 5	
1893 Jan	Gregory	1900	plate 10	sketch, present Map 4.3:3.
1899 Aug–Sept	Mountain Club of Kenya, archive			by Hausburg
1920s	MCK archive			Curling Pond
1926 Feb	Dutton	1929	figs. 20, 32	fig. 32 is present Photo 4.3:11.
1926 Feb	Dutton	1929	figs. 19, 22, 31	Curling Pond
1934 Apr–May	Troll and Wien	1934	figs. 3, 4, 6	fig. 6 is present Photo 4.3:12.
1938 Jan 13	Light	1941	photos 183, 184, 185	aerial
1941 June	Hodgkin	1941	second fig. following p. 312	present Photo 4.3:13.
1941 Feb	Douglas-Hamilton	1941–2	facing p. 217, bottom	Curling Pond
1945 Feb	Spink	1945	facing p. 217, bottom	by David, Curling Pond; present Photo 4.3:18.
1945	Spink	1949	p. 281, bottom	aerial, by RAF
1947 Feb	Appendix 1			vertical air photography
1940 Feb–March	Cameron and Reade	1950		
1957 Dec	Charnley	1959	figs. 6, 9	aerial; fig. 6 is present Photo 4.3:14;
1974 Feb	Caukwell and Hastenrath	1977		map (present Map 4.3:9.*)
1978 Feb	Hastenrath and Caukwell	1979		map (present Map 4.3:10.*)
1982 March	Caukwell and Hastenrath	1982		map (present Map 4.3:11.*)

Melhuish (5)

Date	Source			Remarks
1899	Mountain Club of Kenya, archive			by Hausburg

Date	Source			Remarks
1919	Royal Geographical Society; archive; Mountain Club of Kenya, archive			
1930	Mittelholzer	1930	fig. 89	aerial
1934 April–May	Troll and Wien	1949	figs. 3, 4, 6	fig. 6 is present Photo 4.3:12.
1938 Jan 13	Light	1941	photos 183, 184	aerial
1958 Jan	Charnley	1959	fig. 9	

Darwin (6)

Date	Source			Remarks
1899	Mountain Club of Kenya, archive			by Hausburg
1913	Royal Geographical Society, archive			
1919	Royal Geographical Society, archive; Mountain Club of Kenya, archive			
1930	Mittelholzer	1930	fig. 89	aerial
1938 Jan 13	Light	1941	photo 183	aerial
1940s	Howard	1955	facing p. 273, 274	by Firmin
1941 June	Hodgkin	1941	third fig. following p. 312	
1946 Jan	Firmin	1945–6	facing p. 404	

Diamond (7)

Date	Source			Remarks
1899	Mackinder	1900	figs. on p. 465, 467	
1899	Mackinder	1901	fig. facing p. 102	
1908	McGregor Ross	1911	fig. p. 473	
1919	Arthur	1921	fig. facing p. 17, top	
1940s	Howard	1955	facing p. 273, 274	by Firmin
1946 Jan	Firmin	1945–6	facing p. 404	

Forel (8) as Diamond (7) and Heim (9)

Heim (9) as Diamond (7) and Forel (8)

Date	Source			Remarks
Tyndall (10)				
1893	Gregory	1894	fig. 2	map, present Map 4.3:2.
1893	Gregory	1896	facing p. 180	drawing, present Photo 4.3:19.
1893	Gregory	1900	plate 10	map, present Map 4.3:3.
1899 Aug–Sept	Mackinder	1900	p. 467	by Hausburg
1899 Aug–Sept	Mackinder	1901	facing p. 109	by Hausburg
1899 Aug–Sept	Mountain Club of Kenya, archive			by Hausburg
1908	McGregor Ross	1911	p. 473	
1919 Feb	Arthur	1921	fig. facing p. 17	
1919	Gregory	1921	fig. facing p. 150, top	by Melhuish
1926	Dutton	1929	fig. 33	present Photo 9.3:20.
1938 Jan 13	Light	1941	photos 183, 185	aerial
1940s	Howard	1955	facing p. 274	
1946 Jan	Firmin	1945–46	facing p. 404	
1947 Feb	Appendix 1			vertical air photography
1950 Feb–March	Cameron and Reade	1950		
Barlow (11)				
1908	Royal Geographical Society, archive			
1912	Jeannel	1950	plate 25	by Alluaud
1920	Royal Geographical Society, archive			map, present Map 4.3:5.
1926	Dutton	1929		map, present Map 4.3:6.
Northwest Pigott (12)				
1920	Royal Geographical Society, archive			map, present Map 4.3:5.
1926	Dutton	1929		map, present Map 4.3:6.
Cesar (13)				
1899 Aug–Sept	Mackinder	1900		map, present Map 4.3:4.
1908	McGregor Ross	1911	p. 469	present Photo 4.3:22.

Date	Source			Remarks
1919 Feb	Arthur	1921	text p. 16	text only
1947 Feb	Appendix 1			vertical air photography
1950 Feb–March	Cameron and Reade	1950		

Joseph (14)

1899 Aug–Sept	Mackinder	1900		map, present Map 4.3:4.
1926	Dutton	1929		map, present Map 4.3:6.
1930s early	Tilman	1937	facing p. 64	present Photo 4.3:25.
1938 Jan 13	Light	1941	photo 182	aerial
1945	Spink	1949	p. 281, top	aerial, by RAF
1947 Feb	Appendix 1			vertical air photograph

Peter (15)

1926	Dutton	1929		map, present Map 4.3:6.

Northey (16)

1899	Mackinder	1900		map, present Map 4.3:4.
1926	Dutton	1929		map, present Map 4.3:6.
1930s early	Tilman	1937	facing p. 66	present Photo 4.3:28.
1938 Jan 13	Light	1941	photo 182	aerial
1944 Aug	Spink	1945	facing p. 216, top	by Waddington; present Photo 4.3:3.
1945	Spink	1949	p. 281, top	aerial, by RAF
1945	Appendix 1			vertical air photography

Mackinder (17)

1899	Mackinder	1900		map, present Map 4.3:4.

Arthur (18)

1899	Mackinder	1900		map, present Map 4.3:4.
1920 approx.	Royal Geographical Society, archive			map, present Map 4.3:5.
1926	Dutton	1929		map, present Map 4.3:6.

APPENDIX 3

DATA SUPPLIED TO WORLD GLACIER INVENTORY

List of Tables in Appendix 3

TABLE 1
Identification code of glaciers on Kilimanjaro

TANZANIA	: EAT3H . . . /xxx
draining to Indian Ocean	

Glaciers on KILIMANJARO are counted clockwise, staring from North

KILIMANJARO	: EAT3H . . . /. xx
glacier 1	EAT3H . . . /. . 1
2	/. . 2
3	/. . 3
4	/. . 4
5	/. . 5
6	/. . 6
7	/. . 7
8	/. . 8
9	/. . 9
10	/. 10
11	/. 11
12	/. 12
13	/. 13
14	/. 14
15	/. 15
16	/. 16
17	/. 17
18	/. 18
19	/. 19
20	/. 20

TABLE 2
Identification code of glaciers in the Ruwenzori

UGANDA	: EAU3E . . . /xxx
ZAIRE	CGO3E . . . /xxx
drainage to Nile	

The codification is in part adopted from Whittow *et al.* (1963)

RUWENZORI	: EAU3E . . . /xxx ,	CGO3E . . . /xxx
MOUNT EMIN	: CGO3E . . . /Axx	
glacier 1	/A.1	
2	/A.2	
3	/A.3	
4	/A.4	
5	/A.5	
6	/A.6	
MOUNT GESSI	: EAU3E . . . /Bxx	
glacier 1	EAU3E . . . /B.1	
2	/B.2	
3	/B.3	
4	/B.4	
5	/B.5	
6	/B.6	
7	/B.7	
8	/B.8	
9	/B.9	
MOUNT SPEKE	: EAU3E . . . /Cxx ,	CGO3E . . . /Cxx
glacier 1	CGO3E . . . /C.1	
2	/C.2	
3	/C.3	
4	/C.4	
5	/C.5	
MOUNT STANLEY	: EAU3E . . . /Dxx ,	CGO3E . . . /Dxx
glacier 1	CGO3E . . . /D.1	
2	CGO3E . . . /D.2	
3	EAU3E . . . /D.3	
4	/D.4	
5	/D.5	
6	/D.6	
7	/D.7	
8	/D.8	
9	/D.9	
10	/D10	
11	/D11	
12	/D12	
13	/D13	
14	/D14	
15	/D15	

Table 2, continued

UGANDA	: EAU3E . . . /xxx
ZAIRE	CGO3E . . . /xxx
drainage to Nile	

The codification is in part adopted from Whittow *et al.* (1963)

MOUNT BAKER		: EAU3E . . . /Exx
glacier	1	EAU3E . . . /E.1
	2	/E.2
	3	/E.3
	4	/E.4
	5	/E.5
	6	/E.6
	7	/E.7
MOUNT LUIGI DI SAVOIA		: EAU3E . . . /Fxx
glacier	1	EAU3E . . . /F.1
	2	/F.2

TABLE 3

Identification code of glaciers at Mount Kenya

KENYA	: EAK3H . . . /xxx
draining to Indian Ocean	

Glaciers on MOUNT KENYA are numbered clockwise, starting from North

MOUNT KENYA		: EAK3H . . . /. xx
glacier	1	EAK3H . . . /. . 1
	2	/. . 2
	3	/. . 3
	4	/. . 4
	5	/. . 5
	6	/. . 6
	7	/. . 7
	8	/. . 8
	9	/. . 9
	10	/. 10
	11	/. 11
	12	/. 12
	13	/. 13
	14	/. 14
	15	/. 15
	16	/. 16
	17	/. 17
	18	/. 18

TABLE 4

Characteristic parameters of Kilimanjaro glaciers

Glacier number	Latitude S °	′	Longitude E °	′	Total area [10^4 m^2]	Mean width [10^2 m]	Mean length [10^2 m]	Orientation	Highest elevation [m]	Lowest elevation [m]	Classification
1	3	3.32	37	21.66	7	2	3	NE	5800	5450	290011
2	3	3.70	37	21.88	50	8	5	E	5800	5500	360112
3	3	4.54	37	21.80	6	3	2	SE	5700	5650	060112
4	3	4.63	37	21.13	30	15	2	S	5850	5700	290011
5	3	4.92	37	21.34	12	3	4	SE	5720	5550	403112
6	3	4.95	37	21.14	20	3	7	S	5800	5300	403112
7	3	4.95	37	20.86	40	4	12	S	5850	5200	403112
8	3	4.89	37	20.68	70	5	14	SW	5850	5000	403112
9	3	4.15	37	20.95	12	2	6	W	5690	5660	260010
10	3	4.59	37	20.58	8	2	4	W	5600	5450	690000
11	3	4.85	37	20.46	4	1	4	SW	5200	4900	690000
12	3	4.46	37	20.33	12	1	11	SW	5400	4900	600000
13	3	4.31	37	20.24	20	3	7	SW	5350	4900	600000
14	3	4.25	37	20.09	2	1	2	SW	5100	4900	200002
15	3	3.79	37	20.13	1	1	2	SW	5450	5350	290002
16	3	3.59	37	20.11	25	2	14	W	5750	5200	503112
17	3	3.45	37	20.07	20	2	11	W	5750	5350	503112
18	3	3.36	37	20.04	25	2	14	W	5750	5200	503112
19	3	3.11	37	20.12	65	4	13	NW	5750	5200	503112
20	3	3.32	37	20.94	120	5	20	N	5800	5600	303012

TABLE 5
Characteristic parameters of Ruwenzori glaciers

Glacier number	Latitude S °	′	Longitude E °	′	Total area [10^4 m^2]	Mean width [10^2 m]	Mean length [10^2 m]	Orientation	Highest elevation [m]	Lowest elevation [m]	Classification
MOUNT EMIN											
1	0	26.33	29	53.94	1	1	1	SW	4700	4600	500000
2	0	25.87	29	53.87	1	1	1	NW	4700	4600	500000
3	0	26.43	29	53.94	1	1	1	NW	4750	4700	500000
4	0	26.26	29	54.03	3	1	1	E	4660	4620	790000
5	0	25.96	29	53.98	1	1	1	W	4650	4600	790000
6	0	25.81	29	53.79	1	1	1	S	4650	4600	790000
MOUNT GESSI											
1	0	26.00	29	54.89	1	1	1	NW	4600	4500	500000
2	0	25.87	29	54.87	3	1	2	W	4600	4500	500000
3	0	25.69	29	54.89	1	1	2	W	4600	4500	500000
4	0	25.54	29	54.87	1	1	1	SW	4600	4500	500000
5	0	25.35	29	55.02	3	1	2	SW	4600	4500	500000
6	0	25.41	29	55.17	3	1	2	S	4600	4500	500000
7	0	25.46	29	55.17	1			N	4650	4620	790000
8	0	25.55	29	55.11	4	1	1	E	4650	4550	790000
9	0	25.98	29	54.89	1	1	1	W	4600	4550	790000
MOUNT SPEKE											
1	0	24.52	29	53.35	30	8	3	W	4800	4600	500000
2	0	24.59	29	53.57	1	5	2	E	4800	4600	500000
3	0	23.97	29	53.85	5	2	2	E	4800	4550	500000
4	0	23.76	29	53.78	5	2	2	SE	4800	4550	500000
5	0	23.76	29	53.50	37	3	10	S	4814	4400	533112

MOUNT STANLEY											
1	0	23.34	29	52.07	20	5	4	W	5100	4550	533110
2	0	23.38	29	52.30	1	1	1	N	5100	5000	533110
3	0	23.37	29	52.54	3	2	3	NE	5000	4750	533110
4	0	23.22	29	52.75	6	2	3	E	4800	4600	533110
5	0	22.85	29	52.61	15	5	3	E	4850	4550	000000
6	0	22.52	29	52.85	12	2	7	SE	4800	4600	533112
7	0	22.41	29	52.70	2	1	2	E	4900	4700	533112
8	0	22.17	29	52.78	10	2	5	SE	4800	4650	533112
9	0	22.33	29	52.61	1	1	1	SW	4800	4550	530000
10	0	22.21	29	52.62	1	1	1	W	4800	3700	530000
11	0	22.39	29	52.56	1	1	1	SW	4800	4700	530000
12	0	22.46	29	52.50	1	1	1	W	4800	4700	530000
13	0	22.56	29	52.37	12	2	6	SW	4800	4300	530000
14	0	22.97	29	52.20	18	3	6	W	5100	4500	530000
15	0	22.24	29	56.86	1	1	2	SE	4700	4600	533112
MOUNT BAKER											
1	0	22.09	29	53.87	12	4	3	5	4600	4450	533110
2	0	21.87	29	53.43	1	1	1	SW	4700	4450	533110
3	0	21.62	29	53.61	8	2	4	SW	4700	4600	533110
4	0	21.85	29	53.74	12	3	4	S	4600	4500	533110
5	0	22.24	29	54.18	4	1	4	S	4500	4200	533112
6	0	22.26	29	54.32	1	1	1	SW	4600	4400	533110
7	0	21.70	29	53.41	1	1	1	SW	4700	4500	533110
MOUNT LUIGI DI SAVOIA											
1	0	20.00	29	53.39	2	1	2	E	4550	4450	533110
2	0	19.84	29	53.27	2	1	2	SE	4550	4500	000000

TABLE 6

Characteristic parameters of Mount Kenya glaciers

Glacier number	Latitude S °	′	Longitude E °	′	Total area [10^4 m^2]	Mean width [10^2 m]	Mean length [10^2 m]	Orientation	Highest elevation [m]	Lowest elevation [m]	Classification
1	0	09.02	37	18.60	4	1	4	N	4880	4690	533112
2	0	9.08	37	18.86	8	2	3	N	4900	4690	533112
3	0	9.13	37	19.07					4870		
4	0	9.37	37	18.71	30	3	11	SW	4975	4587	533112
5	0	9.38	37	18.49					4860	4780	
6	0	9.29	37	18.42	4	2	2	SW	5040	4650	533112
7	0	9.15	37	18.47	1	1	2	SW	5140	4090	630200
8	0	9.07	37	18.39	3	3	1	W	5060	4840	630200
9	0	9.10	37	18.32	2	2	1	W	4840	4700	630200
10	0	9.13	37	18.21	9	2	5	SW	4800	4515	533322
11	0	9.08	37	18.00				W			
12	0	8.95	37	18.11				NW			
13	0	8.87	37	18.16	3	1	4	W	4800	4600	533112
14	0	8.78	37	18.27	3	1	3	W	4800	4630	533112
15	0	8.67	37	18.32							
16	0	08.88	37	18.41	3	1	3	N	5040	4700	533522
17	0	08.91	37	18.49				NE			
18	0	08.97	37	18.50				NE			

TABLE 7

Classification according to World Glacier Inventory

	Digit 1 Primary classification	Digit 2 Form	Digit 3 Frontal characteristic	Digit 4 Longitudinal profile	Digit 5 Major source of nourishment	Digit 6 Activity of tongue
0	Uncertain or misc.	Uncertain or misc.	Normal or misc.	Uncertain or misc.	Uncertain or misc.	Uncertain
1	Continental ice sheet	Compound basins	Piedmont	Even; regular	Snow and/or drift snow	Marked retreat
2	Ice-field	Compound basin	Expanded foot	Hanging	Avalanche ice and/or avalanche snow	Slight retreat
3	Ice cap	Simple basin	Lobed	Cascading	Superimposed ice	Stationary
4	Outlet glacier	Cirque	Calving	Ice-fall		Slight advance
5	Valley glacier	Niche	Confluent	Interrupted		Marked advance
6	Mountain glacier	Crater				Possible surge
7	Glacieret and snowfield	Ice apron				Known surge
8	Ice shelf	Group				Oscillating
9	Rock glacier	Remnant				

BIBLIOGRAPHY

(Asterisks indicate the more important items)

Abruzzi, S. A. R. il Duca degli, 1907a. 'Esplorazione nella catena del Ruwenzori'. *Bolletino della Società Geografica Italiana,* Series 4, vol. 8, part 2, 99–127.

Abruzzi, S. A. R. il Duca degli, 1907b: 'The snows of the Nile'. *Geogr. J.*, **29**, 121–147.

*Abruzzi, S. A. R. il Duca degli, 1909a: *Il Ruwenzori, parte scientifica* vol. 1. *Zoologia, botanica* (Camerano, Chiovenda e Cortesi, *et al.*), 603 pp. vol. 2. *Geologia, petrografia, mineralogia* (Roccati *et al.*), 286 pp. Hoepli, Milano.

Abruzzi, S. A. R. il Duca degli, 1909b: 'The Mountains of the Moon'. *Blackwood's Magazine*, **185**, 61–69.

Academy of Sciences USSR, 1964: *Physical-geographic atlas of the World* (in Russian). Moscow, 298 pp.

Adamson, D. A., Gasse, F., Street, F. A., Williams, M. A. J., 1980: 'Late Quaternary history of the Nile'. *Nature*, **288**, 50–55.

*Allison, I., Kruss, P., 1977: 'Estimation of recent climatic change in Irian Jaya by numerical modelling of its tropical glaciers'. *Arctic and Alpine Research*, **9**, 49–60.

Alluaud, Ch., 1910: 'Mission scientifique aux monts Ruwenzori'. *La Géographie*, Paris, **21**, 291–292.

Almagià, R., 1908: 'I resultati geologici della spedizione di. S. A. R. il Duca degli Abruzzi al Ruwenzori'. *Bolletino della Società* Geografica Italiana, Series 4, vol. 9, part 1, 257–263.

*Ambach, W., H. Eisner, K. Pessl, 1972: 'Isotopic oxygen composition of firn, old snow, and precipitation in Alpine regions'. *Zeitschrift für Gletscherkunde und Glazialgeologie*, **8**, 125–135.

Anonymous, 1900: review of "Meyer, H., 1900: Der Kilimanjaro". *Geogr. J.*, **16**, 352.

Anonymous, 1904: 'Scientific researches in the Ruwenzori region'. *Geogr. J.*, **24**, 348–349.

Anonymous, 1906: 'A visit to Mount Kenya'. *Scottish Geogr. Mag.*, **25**, 346–352.

Anonymous, 1912: 'Africa'. *Geogr. J.*, **39**, 490–491.

Anonymous, 1929: 'Mount Kenya climbed'. *The Times*, London, 15 January 1929.

Anonymous, 1931: 'Africa's two highest peaks claimed from the air: Mounts Kilimanjaro and Kenya'. *The Illustrated London News*, 24 January 1931.

Anonymous, 1932: 'Flying over Africa's highest peak: Kilimanjaro, other craters'. *The Illustrated London News*, 27 August 1932, p. 306–309.

Anonymous, 1937: review of "G. A. Geilinger, Der Kilimanjaro". *Geogr. J.*, **89**, 391–392.

*Arthur, J. W., 1921: 'Mount Kenya'. *Geogr. J.*, **58**, 8–25.

*Atkinson, G., Sadler, J.C., 1970: 'Mean cloudiness and gradient-level wind charts over the tropics'. USAF, Air Weather Service, Technical Report No. 215.

*Baker, B. H., 1967: 'Geology of the Mount Kenya area'. Geological Survey of Kenya, Report No. 79, Nairobi, 78 pp.

Baker, B. H., Mohr, P. A., Williams, L. A. J., 1972: 'Geology of the Eastern Rift system of Africa'. Geological Society of America, Special Paper no. 136, Boulder, Colo., 67 pp.

Barth, H., 1862: 'Dr. August Petermann und die Schneeberge'. *Zeitschrift für allgemeine Erdkunde*, N. F., **13**, 342–347.

Behrens, T. T., 1906: 'The snow peaks of Ruwenzori'. *Geogr. J.*, **28**, 42–50.

Benuzzi, F., 1947: *Fuga sul Kenya*. L'Eroica, Milano, 421 pp.

Bere, M., Hicks, P. H., 1946: 'Ruwenzori'. *Uganda Journal*, **10**, 84–96.

Berg, H., 1949: *Einführung in die Physik der festen Erde*. Hirzel, Stuttgart, 296 pp.

Bergström, E., 1953: 'Som glaciolog på Ruwenzori'. *Ymer*, **73**, 1–23.

Bergström, R., 1955: 'British Ruwenzori Expedition, 1952, glaciological observations – preliminary report'. *J. Glaciol.*, **2**, 469–476.

Bernhardt, F., Philipps, H., 1958: 'Die räumliche und zeitliche Verteilung der Einstrahlung, der Ausstrahlung, und der Strahlungsbilanz im Meeresniveau. I. Die Einstrahlung'. *Abhandlungen Meteorol. Hydrol. Dienst DDR*, No. 45, Akademie-Verlag, Berlin, 227 pp.

*Bhatt, N. V., Hastenrath, S., Kruss, P., 1980: 'Ice thickness determination at Lewis Glacier, Mount Kenya: seismology, gravimetry, dynamics'. *Zeitschrift für Gletscherkunde und Glazialgeologie*, **16**, 213–228.

Bremer, H., 1965: 'Musterböden in tropisch-subtropischen Gebieten und Frostmusterböden'. *Zeitschrift für Geomorphologie*, **9**, 222–236.

Bright, R. G. T., 1908: 'The Uganda-Congo boundary commission'. *Geogr. J.*, **32**, 488–493.

*Brinkman, S. E., Wurzel, P., Jaetzold, R., 1968: 'Meteorological observations on Mount Kenya'. *East African Meteor. Dept. Memoirs*, vol. 4, No. 5, 42 pp.

British Museum, National History, 1939: *Ruwenzori Expedition 1934–35,* vol. 1 Trustees of British Museum, London, 127 pp.

Brown, L. H. Cochemè, J., 1969: *A study of the agroclimatology of the highlands of Eastern Africa*. FAO-UNESCO-WMO Interagency Agroclimatology Project, FAO, Rome; 330 pp.

Budd, W. F., 1969: *The dynamics of ice masses*. ANARE scientific reports, series A(IV) Glaciology Publ. No. 108, 216 pp.

*Budd, W. F., 1970: 'The longitudinal stress and strain-rate gradients in ice masses'. *J. Glaciol.*, **9**, 19–27.

*Budd, W. F., 1975: 'A first simple model for periodically self-surging glaciers'. *J. Glaciol.*, **14**, 3–21.

*Budd, W. F., Jenssen, D., 1975: 'Numerical modelling of glacier systems'. in: *Proceedings of the Moscow Symposium on snow and ice in mountainous regions*, August 1971. IAHS-AISH Publ. No. 104, 257–291.

*Budyko, M. I., 1974: *Climate and life*. International Geophysics Series, vol. 18. Academic Press, New York and London, 508 pp.

Busk, D. L., 1954a: 'The Royal peaks of Ruwenzori'. p. 137–146 in: *The Mountain World*, Harper and Brothers, New York. 224 pp.

Busk, D. L., 1954b: 'The southern glaciers of the Stanley group of the Ruwenzori'. *Geogr. J.*, **120**, 131–145.

*Butzer, K. W., 1971: *Recent history of an Ethiopian delta*. University of Chicago, Dept. Geography, Res. Paper No. 136, 184 pp.

*Butzer, K. W., Isaac, G. C., Richardson, J. L., Washbourn-Kamau, C., 1972: 'Radiocarbon dating of East African lake levels'. *Science*, **175**, 1069–1076.

*Butzer, K. W. B., 1980: 'The Holocene lake plain of North Rudolf, East Africa'. *Physical Geogr.*, **1**, 42–58.

Callendar, G. S., 1961: 'Temperature fluctuations and trends over the earth'. *Quart. J. Roy. Meteor. Soc.*, **87**, 1–12.

Cameron, W. O. and N. Reade, 1950: *Mount Kenya investigation (20th February to 4th March 1950)*. unpublished report, Water Department, Nairobi.

*Caukwell, R. A., Hastenrath, S., 1977: 'A new map of Lewis Glacier, Mount Kenya'. *Erdkunde*, **31**, 85–87.

*Caukwell, R. A., Hastenrath, S., 1983: 'Variations of Lewis Glacier, Mount Kenya, 1978–82'. *Erdkunde,* **36**, 299–303.

Caviedes, C. N., Paskoff, R., 1975: 'Quaternary glaciations in the Andes of North-Central Chile'. *J. Glaciol.*, **14**, 155–170.

Chanel, J., 1899: 'Voyage au Kilima Ndjaro'. *Tour du Monde*, Paris, n.s., 5 385, 397, 409, 421, 433.

Chapin, J. P., 1927: 'Ruwenzori from the West'. *Natural History*, **27**, 615–627.

*Charnley, F. E., 1959: 'Some observations on the glaciers of Mount Kenya'. *J. Glaciol.*, **3**, 483–492.

Clapperton, C. M., 1972: 'The pleistocene moraine stages of West-Central Peru'. *J. Glaciol.*, **11**, 255–263.

Coetzee, J. A., 1964: 'Evidence for a considerable depression of the vegetation belts during the upper Pleistocene on the East African mountains'. *Nature*, **204**, 564–566.

*Coetzee, J. A., 1967: 'Pollen analytical studies in East and Southern Africa'. *Palaeoecology of Africa*, **3**, 1–146.

Coetzee, J. A., 1978: 'Phytogeographical aspects of the montane forests of the chain of mountains on the Eastern side of Afirca'. *Erdwissenschaftliche Forschung*, vol. 11, p. 482–494.

Cohen, A., 1981: 'Palaeolimnological research at Lake Turkana, Kenya'. *Palaeoecology of Africa*, **13**, 61–82.

Cora, G., 1906: 'La spedizione del Duca degli Abruzzi al Ruwenzori o Runssoro (Africa Centrale)'. *Nuova Antologia*, **122**, 323–329.

*Coutts, H. H., 1969: 'Rainfall of the Kilimanjaro area'. *Weather*, **24**, 66–69.

Dansgaard, W., 1964: 'Stable isotopes in precipitation'. *Tellus*, **16**, 436–468.

Dansgaard, W., Johnsen, S. J., Moller, J., Langway, C. C., Jr., 1969: 'One thousand centuries of climatic record from Camp Century on the Greenland ice sheet'. *Science*, **166**, 377–381.

David, J. J., 1904: 'Scientific researches in the Ruwenzori region'. *Geogr. J.*, **24**, 348.

David, J. J., 1905: 'Dr. David's journey around Ruwenzori'. *Geogr. J.*, **25**, 93.

*Davies, T. D., Brimblecombe, P., Vincent, C. E., 1977a: 'The first ice core from East Africa'. *Weather*, **32**, 386–390.

*Davies, T. D., Brimblecombe, P., Vincent, C. E., 1977b: 'The daily cycle of weather on Mount Kenya'. *Weather*, **32**, 406–417.

Davies, T. D., Vincent, C. E., Brimblecombe, P., 1979b: 'Condensation nuclei and weather on Mount Kenya'. *J. Appl. Meteor.*, **18**, 1239–1243.

Dawe, M. T., 1906: 'An ascent of Ruwenzori'. *African Affairs*, **5**, 182–186.

Decken, C., von der, 1869: *Reisen in Ostafrika*. vol. 3, 'Wissenschaftliche Ergebnisse'. Winter, Leipzig and Heidelberg, 169 pp.

Degens, E. T., Hecky, R. E., 1974: 'Palaeoclimatic reconstruction of late Pleistocene and Holocene based on biogenic sediments from the Black Sea and a tropical lake'. p. 13–24 in: *Colloques Internationaux du Centre National de la Recherche Scientifique No. 219, Les méthodes quantitatives d'étude des variations du climat au cours du pleistocène*. June 1973, Paris, 317 pp.

Dobrin, M. B., 1960: Introduction to geophysical prospecting. McGraw Hill, New York, 446 pp.

Douglas-Hamilton, M., 1941–42: 'A climb on Mount Kenya – February 1942'. *Alpine J.*, **53**, 215–225.

Downie, C., Humphries, D. W., Wilcockson, W. H., Wilkinson, P., 1956: 'Geology of Kilimanjaro'. *Nature*, *178*, 828–830.

Downie, C., 1964: 'Glaciations of Mount Kilimanjaro, Northeast Tanganyika'. *Bull. Amer. Geol. Soc.*, **75**, 1–16.

*Downie, C., Wilkinson, P., 1972: *The geology of Kilimanjaro*. Geology Department, University of Sheffield, 253 pp.

Dronia, H., 1967: 'Der Stadteinfluss auf den weltweiten Temperaturtrend'. *Meteor. Abhandl., Freie Universität Berlin*, **74**, 4, 65 pp.

Drummond, H., 1980: *Tropical Africa*. Alden, New York, 132 pp.

*Dutton, E. A. T., 1929: *Kenya mountain*. Johathan Cape, London, 218 pp.

*East African Meteorological Department, 1958: *Climate, mean monthly rainfall and humidity*. Survey of Kenya, Nairobi.

*East African Meteorological Department, 1966a: *Monthly and annual rainfall in Kenya, during the 30 years 1931 to 1960*. Nairobi, 172 pp.

*East African Meteorological Department, 1966b: *Monthly and annual rainfall in Tanganyika and Zanzibar, during the 30 years 1931 to 1960*. Nairobi.

*East African Meteorological Department, 1966c: *Monthly and annual rainfall in Uganda, during the 30 years 1931 to 1960*. Nairobi.

East African Meteorological Department, 1968: *Tables showing diurnal variation of precipitation in East Africa and Seychelles*. East African Meteor. Dept. Technical Memoirs No. 10, 49 pp.

*East African Meteorological Department, 1970: *Temperature data for stations in East Africa. Part 1. Kenya, Part 2. Tanzania, Part 3. Uganda*. Nairobi.

*East African Meteorological Department, 1971: *Mean annual rainfall map of East Africa (based on all available data at 1966)*. Scale 1 : 200 000. Survey of Kenya, Nairobi.

Eberstein, von. 1887: Die Besteigung des Kilima-Ndjaro durch Herrn Dr. Hans Meyer and Herrn Lieutenant von Eberstein'. *Kolonial-Politische Korrespondenz*, **3**, no. 48, 377–380.

Ehlers, O. E., 1889: 'Meine Besteigung des Kilima Ndscharo'. *Petermann's Mitteilungen*, **35**, 68–71.
*Eisenmann, E., 1939a: 'Die Ruwenzori-Kundfahrt 1937/38 des Zweiges Stuttgart'. *Zeitschrift des Deutschen Alpenvereins*, **70**, 40–49.
Eisenmann, E., 1939b: *Schwarze Menschen, weisse Berge*. Kosmos, Frankh, Stuttgart, 92 pp.
*Electrotechnical Labs, Mandrel, 1960s: *Operator's manual, recording interval timer, model ER-75A*. Ampex Corp., Houston, Tex., 55 pp.
Fantin, M., 1968: *Sui ghiacciai dell' Africa*. Cappelli, Bologna, 450 pp.
Filippi, F. de, 1908a: *Il Ruwenzori, viaggio di esplorazione e prime ascensioni*. Hoepli, Milano, 358 pp.
*Filippi, F. de, 1908b: *Ruwenzori*. Dutton and Co., New York, 403 pp.
Filippi, F. de, 1909a: *Le Ruwenzori*. Plon-Nourrit, Paris, 352 pp.
Filippi, F. de, 1909b: *Der Ruwenzori*. Brockhaus, Leipzig, 471 pp.
Firmin, K., 1945: 'Ruwenzori, the Mountains of the Moon'. *East African Annual*, 1944/45, 25–35.
Firmin, A. J., 1945–56: 'The first ascent of the South face of Mount Kenya'. *Alpine, J.*, **55**, 400–405.
*Flenley, J. R., 1979a: *The equatorial rain forest: a geological history*. Butterworth, London, 162 pp.
Flenley, J. R., 1979b: 'The late Quaternary vegetational history of the equatorial mountains'. *Progress in Physical Geography*, **4**, 488–509.
Flenley, J. R., Morley, R. J., 1978: 'A minimum age for the deglaciation of Mt. Kinabalu, East Malaysia'. *Modern Quaternary Research in Southeast Asia*, **4**, 57–61.
Flohn, H., Fraedrich, K., 1966: 'Tagesperiodische Zirkulation und Niederschlagsverteilung am Victoria-See (Ostafrika)'. *Meteor. Rundschau*, **19**, 157–165.
Flückinger, O., 1934: 'Schuttstrukturen am Kilimandscharo'. *Petermann's Mitteilungen*, **80**, 321–324, 357–359.
*Forschungsunternehmen Nepal Himalaya, 1963: *Mount Kenya* 1 : 5 000. Kartographische Anstalt Freytag-Berndt und Artaria, Wien.
Forster, B., 1901: 'Das Tanganikaproblem und das Runsorogebirge'. *Globus*, **79**, p. 131.
Forster, B., 1902: 'Das Runssorogebirge'. *Globus*, **81**, p. 78–79.
Forster, B., 1907: 'Die Ruwenzori-Ferner'. *Globus*, **91**, 245–249.
Fraedrich, K., 1972: 'A simple climatological model of the dynamics and energetics of the nocturnal circulation at Lake Victoria'. *Quart. J. Roy. Meteor. Soc.*, **98**, 322–335.
Freiberg, H., 1929: *Das Ruwenzorigebirge, Zentralafrika*. Inaugural Dissertation, Universität Leipzig, 111 pp.
Freshfield, D. W., 1906: 'A note on the Ruwenzori group'. *Geogr. J.*, **27**, 481–486.
Furon, R., 1963: *Geology of Africa*. Oliver and Boyd, Edinburgh and London, 377 pp.
Furrer, G., Freund, R., 1973: 'Beobachtungen zum subnivalen Formenschatz am Kilimandjaro'. *Zeitschrift für Geomorphologie*, N. F., Suppl. vol. 16, 180–203.
Furrer, G., Graf, K., 1978: 'Die subnivale Höhenstufe am Kilimandjaro und in den Anden Boliviens und Ecuadors'. *Erdwissenschaftliche Forschung*, vol. 11, p. 441–457.
Galloway, R. W., Hope, G. S., Löffler, E., Peterson, J. A., 1972: 'Late Quaternary glaciation and periglacial phenomena in Australia and New Guinea'. *Palaeoecology of Africa*, **8**, 125–138.
Gansser, A., 1955: 'Ein Beitrag zur Geologie und Petrographie der Sierra Nevada de Santa Marta (Kolumbien, Südamerika)'. *Schweizerische Mineralogische und Petrographische Mitteilungen*, **35**, 209–278.
Gasse, F., 1977: 'Evolution of Lake Abhé [Ethiopia and TFAI), from 70 000 b.p.' *Nature*, **265**, 42–45.
Gasse, F., 1978: 'Les diatomées holocènes d'une tourbière (4040 m) d'une montagne èthiopienne: Le Mont Badda'. *Rev. Algol.*, N.S., **13**(2), 105–149.
*Gasse, F., 1980: 'Late quaternary changes in lake-levels and diatom assemblages on the Southeastern margin of the Sahara'. *Palaeoecology of Africa*, **12**, 333–350.
*Gasse, F., Rognon, P., Street, F. A., 1980: 'Quaternary history of the Afar and Ethiopian Rift lakes'. p. 361–400. in: Williams, M. A. J., Faure, H., eds: *The Sahara and the Nile*, Balkema, Rotterdam, 607 pp.
Gasse, F., Street, F. A., 1978: 'Late quaternary lake level fluctuations and environments of the Northern Rift Valley and Afar region (Ethiopia and Djibouti)'. *Paleogeography, Palaeoclimatology, Palaeoecology*, **24**, 279–325.
*Gasse, F., Descourtieux, C., 1979: 'Diatomées et évolution de trois milieux Éthiopiens d'altitude différente, au cours du quaternaire supérieur'. *Palaeoecology of Africa*, vol. 11, p. 117–134.

*Geilinger, W., 1930: *Der Kilimandjaro*. Huber, Bern-Berlin, 182 pp.

*Geilinger, W., 1936: 'The retreat of the Kilimanjaro glaciers'. *Tanganyika Notes and Records,* **2**, 7–20.

*Geilinger, W., 1937: 'Recession of Kilimanjaro ice cap'. *Geogr. J.,* **89**, 391–392.

Gillman, D., 1923a: 'Dr. Klute's map of Kilimanjaro'. *Geogr. J.,* **61**, p. 70.

Gillman, C., 1923b: 'An ascent of Kilimanjaro'. *Geogr. J.,* **61**, 1–27.

Gillman, C., 1944: 'A bibliography of Kilimanjaro'. *Tanganyika Notes and Records,* **18**, 60–68.

*Glen, J. W., Homer, D. R., Paren, J. G., 1977: 'Water at grain boundaries: its role in the purification of temperate glacier ice'. *International Association of Hydrological Sciences*, Publ. 118, 263–271.

*Gonfiantini, R., 1970: 'Discussion'. *Isotope Hydrology,* **56**, IAEA, Vienna.

Gonzalez, E., Van der Hammen, Th., Flint, R. F., 1965: 'Late quaternary glacial and vegetational sequence in Valle de Lagunillas, Sierra Nevada del Cocuy, Colombia'. *Leidse Geologische Mededelingen,* **32**, 157–182.

*Gregory, J. W., 1893: 'Mr. Gregory's expedition to Mount Kenya'. *Geogr. J.,* **1**, 456–457, and **2**, 326–327.

*Gregory, J. W., 1894: 'The glacial geology of Mount Kenya'. *Quart. J. Geol. Soc.,* **50**, 515–530.

*Gregory, W. J., 1896: *The Great Rift Valley*. Murray, London, 422 pp., reprinted by Cass and Co., London 1968.

*Gregory, W. J., 1900: 'The geology of Mount Kenya'. *Quart. J. Geol. Soc.,* **56**, 205–222.

*Gregory, W. J., 1921: *The Rift Valley and the geology of East Africa*. Seeley, Service and Co., London, 479 pp.

*Griffiths, J. F., ed., 1972: *Climates of Africa*. World Survey of Climatology vol. 10. Elsevier, Amsterdam–London–New York, 604 pp.

*Grove, A. T., 1973: 'Desertification in the African environment'. p. 33–45 in: Dalby, B., Church, R. J. H.: *Drought in Africa*, report of the 1973 Symposium, Centre for African Studies, University of London, 124 pp.

*Grove, A. T., Street, F. A., Goudie, A. S., 1975: 'Former lake levels and climatic change in the Rift Valley of Southern Ethiopia'. *Geogr. J.,* **141**, 177–202.

Grunne, V. de, 1933: 'Ruwenzori from the West'. *Alpine J.,* **45**, p. 275–289.

*Grunne, X. de, Hauman, L., Burgeon, L., Michot, P., 1937: *Vers les glaciers de l'Équateur, le Ruwenzori*. Dupriez, Bruxelles, 300 pp.

Gumprecht, T. E., 1849: 'Die von Rebmann im östlichen Süd-Africa in der Nähe des Aequators entdeckten Schneeberge'. *Monatsberichte der Gesellschaft für Erdkunde zu Berlin,* **6**, 285–297.

Gumprecht, T. E., 1853: 'Schnee und neue Schneeberge im tropischen Afrika'. *Zeitschrift für allgemeine Erdkunde,* **1**, 230–240.

Hamilton, A. C., 1972: 'The interpretation of pollen diagrams from highland Uganda'. *Palaeoecology of Africa,* **7**, 45–149.

Hamilton, A. C., 1973: 'The history of vegetation'. p. 188–209 in: Lind, E. M., Morrison, M. E. S., ed., *The vegetation of East Africa*, Longman, London, 257 pp.

Hamilton, A. C., 1974: 'Distribution patterns of forest trees in Uganda and their historic significance'. *Vegetatio,* **29** 21–35.

Hamilton, A. C., 1982: *Environmental history of East Africa*. Academic Press, London–New York, 328 pp.

*Hamilton, A., Perrott, A., 1978: 'Date of deglacierisation of Mount Elgon'. *Nature,* **273**, 49.

*Hamilton, A., Perrott, A., 1979: 'Aspects of the glaciation of Mt. Elgon, East Africa'. *Palaeoecology of Africa,* **11**, 153–161.

*Hammer, S., 1939: 'Terrain corrections for gravimeter stations'. *Geophysics,* **4**, 184–194.

*Hammer, S., Heck, E. T., 1941: 'A gravity profile across the Central Appalachians, Buckhannon, West Virginia, to Swift Run Gap, Virginia'. *Transactions Amer. Geophys. Union, Reports and Papers, Geodesy*, p. 353–362.

Harper, G., 1969: 'Periglacial evidence in Southern Africa during the pleistocene epoch'. *Palaeoecology of Africa,* **4**, 71–101.

Harris, W., 1929: 'Mount Kenya'. *Alpine J.,* **41**, 362–367.

Hastenrath, S., 1967: 'Rainfall distribution and regime in Central America'. *Archiv Meteor. Geophys. Bioklim.*, Ser. 8, **15**, 201–241.

Hastenrath, S., 1970: 'Diurnal variation of rainfall over Southern Africa'. *Notos,* **19**, 85–94.

Hastenrath, S., 1971: 'On snowline depression and atmospheric circulation in the tropical Americas during the Pleistocene'. *South African Geogr. J.*, **53**, 53–69.

Hastenrath, S., 1972: 'A note on recent and pleistocene altitudinal zonation in Southern Africa'. *South African J. of Science,* **68**, 96–102.

Hastenrath, S., 1973a: 'Observations on the periglacial morphology of Mts. Kenya and Kilimanjaro, East Africa'. *Zeitschrift für Geomorphologie*, N. F., Suppl. vol. 16, 161–179.

Hastenrath, S., 1973b: 'On the pleistocene glaciation of the Cordillera de Talamanca, Costa Rica'. *Zeitschrift für Gletscherkunde und Glazialgeologie,* **9**, 105–121.

Hastenrath, S., 1974a: 'Spuren pleistozäner Vereisung in den Altos de Cuchumatanes, Guatemala'. *Eiszeitalter und Gegenwart,* **25**, 25–34.

Hastenrath, S., 1974b: 'Glaziale und periglaziale Formbildung in Hoch-Semyen, Nord Äthiopien'. *Erdkunde,* **28**, 176–186.

Hastenrath, S., 1974c: 'Soil frost phenomena in the mountains of East Africa'. pp. 81–85 in: *Agroclimatology of the Highlands of Eastern Africa*. Proceedings of Technical Conference, 1–5 October 1973, Nairobi, Kenya. WMO-No. 389, Geneva, 242 pp.

*Hastenrath, S., 1975: 'Glacier recession in East Africa'. pp. 135–142 in *Proceedings of WMO/IAMAP Symposium on Long-term Climatic Fluctuations*, 18–23 August 1975, Norwich, England, WHO-No. 421.

Hastenrath, S., 1976: 'Weitere Beobachtungen zu Bodenfrosterscheinungen am Mount Kenya'. *Zeitschrift für Geomorphologie,* **20**, 235–239.

Hastenrath, S., 1977: 'Pleistocene mountain glaciation in Ethiopia'. *J. Glaciol.*, **18**, 309–313.

*Hastenrath, S., 1978a: 'Heat budget measurements on the Quelccaya Ice Cap, Peruvian Andes'. *J. Glaciol.*, **20**, 85–97.

*Hastenrath, S., 1978b: 'On the three-dimensional distribution of subnival soil patterns in the high mountains of East Africa'. *Erdwissenschaftliche Forschung*, vol. 11, p. 458–481.

Hastenrath, S., 1979: Clima y sistemas glaciales tropicales. p. 47–53 in M. S. Salgado-Labouriau, ed.: El Medio ambiente páramo, Actas del seminario de Mérida, Venezuela, Nov. 1978. CIFCA-CIET-UNESCO-IVIC, Ediciones Centro de Estudios Avanzados.

*Hastenrath, S., 1981: *The Glaciation of the Ecuadorian Andes.* Balkema, Rotterdam, 160 pp.

*Hastenrath, S., 1983a: 'Net balance, surface lowering, and ice flow pattern in the interior of Lewis Glacier, Mount Kenya'. *J. Glaciol.*, **29**, in press.

Hastenrath, S., 1983b: Diurnal thermal forcing and hydrological response of Lewis Glacier, Mount Kenya. *Archiv Meteor. Geophys. Bioklim., Ser. A,* **32**, 361–373.

Hastenrath, S., Caukwell, R. A., 1979: 'Variations of Lewis Glacier, Mount Kenya, 1974–78'. *Erdkunde,* **33**, 292–297.

*Hastenrath, S., Kruss, P., 1979: 'Dynamics of crevasse pattern at Lewis Glacier, Mount Kenya'. *Zeitschrift für Gletscherkunde und Glazialgeologie,* **15**, 201–207.

*Hastenrath, S., Kruss, 1982: 'On the secular variation of ice flow velocity at Lewis Glacier, Mount Kenya'. *J. Glaciol.,* **28**, 333–339.

*Hastenrath, S., and Kutzbach, J. E., 1983: 'Paleoclimatic estimates from water and energy budgets of East African lakes'. *Quaternary Research,* **19**, 141–153.

*Hastenrath, S., Lamb, P., 1979: *Climatic atlas of the Indian Ocean*. Vol. 1. Surface climate and atmospheric circulation. Vol. 2. The oceanic heat budget. University of Wisconsin Press.

*Hastenrath, S., Patnaik, J. K., 1980: 'Radiation measurements on Lewis Glacier, Mount Kenya'. *J. Glaciol.,* **25** 439–444.

*Hastenrath, S., Wilkinson, J., 1973: 'A contribution to the periglacial morphology of Lesotho, Southern Africa'. *Biuletyn Periglacjalny,* **22**, 157–167.

Hecky, R. E., 1979: 'The Kivu-Tanganyika basin; the last 14 000 years'. *Pol. Arch. Hydrobiol.,* **25**, 159–165.

*Hedberg, O., 1951: 'Vegetation belts of East African mountains'. *Svensk Botanisk Tidskrift,* **45**, 140–195.

Hedberg, O., 1964: 'Features of afro-alpine plant ecology'. *Acta phytogeogr. suecia,* **49**, 1–144.

Heine, K., 1975: *Studien zur jungquartären Glazialmorphologie Mexikanischer Vulkane. Mexiko Projekt DFG*, vol. 7, Steiner Verlag, Wiesbaden, 178 pp.

*Heinzelin, J. de, 1951: 'Le retrait des glaciers du flanc Oüest du Massif Stanley (Ruwenzori)'. *Union Geod. Geophys. Int. Assemblée Générale*, vol. 1, 203–205.

Heinzelin, J. de, 1952: 'Glacier recession and periglacial phenomena in the Ruwenzori range (Belgian Congo)'. *J. Glaciol.*, **2**, 137–140.

*Heinzelin, J. de, 1953: *Les stades de récession du glacier Stanley Occidental (Ruwenzori, Congo Belge)*. Institut des Parcs Nationaux du Congo Belge, Parc National Albert, 2nd series, fasc. 1., p. 1–25.

*Heinzelin, J. de, 1962: 'Carte des extensions glaciaires du Ruwenzori (versant Congolais)'. *Biuletyn Periglacjalny*, **11**, 133–139.

Henrici, E. O., 1911–1912: 'The height of Ruwenzori'. *Geogr. J.*, **38**, 607–608; **39**, 490.

Herd, D. G., Naeser, C. W., 1974: 'Radiometric evidence for Pre-Wisconsin glaciation in the Northern Andes'. *Geology*, **2** 603–604.

Hicks, P. H., 1945–46: 'Mount Kenya – first ascent of the North face'. *Alpine J.*, **55**, 76–81.

Hicks, P. H., 1946: 'The Portal Peaks of Ruwenzori'. *Geogr. J.*, **107**, 210–220.

Hicks, P. H., 1947: 'Mount Gessi of the Ruwenzori Range'. *East African Annual*, 1946–47, 14–21.

Hildebrandt, J. M., 1878a: 'Meine zweite Reise in Ostafrika'. *Globus*, **33**, 269–271, 279–281, 296–298.

Hildebrandt, J. M., 1878b: 'On his travels in East Africa'. *Proceedings Roy. Geogr. Soc.*, **22**, 446–453.

Hobley, C. W., 1914: 'The alleged desiccation of East Africa'. *Geogr. J.*, **44**, 467–477.

Hobson, G. D., Jobin, C., 1975: 'A seismic investigation – Peyto Glacier, Banff National Park, and Woolsey Glacier, Mt. Revelstoke National Paark'. *Geoexploration*, **13**, 117–127.

*Hope, G. S., Peterson, J. A., Radok, U., Allison, I., eds., 1976: *The equatorial glaciers of New Guinea; results of the Australian Universities expeditions to Irian Jaya, 1971–73*. Balkema, Rotterdam, 244 pp.

Hodgkin, R. A., 1941: 'Kenya and Ruwenzori'. *Alpine J.*, **53**, 265, pp. 309–318.

Hodgkin, R. A., 1944: 'The Mountains of the Moon'. *Geogr. Mag.*, **16**, 304–311.

Howard, J. W., 1955: 'Mount Kenya, 1939–1952'. *Alpine J.*, **60**, 270–275.

Humphreys, G. N., 1927: 'New routes on Ruwenzori'. *Geogr. J.*, **69**, 516–531.

Humphreys, G. N., 1933: 'Ruwenzori, flights and further exploration'. *Geogr. J.*, **82**, 481–514.

Humphries, D. W., 1959: 'Preliminary notes on the glaciology of Kilimanjaro'. *J. Glaciol.*, **3**, 475–479.

*Humphries, D. W., 1972: 'Glaciology and glacial history'. pp. 31–71 in: Downie, C., Wilkinson, P., 1972: *The geology of Kilimanjaro*, Dept. of Geology, University of Sheffield, 253 pp.

Hunt, T. C., 1947: 'Weather conditions on Kilimanjaro; an ascent of the Western slopes'. *Weather*, **2**, 338–343.

*Huntec Ltd., 1958: *FS-3 Portable Facsimile Seismograph, Manual*. Toronto, 140 pp.

Hurni, H., 1982: *Hochgebirge von Semien, Äthiopien*, vol. II: Klima und Dynamik der Höhenstufung von der letzten Kaltzeit bis zur Gegenwart. Geographia Bernensia, G13, Bern, 196 pp. plus maps.

*International Association of Hydrological Sciences – UNESCO, 1977: *Fluctuations of glaciers, 1970–75*. Paris, 269 pp. plus maps.

Jackson, S. P., 1961: *Climatological atlas of Africa*. Government Printer, Pretoria.

*Jaeger, F., 1909: 'Forschungen im den Hochregionen des Kilimandscharo'. (with sketch of Western Kibo at scale 1:40 000) *Mitteilungen aus den Deutschen Schutzgebieten*, vol. 22, 113–197.

Jaeger, F., 1931: 'Veränderungen der Kilimandjaro–Gletscher'. *Zeitschrift für Gletscherkunde*, **19**, 285–299.

Jaeger, F., 1968: 'Weiterer Rückgang der Gletscher am Kilimandjaro'. *Zeitschrift für Gletscherkunde und Glazialgeologie*, **5**, 99–101.

Jeannel, R., 1950: *Hautes montagnes d'Afrique; vers les neiges éternelles sous l'Équateur*. Éditions du Museum, Paris, 253 pp.

Johannes, K., 1899: 'Besteigung des Kilimandscharo'. *Mutter Erde*, **2**, 267–272.

Johannes, K., 1899: 'Letter to Hans Meyer'. *Globus*, **75**, p. 35.

Johnsen, S. J., Dansgaard, W., Clausen, H. B., Langway, C. C. Jr., 1972: 'Oxygen isotope profiles through the Antarctic and Greenland ice sheets'. *Nature*, **235**, 429–434.

Johnson, D. H., Mörth, H. T., 1960: 'Forecasting research in East Africa'. pp. 56–137 in D. J. Bargman, ed., *Symposium on Tropical Meteorology in East Africa*. Munitalp Foundation, Nairobi.

Johnson, T. B., 1912: *Tramps round the mountains of the moon*. Fisher Unwin, London–Leipzig, 316 pp.
Kassner, T., 1911: *My journey from Rhodesia to Egypt, including an ascent of Ruwenzori*. Hutchinson and Co., London, 310 pp.
Kendall, R. L., 1969: 'An ecological history of the Lake Victoria basin'. *Ecol. Monogr.*, **39**, 121–176.
Kendall, R. L., Livingstone, D. A., 1967: 'Paleoecological studies on the East Africa plateau'. *Congrés panafricain de Prêhistoire*, Dakar, p. 386–388.
Kennedy, W. Q., 1953: 'The mountains of the Moon'. *Illustrated London News*, **223**, 2–3.
Kersten, O., 1863: 'Briefliche Mitteilung von Dr. Kersten über seine Besteigung des Kilimandjaro in der Gesellschaft des Herrn von der Decken'. *Zeitschrift für allgemeine Erdkunde*, N.F., **15**, 141–149.
*King, H. W., Brater, E. F., 1963: *Handbook of hydraulics*. McGraw-Hill, New York, 5th ed., 565 pp.
Klute, F., 1914: 'Forschungen am Kilimandscharo im Jahre 1912'. *Geogr. Zeitschrift*, **20**, 496–505.
*Klute, F., 1920: *Ergebnisse der Forschungen am Kilimandscharo 1912*. Reimer-Vohsen, Berlin, 135 pp.
Klute, F., 1921: 'Die stereophotogrammetrische Aufnahme der Hochregionen des Kilimandscharo'. *Zeitschrift der Gesellschaft für Erdkunde zu Berlin*, 00, p. 144–151.
*Klute, F., 1929: 'Der Kilimandscharo'. *Geologische Charakterbilder*, **36**, 29 pp.
Klute, F., Oehler, E., 1920: 'Karte der Hochregionen des Kilimandscharo-Gebirges nach stereophotogrammetrischen Aufnahmen 1912'. (konstruiert von F. Schroeder und F. Klute), 1:50 000. (in Klute: *Die Ergebnisse der Forschungen am Kilimandscharo*). reproduced in *Zeitschrift für Vulkanologie*, 1922, vol. 6, p. 198.
Kmunke, R., 1913: *Ouer durch Uganda*. Reimer, Berlin, 186 pp.
Koopmans, B. N., Stauffer, P. H., 1968: 'Glacial phenomena on Mount Kinabalu, Sabah'. *Malaysian Geol. Survey, Bull.*, 8, 25–35.
Krapf, L., 1849: 'Von der afrikanischen Ostküste'. *Zeitschrift der Deutschen Morgenländischen Gesellschaft*, **3**, 310–321.
Krapf, L., 1858: *Reisen in Ostafrika, ausgeführt in den Jahren 1837–1855*. 2 vols. Stroh, Kronthal and Stuttgart; vol. 1: 506 pp., vol. 2: 522 pp.
Krapf, L., 1860: *Travels, researches, and missionary labors, during an eighteen years' residence in Eastern Africa*. Trübner, London, 464 pp.
Kraus, E. B., 1955a: 'Secular changes of tropical rainfall regimes'. *Quart. J. Roy. Meteor. Soc.*, **81**, 198–210.
Kraus, E. B., 1955b: 'Secular changes of east-coast rainfall regimes'. *Quart. J. Roy. Meteor. Soc.*, **81**, 430–439.
Krenkel, E., 1925: *Geologie Afrikas*, part 1. Borntraeger, Berlin, 461 pp.
*Kruss, P. D., 1981: *Numerical modelling of climatic change from the terminus record of Lewis Glacier, Mount Kenya*. Ph.D. Diss., Dept. Meteorology, University of Wisconsin, Madison, 128 pp.
Kruss, P. D., 1983a: 'Terminus response of Lewis Glacier, Mount Kenya, to sinusoidal net balance forcing'. *J. Glaciol.*, **29**, in press.
Kruss, P. D., 1983b: 'Climatic change in East Africa: numerical modeling from the 100 years' terminus record of Lewis Glacier, Mount Kenya'. *Zeitschrift für Gletscherkunde und Glazialgeologie*, in press.
*Kruss, P. D., Hastenrath, S., 1983: 'Variation of ice velocity at Lewis Glacier, Mount Kenya: verification midway into a forecast'. *J. Glaciol.*, **29**, 48–54.
Kurz, M., 1948: 'Fremde Berge – ferne Ziele'. *Berge der Welt*, vol. 3, 526 pp. Verbandsdruckerei AG, Bern.
*Lamb, H. H., 1966: 'Climate in the 1960's; changes in the World's wind circulation reflected in prevailing temperatures, rainfall patterns and the levels of African lakes'. *Geogr. J.*, **132**, 183–213.
Lange, M., 1912: 'Eine Kibo-Besteigung'. *Zeitschrift Ges. f. Erdkunde*, Berlin, **47**, 513–522.
Latham, D. V., 1926: 'Kilimanjaro and some observation on the physiology of high altitudes in the tropics'. *Geogr. J.*, **68**, 492–505.
La Vallée Poussin, J. de, 1933 'Les glaciers du Ruwenzori'. *Annales Sociéte Scientifique de Bruxelles, Serie B, Sciences Physiques et Naturelles*, **53**, 45–47.
Lent, C., 1894a: 'Bericht über die wissenschaftliche Station am Kilima-Ndjaro'. *Mitteilungen aus den Deutschen Schutzgebieten*, **7**, 61–66.

Lent, C., 1894b: 'Bericht über topographische Aufnahmen am Kilima-Ndjaro'. *Mitteilungen aus den Deutschen Schutzgebieten,* 7, 243–246.

*Light, R. U., 1941: *Focus on Africa.* Amer. Geogr. Soc. Special Publications No. 25, New York, 288 pp.

Lind, E. M., Morrison, M. E. S., 1973: *The vegetation of East Africa.* Longman, London, 257 pp.

*List, R. J., 1968: *Smithsonian meteorological tables.* Smithsonian Meteorological Collections, vol. 114, sixth revised edition. Washington, D.C., 527 pp.

*Livingstone, D. A., 1962: 'Age of deglaciation in the Ruwenzori range, Uganda'. *Nature,* **194**, 859–860.

Livingstone, D. A., 1967: 'Postglacial vegetation of the Ruwenzori mountains in Equatorial Africa'. *Ecol. Monogr.,* **37**, 25–52.

Livingstone, D. A., 1975: 'Late quaternary climatic change in Africa'. *Annual Review of Ecology and Systematics,* **6**, 249–280.

Livingstone, D. A., 1976a: 'The Nile – palaeolimnology of headwaters', pp. 21–30 in: J. Rzoska, ed., *The Nile, biology of an ancient river.* W. Junk, The Hague, 417 pp.

Livingstone, D. A., 1976b: 'Limnology, paleolimnology and palynology of Tropical Africa'. *Palaeoecology of Africa,* vol. 9, 128–130.

*Livingstone, D. A., 1980: 'Environmental changes in the Nile headwaters'. pp. 339–359 in: Williams, M. A. J., Faure, H., eds.: *The Sahara and the Nile,* Balkema, Rotterdam, 607 pp.

Livingstone, D. A., Clayton, W. D., 1980: 'An altitudinal cline in tropical African grass floras and its paleoecological significance'. *Quaternary Research,* **13**, 392–402.

Livingstone, D. A., Kendall, R. L., 1969: 'Stratigraphic studies of East African Lakes'. *Mitt. Internat. Verein. Limnol.,* **17**, 147–153.

Livingstone, D. A., Van der Hammen, T., 1978: 'Palaeogeography and palaeoclimatology'. pp. 61–90 in: *Tropical forest ecosystems,* UNESCO/UNEP/FAO, Natural Resources Research 15, Chapter 3. UNESCO, Paris.

Lliboutry, L., Morales Arnao, B., Schneider, B., 1977: 'Glaciological problems set by the control of dangerous lakes in Cordillera Blanca, Peru, III. study of moraines and mass balance at Safuna'. *J. Glaciol.,* **18**, 275–290.

Loeffler, H., 1968: 'Die Hochgebirgsseen Ostafrikas'. *Hochgebirgsforschung,* **1**, 1–61.

Löffler, E., 1970: 'Untersuchungen zum eiszeitlichen und rezenten klimagenetischen Formenschatz in den Gebirgen Nordostanatoliens'. *Heidelberger Geogr. Arbeiten,* Heft 27, 162 pp.

Löffler, E., 1972: 'Pleistocene glaciation in Papua and New Guinea'. *Zeitschrift für Geomorphologie,* suppl. vol. 13, 32–58.

Löffler, E., 1975: 'Beobachtungen zur periglazialen Höhenstufe in den Hochgebirgen von Papua New Guinea'. *Erdkunde,* **29**, 285–292.

*Löffler, E., 1976: 'Potassium-Argon dates and pre-Würm glaciations of Mount Giluwe volcano, Papua New Guinea'. *Zeitschrift für Gletscherkunde und Glazialgeologie,* **12**, 55–62.

Löffler, E., 1977: *Geomorphology of Papua New Guinea.* CSIRO and Australian National University Press, Canberra, 196 pp.

Löffler, E., 1982: 'Pleistocene and present-day glaciations'. pp. 39–55, in: Gressit, J. L., ed., *Biogeography and ecology of New Guinea,* Monographiae Biologicae, vol. 42, Junk, The Hague, 983 pp.

*Mackinder, J. H., 1900: 'A journey to the summit of Mount Kenya, British East Africa'. *Geogr. J.,* **15**, 453–486.

Mackinder, J. H., 1901: 'The ascent of Mount Kenya'. *Alpine J.,* **20**, 102–111.

Mackinder, J. H., 1930: 'Mount Kenya in 1899'. *Geogr. J.,* **76**, 529–534.

Mahaney, W. C., 1972: 'Late quarternary history of the Mount Kenya afroalpine area, East Africa'. *Paleoecology of Africa,* vol. 6, Balkema, pp. 139–149.

Mahaney, W. C., 1976: 'Quaternary stratigraphy of Mount Kenya'. *Palaeoecology of Africa,* vol. 9, p. 130.

Mahaney, W. C., 1979: 'Quaternary stratigraphy of Mount Kenya: a reconnaissance'. *Palaeoecology of Africa,* vol. 11, pp. 163–170.

Mahaney, W. C., 1980: 'Late quaternary rock glaciers, Mount Kenya'. *J. Glaciol.,* **25**, 487–497.

Mahaney, W. C., 1982: 'Chronology of glacial and periglacial deposits, Mount Kenya, East Africa: description of type sections'. *Palaeoecology of Africa,* **14**, 25–43.

Maury, M. J., 1912: 'La description de la frontière oriental du Congo Belge'. *Le Mouvement Géographique,* **29**, 253–260.
*McGregor Ross, W., 1911: 'The snowfields and glaciers of Kenya'. *Pall Mall Magazine*, 197–208, 463–475.
Meersch, Ganshof van der, 1933: 'Une mission scientifique Belge dans le massif du Ruwenzori'. *L'Illustration,* **184**, pp. 103–110, 134–140.
*Meidav, T., 1969: 'Hammer reflection seismics in engineering geophysics'. *Geophysics,* **34**, 383–395.
*Menzies, I. R., 1951a: 'Some observations on the glaciology of the Ruwenzori range'. *J. Glaciol.* **1**, 511–512.
*Menzies, I. R., 1915b: 'The glaciers of Ruwenzori'. *Uganda Journal,* **15**, 177–181.
Mercer, J. H., Palacios, O., 1977: 'Radiocarbon dating of the last glaciation in Peru'. *Geology,* **5**, 600–604.
Messerli, B., 1967: 'Die eiszeitliche und die gegenwärtige Vergletscherung im Mittelmeerraum'. *Geographica Helvetica,* **3**, 105–228.
Messerli, B., Baumgartner, R., 1978: 'Kamerun, Bericht über die Exkursion 1978'. *Geographia Bernensica*, G 9, 315 pp.
*Messerli, B., Winiger, M., Rognon, P., 1980: 'The Saharan and East African uplands during the quaternary'. pp. 87–132 in M. A. J. Williams and H. Faure, editors; *The Sahara and the Nile*, Balkema, Rotterdam, 640 pp.
Methner, W., 1932: 'The ascent of Mount Kilimanjaro'. *Ice Cap*, pp. 9–21.
Meyer, H., 1887a 'Über seine Besteigung des Kilimandscharo'. *Verhandlungen Ges. Erdkunde Berlin,* **14**, 446–454.
Meyer, H., 1887b. 'Über seine neue Kilima-Ndscharo – Expedition'. *Verhandlungen Ges. Erdkunde Berlin,* **26**, 88–101.
Meyer, H., 1887c: Vorläufiger Bericht über meine Besteigung des Kilimandscharo, Juli 1887'. *Petermann's Mitteilungen,* **33**, 353–355.
*Meyer, H., 1888a: *Zum Schneedom des Kilimandscharo*. Meidinger, Berlin (40 photographs with text).
Meyer, H., 1888b: 'Die Schneeverhältnisse am Kilimandscharo im Sommer 1887'. *Wissenschaftliche Veröffentlichungen des Vereins f. Erdkunde Leipzig*, vol. 1, 277–282. (also pp. XXIV–XXVIII).
Meyer, H., 1890a: 'Die Besteigung des Kilimandscharo'. (with sketch of Kibo crater at scale 1 : 20 000) *Petermann's Mitteilungen,* **36**, 15–22.
Meyer, H., 1890b: 'Die Besteigung des Kilimandscharo'. *Petermann's Mitteilungen,* **36**, 15–22.
Meyer, H., 1890c: 'Das Bergland Ugueno und der Westliche Kilimandscharo'. *Petermann's Mitteilungen,* **36**, pp. 46–48.
*Meyer, H., 1890d: *Ostafrikanische Gletscherfahrten*. Duncker und Humblot, Leipzig, 376 pp.
Meyer, H., 1890e: 'Ascent to the summit of Kilima-njaro'. *Proceedings Roy Geogr. Soc.*, **12**, 321–345.
Meyer, H., 1891a: 'Zur Kenntnis von Eis und Schnee des Kilimandscharo'. *Wissenschaftliche Veröffentlichungen d. Vereins f. Erdkunde Leipzig*. vol. 1, 289–294.
Meyer, H., 1891b: *Across East African glaciers*. Philip and Son, London, 397 pp.
*Meyer, H., 1900a: *Der Kilimandjaro*. Reimer-Vohsen, Berlin, 436 pp.
Meyer, H., 1900b: 'Mount Kilimanjaro'. *Geogr. J.,* **16**, p. 352.
Meyer, H., 1901: 'Heutige und einstige Vergletscherung im tropischen Ostafrika'. *Verhandlungen des 7. Internationalen Geographen-Kongresses, Berlin*, 1899, **11**, 767–773.
Meyer, H., 1928: *Hochtouren im tropischen Afrika*. Brockhaus, Leipzig, 153 pp.
Michot, P., 1933: 'Les traits caractéristiques de la morphologie du Ruwenzori dans leur relations avec la tectonique du massif'. *Bull. Soc. Royale Belge de Geographie,* **47**, 5–13.
Michot, P., 1937 'Géologie et géographie physique'. pp. 207–250, in: de Grunne, X., Haumann, L., Burgeon, L., Michot, P., eds., *Vers les glaciers de l'équateur, le Ruwenzori*, Dupriez, Bruxelles, 300 pp.
Minucci, E., 1938a: 'Ricerche geologiche nella regione del Tana'. pp. 19–35 in: *Missione di studio al Lago Tana*, vol. 1, Reale Accademia d'Italia, Roma, 175 pp.
*Minucci, E., 1938b: 'Ricerche geologiche nella regione del Semien'. pp. 37–46 in: *Missione di studio al Lago Tana*, vol. 1, Roma, Reale Accademia d'Italia, 175 pp.

Mitchell, J. M. Jr., 1963: 'On the world-wide pattern of secular temperature change', in: *Changes of climate, Arid Zone Research*, vol. 20, UNESCO, Paris, pp. 161–181.
*Mittelholzer, W., 1930: *Kilimandjaro Flug*. Orell-Füssli, Zurich-Leipzig, 2nd ed., 102 pp.
Mittelholzer, W., 1932: 'Flying over Africa's highest peak: Kilimanjaro; other craters'. *The Illustrated London News*, 27 Aug. 1932.
Mohaupt, W., 1932: *Beobachtungen über Bodenversetzungen und Kammeisbildungen aus dem Stubai und dem Grödener Tal*. Diss. Hamburg, 1932, 69 pp.
Mohr, P. A., 1963: 'General report of an expedition to the Semien Mountains'. *Bulletin of the Geophysical Observatory*, Addis Abeba, vol. 3, no. 2, 155–167.
Mohr, P. A., 1967: 'Review of the geology of the Semien Mountains'. *Bulletin of the Geophysical Observatory*, Addis Abeba, vol. 10, 79–93.
Mohr, P. A., 1971: *The geology of Ethiopia*. Haile Sellasie I University Press, Addis Abeba, reprinted, 268 pp.
Mooney, H. M., 1974: 'Seismic shear waves in engineering'. *J. of Geotech. Engineering Division* vol. 100, No. GT 8, pp. 905–923.
*Moore, J. E. S., 1901: *To the Mountains of the Moon*. Hurst and Blackett Ltd., London, 350 pp.
Morgan, W. T. W., 1973: *East Africa*. Longman, London, 410 pp.
Morrison, M. E. S., 1961: 'Pollen analysis in Uganda'. *Nature*, **190**, 483–486.
Morrison, M. E. S., 1968: 'Vegetation and climate in the uplands of Southwestern Uganda during the later pleistocene period. I. Muchoya Swamp, Kigezi District'. *J. Ecol.*, **56**, 363–384.
Morrison, M. E. S., Hamilton, A. C., 1974: 'Vegetation and climate in the uplands of Southwestern Uganda during the later pleistocene period. II. Forest clearance and other vegetational changes in the Rukiga Highlands during the past 8000 years'. *J. Ecol.*, **62**, 1–32.
Mörth, H. T., 1967: 'Investigation into the meteorological aspects of the variations in the level of Lake Victoria'. *East African Meteorological Department, Memoirs*, vol. 4, no. 2, 23 pp.
Mörth, H. T., 1970: *Rainfall measured on the slopes of Kilimanjaro*. East African Meteor. Dept., mimeographed, 3 pp., Nairobi.
*Mostertz, H., 1929/30: 'Am Krater des Kibo (Kilimandjara)'. *Zeitschrift für Vulkanologie*, **12**, 299–304.
*Mountain Club of East Africa, 1932: *The Icecap*. vol. 1.
*Mountain Club of Kenya, 1971: *Guide book to Mount Kenya and Kilimanjaro*. Nairobi, 3rd ed., 240 pp.
Mountain Club of Kenya, 1981: *Guide to Mount Kenya and Kilimanjaro*. Nairobi, 4th ed., 284 pp.
Nangeroni, G., 1952: 'I fenomeni di morfologia periglaciale in Italia'. pp. 213–220 in *Proc. 8th gen. assembly and 17th Internat. Congr., Internat. Geogr. Union. Washington, D.C., Aug. 8–15, 1952*. Commiss. on periglacial morphology. pp. 207–226; U.S. Nat. Comm. of IGU.
Nicol, H., 1954: 'The North face of Mount Kenya and other climbs'. *Alpine J.*, **59**, 81–87.
Nicol, H. G., 1964: 'North face of Mount Kenya'. *Alpine J.*, **69**, 25–27.
*Nilsson, E., 1931: 'Quaternary glaciations and pluvial lakes in British East Africa'. *Geografiska Annaler*, **13**, 249–349.
*Nilsson, E., 1935: 'Traces of ancient changes of climate in East Africa'. *Geografiska Annaler*, **1/2**, 1–21.
*Nilsson, E., 1940: 'Ancient changes of climate in British East Africa and Abyssinia'. *Geografiska Annaler*, **22**, 1–79.
*Nilsson, E., 1949 'The pluvials of East Africa'. *Geografiska Annaler*, **31**, 204–211.
*Nye, J. F., 1952: 'The mechanics of glacier flow'. *J. Glaciol.*, **2**, 82–93.
Nye, J. F., 1965: 'The flow of a glacier in a channel of rectangular, elliptic or parabolic cross-section'. *J. Glaciol.*, **5**, 661–690.
Osmaston, H. A., 1961: 'Notes on the Ruwenzori glaciers'. *Uganda Journal*, **25**, 99–104.
*Osmaston, H. A., 1965: *The past and present climate and vegetation of Ruwenzori and its neighborhood*. Thesis, Oxford, 238 pp., cyclostyled.
*Osmaston, H. A., Pasteur, D., 1972: *Guide to the Ruwenzori*. Alden Press, Oxford, 200 pp.
Owen, R. B., Barthelme, J. W. Renaut, R. W., Vincens, A., 1982: 'Paleolimnology and archaeology of Lake Turkana, Kenya'. *Nature*, **298**, 523–529.
*Parasnis, D. S., 1972: *Principles of applied geophysics*. Chapman and Hall Ltd., London, 214 pp.

Paterson, W. S. B., 1969: *The physics of glaciers*. Pergamon Press, 250 pp.
*Perrott, A. R., 1982a: 'A postglacial pollen record from Mt. Satima, Aberdare Range, Kenya'. p. 153 in *Amer. Quat. Assoc., seventh Biennial Conference, Seattle, June 1982*, Program and Abstracts, 188 pp.
Perrott, R. A., 1982b: 'A high attitude pollen diagram from Mount Kenya: its implications for the history of glaciation'. *Palaeoecology of Africa,* **14**, 77–83.
Petermann, A., 1853: 'The snowy mountains of Eastern Africa'. *The Athenaeum Journal of Literature, Science, and the Fine Arts*. pp. 1014–1015.
Peterson, J. A., Hope, G., 1972: Lower limit and maximum age for the last major advance of the Carstensz Glacier, West Irian. *Nature* **240**, 36–37.
*Platt, C. M., 1966: 'Some observations on the climate of Lewis Glacier, Mount Kenya, during the rainy season'. *J. Glaciol.,* **5**, 661–690.
Potter, E. C., 1976: 'Pleistocene glaciation in Ethiopia'. *J. Glaciol.,* **17**, 148–150.
Priestley, C. H. B., 1959: *Turbulent transfer in the lower atmosphere*. University of Chicago Press, 130 pp.
Raasveldt, H. C., 1957: 'Las glaciaciones de la Sierra Nevada de Santa Marta'. *Revista de la Academia Colombiana de Ciencias exactas, físicas, y matemáticas,* **9**, 469–482.
Ravenstein, E. G., 1901: 'The lake-level of the Victoria Nyanza'. *Geogr. J.,* **18**, 403–406.
Rebmann, J., 1849a: 'Narrative of a journey to Jagga, the snow country of Eastern Africa'. *Church Missionary Review,* **1**, 12–23.
Rebmann, J., 1849b: 'Journal d'une excursion au Djagga'. *Nouvelles Annales des Voyages et des Sciences Géographiques,* **2**, 257–307.
Rebmann, J., 1855: 'Extracts of letter to Rev. H. Venn'. *Report of 24th meeting of the British Association for the Advancement of Science, London*, 123–124.
Reck, H., 1921/22: 'F. Klute, Ergebnisse der Forschungen am Kilimandjaro 1912'. (F. Klute). *Zeitschrift für Vulkanologie,* **6**, 198–206.
Reischel, G., 1899: 'Die Ergebnisse von Hans Meyer's Forschungsreise am Kilimandscharo'. *Die Natur,* **48**, 27–29.
Revelli, P., 1906 'Il Runssóro (Ruwenzori) secondo le esplorazioni (1904) del dott. J. J. David'. *Bolletino della Società Geografica Italiana*, ser. 4, vol. 7, part 1, 345–365.
Rice, N. R., 1932a: *Sketch map of the Crater of Kibo* 1:10 000. Ice cap, 1, plate 1.
Rice, N. R., 1932b: *Mawenzi, sketch map of its peaks*, 1:5000. Ice cap 1, plate 2.
Richardson, J. L., 1966: 'Changes in level of Lake Naivasha, Kenya, during post-glacial time'. *Nature,* **209**, 290–291.
Richardson, J. L. and A. E., 1972: 'The history of an African rift lake and its climatic implications'. *Ecological Monographs,* **42**, 499–534.
Roccati, A., 1907: 'Nell'Uganda e nella catena del Ruwenzori'. *Bolletino della Società Geologica Italiana,* **26**, 127–158.
Roccati, A., 1909: 'Geologia'. in L. A. S. Duca degli Abruzzi, *Il Ruwenzori*, parte scientifica, vol. 2. Hoepli, Milano, 286 pp.
*Root, A., 1974: 'Sailing over Kilimanjaro'. *Daily Nation*, Friday 26 March., pp. 16–17.
Ross, R., 1955a: 'Some aspects of the vegetation of Ruwenzori'. *Webbia,* **11**, 451–457.
Ross, R., 1955b: 'Some aspects of the sub-alpine zone on Ruwenzori'. *Proc. Linn. Soc. London,* **165**, 2, 135–140.
Rzoska, J., ed., 1976: *The Nile, biology of an ancient river*. W. Junk, The Hague, 417 pp.
Salgado-Labouriau, M. L., Schubert, C., Valastro, S., 1977 'Paleoecologic analysis of a Late-Quaternary terrace from Mucubají, Venezuelan Andes'. *J. Biogeogr.,* **4**, 313–325.
Salis, C. von 1926: 'Kilimandjaro'. *Alpen,* **2**, 281–286.
Salt, G., 1951: 'The Shira Plateau of Kilimanjaro'. *Geogr. J.,* **117**, 150–166.
Salt, G., 1954: 'A contribution to the ecology of the upper Kilimanjaro'. *J. Ecol.,* **42**, 375–423.
Salt, G., 1974: 'Rainfall on Kilimanjaro'. *Geogr. J.,* **140**, 344–346.
Sampson, D. H., 1974: 'The geology, volcanology and glaciology of Mt. Kilimanjaro'. *Tanganyika Notes and Records,* **64**, 118–125.
Scott Elliot, G. F., 1895: 'Expedition to Ruwenzori and Tanganyika'. *Geogr. J.,* **6**, 301–324.
Schenk, E., 1955: 'Die Mechanik der periglazialen Strukturböden'. *Hessisches Landesamt für Bodenforschung, Abhandlungen* No. 13, 92 pp.

Schnell, R. G., Odh, S. A., 1977: *Mount Kenya baseline station feasibility study*. (S. N. Tan-Schnell, ed.) WHO/UNEP Mount Kenya Project, c.o. UNDP, P.O. Box 30218, Nairobi, Kenya.

Schubert, C., 1972a: 'Geomorphology and glacier retreat in the Pico Bolivar area, Sierra Nevada de Mérida, Venezuela'. *Zeitschrift für Gletscherkunde und Glazialgeologie*, 8, 189–202.

Schubert, C., 1972b: 'Late glacial chronology in the Northeastern Venezuelan Andes'. *24th International Geol. Congr. Montreal*, Sect. 12, pp. 103–109, Montreal.

Schubert, C., 1974: 'Late pleistocene Mérida glaciation, Venezuelan Andes'. *Boreas*, **3**, 147–152.

Schubert, C., 1975. 'Glaciation and periglacial morphology in the Northwestern Venezuelan Andes'. *Eiszeitalter und Gegenwart*, **26** 196–211.

Sellers, W. O., 1965: *Physical climatology*. University of Chicago Press, 272 pp.

Serviço Meteorológico Nacional (Portugal) 1965: *Climatologia dinâmica da Africa meridional*. Lisboa, 207 pp.

*Sharp, R., Epstein, S., Vidziunas, I., 1960: 'Oxygen-isotope ratios in the Blue Glacier, Olympic Mountains, Washington, USA'. *J. Geophys. Res.*, **65**, 4043–4059.

Shell, 1968: *Road map of Kenya* 1 : 1 000 000. George Philip and Son Ltd., London.

Shipton, E., 1931: 'The first traverse of the twin peaks of Mount Kenya'. *Alpine J.*, **43**, 138–146.

Shipton, E. E., 1932: 'Mountains of the Moon'. *Alpine Journal*, **44**, 88–96.

*Shipton, E. E., 1956: *Upon that mountain*. Pan Books Ltd., London, 254 pp.

Skinner, N. J., 1977: 'Recent geophysical studies of the Kenya Rift Valley'. *Contemp. Phys.*, **18**, 455–470.

Skinner, N. J., Bhatt, N. V., Hastenrath, S., 1974: 'Negative magnetic anomaly associated with Mount Kenya'. *Nature*, **250**, 561–562.

Spink, P. C., 1943: 'Some effects of vertical insolation upon the Kilimanjaro glaciers'. *Quart. J. Roy. Meteor. Soc.*, **69**, 261–264.

*Spink, P. C., 1944: 'Weather and volcanic activity on Kilimanjaro'. *Geogr. J.*, **103**, 226–229.

*Spink, P. C., 1945: 'Further notes on the Kibo inner crater and glaciers of Kilimanjaro and Mount Kenya'. *Geogr. J.*, **106**, 210–216.

Spink, P. C., 1947: 'World-wide climate and the glaciers and lake levels of East Africa'. *Weather*, **2**, 329–337.

*Spink, P. C., 1949: 'The equatorial glaciers of East Africa'. *J. Glaciol.*, **1**, 277–283.

Spink, P. C., 1952a: 'Recession of African glaciers'. *J. Glaciol.*, **2**, 149–150.

*Spink, P. C., 1952b: 'Decay of the Kilimanjaro glaciers'. *Weather*, **7**, 284–285.

Stanley, H. M., 1878: *Through the dark continent*. Sampson Low, London, 2 vols., 547 pp., 540 pp., map.

Stanley, H. M., 1890: *In darkest Africa*. Sampson Low, London, 2 vols., 547 pp. 540 pp., map.

Street, F. A., 1979: 'Late quaternary precipitation estimates for the Ziway-Shala basin, Southern Ethiopia'. *Palaeoecology of Africa*, **11** 135–143.

Street, F. A., 1980: 'The relative importance of climate and hydrogeological factors in influencing lake level fluctuations'. *Palaeoecology of Africa*, **12**, 137–158.

Street, F. A., 1981: 'Tropical palaeoenvironments'. *Progress in Physical Geography*, **5**, 157–185.

Street, F. A., and Grove, A. T., 1976: 'Environmental and climatic implications of late Quaternary lake-level fluctuations in Africa'. *Nature*, **261**, 385–390.

*Street, F. A., and Grove, A. T., 1979: 'Global maps of lake-level fluctuations since 30 000 B.P.'. *Quaternary Research*, **12**, 83–118.

*Stuhlmann, F., 1894: *Mit Emin Pascha ins Herz von Afrika*. Reimer, Berlin, 902 pp.

*Stumpp, A., 1952: 'Kartierungsarbeiten in Ruwensorigebirge'. *Allgemeine Vermessungsnachrichten*, **6**, 142–147.

Survey of Kenya, 1955–72: Topographic maps 1:50 000, sheets 107/3 *Nanyuki*, 107/4 *Marania*, 108/3 *Meru*, 121/1 *Naro Moru*, 121/2 *Mount Kenya*, 122/1 *Nkubu*, 121/3 *Karatina*, 121/4 *Embu North*, 122/3 *Chuka*. Nairobi.

Survey of Kenya, 1971: *Mount Kenya*, 1 : 25 000. Series SK 75, Edition 4-SK. Nairobi.

Synge, P. M., 1938: *Mountains of the Moon*. Dutton and Co., New York, 210 pp.

Telford, W. M., Geldard, L. P., Sheriff, R. E., Keys, D. A., 1976: *Applied geophysics*. Cambridge University Press, 860 pp.

Temple, P. H., 1966: 'Lake Victoria levels'. *Proceedings East African Academy*, vol. 2, pp. 50–58.

*Temple, P. H., 1968: 'Further observations on the glaciers of the Ruwenzori'. *Geografiska Annaler*, Ser. A., **50**, no. 3, 136–150.

*Temporary Technical Secretariat for World Glacier Inventory of UNESCO–UNEP–IUGG–IASH–ICSI, 1977: *Instructions for compilation and assemblage of data for a World Glacier Inventory.* ETH, Zürich, 29 pp.

*Texas Instruments Inc., 1960 *Worden gravity meter, operating instruction manual.* Houston, Texas, 111 pp.

Thomas, H. B., 1947: 'An early visit to Ruwenzori'. *Uganda Journal*, **11**, 60–63.

Thompson, B. W., 1957: 'Diurnal variation of precipitation in British East Africa'. *E. A. Meteor. Dept. Technical Memoirs* No. 8, 70 pp.

*Thompson, B. W., 1965: *The climate of Africa.* Oxford University Press, Nairobi–London–New York, 132 pp.

Thompson, B. W., 1966: 'The mean annual rainfall of Mount Kenya'. *Weather*, **21**, 48–49.

*Thompson, L. G., Hastenrath, S., Morales Arnao, B., 1979: 'Climatic ice core records from the tropical Quelccaya Ice Cap'. *Science*, **203**, 1240–1243.

*Thompson, L. G., Hastenrath, S., 1980: 'Climatic ice core studies at Lewis Glacier, Mount Kenya'. *Zeitschrift für Gletscherkunde und Glazialgeologie*, **16**, 213–228.

Tiercelin, J. J., Renaut, R. W., Delibrias, G., LeFournier, J., Bieda, S., 1981: 'Late Pleistocene and Holocene lake level fluctuations in the Lake Bogoria basin, northern Kenya Rift Valley'. *Palaeoecology of Africa*, **13**, 105–120.

*Tilman, H. W., 1937: *Snow on the Equator*. G. Bell and Sons Ltd., London, 265 pp.

*Toucheboeuf de Lussigny, P., 1969: *Monographie hydrologique du Lac Tchad.* Office de la Recherche Scientifique et Technique Outre-Mer, Sécrétariat d'État aux Affaires Étrangères chargé de la Cooperation, Paris, 169 pp.

Troll, C., 1944: 'Strukturböden, Solifluktion und Frostklimate der Erde'. *Geol. Rundschau*, **34**, 545–694.

Troll, C., 1958: *Structure soils, solifluction, and frost climates of the Earth.* U.S. Army Snow, Ice and Permafrost Establishment, Translation No. 43, 121 pp.

*Troll, C., Wien, K., 1949: 'Der Lewisgletscher am Mount Kenya'. *Geografiska Annaler*, **31**, 257–274.

Tuckett, R., 1931: 'Africa's two highest peaks'. *The Illustrated London News*, 24 Jan. 1931.

*Uhlig, C., 1904: 'Vom Kilimandscharo zum Meru'. *Zeitschrift Ges. Erdkunde Berlin*, 627–650, 692–718.

Uhlig, C., 1905: 'The glacial features of Kilimanjaro'. *Geogr. J.*, **25**, 213–214.

Uhlig, C., 1908: 'Die Ostafrikanische Expedition der Otto Winter-Stiftung'. *Zeitschrift Ges. Erdkunde Berlin*, 75–94.

*UNESCO, 1970: 'Perennial ice and snow masses. A guide for compilation of data for a World Inventory'. *Technical Papers in Hydrology*, No. 1.

U.S. Weather Bureau, 1958–63: *Monthly climatic data for the World, 1958–62.* Asheville, N.C.

U.S. Weather Burea, 1967: *World weather records*, vol. 5, Africa, 1951–60. Washington, D.C., 545 pp.

Van Zinderen Bakker, E. M., 1962: 'A late-glacial and post-glacial climatic correlation between East Africa and Europe'. *Nature*, **194**, 201–303.

Van Zinderen Bakker, E. M., 1964 'A pollen diagram from Equatorial Africa; Cherangani, Kenya'. *Geol. Mijnbouw*, **43**, 123–128.

Van Zinderen Bakker, E. M., Coetzee, J. A., 1972: 'A re-appraisal of late-quaternary climatic evidence from tropical Africa'. *Palaeoecology of Africa*, **7**, 151–181.

Volkens, G., 1894: 'Bericht über seine Tätigkeit am Kilima-Ndjaro, September 1893'. *Mitteilungen von Forschungsreisenden und Gelehrten aus den Deutschen Schutzgebieten*, vol. 7, pp. 59–61.

Volkens, G., 1897: *Der Kilimandscharo*. Dietrick Reimer, Berlin, 388 pp.

Washbourn-Kamau, C., 1967: 'Lake levels and quaternary climates in the Eastern Rift Valley of Kenya'. *Nature*, **216** 672–673.

*Washbourn-Kamau, C. K., 1971: 'Late quaternary lakes in the Nakuru-Elmenteita basin, Kenya'. *Geogr. J.*, **137**, 522–535.

Washburn, A. L., 1973: *Periglacial processes and environments*. Arnold, London, 320 pp.
Wauters, A. J., 1905: 'Le massif neigeux du Ruwenzori'. *Le Mouvement Géographique,* **22**, 523–527.
Wauters, A. J., 1909: 'La frontière anglo-belge dans le massif du Ruwenzori'. *Le Mouvement Géographique,* **26**, 493–499.
Weiss, M., 1917: 'Am Ruwenzori', *Mitteilungen, Geographische Gesellschaft zu Rostock,* **2**, 120–138.
West, W. C., 1915: 'An ascent of Mount Kibo, Kilimanjaro'. *Mountain Club of South Africa, Journal*, pp. 5–8.
*Whittow, J. B., 1959: 'The glaciers of Mount Baker, Ruwenzori'. *Geogr. J.,* **125**, 370–379.
*Whittow, J. B., 1960: 'Some observations on the snowfall of Ruwenzori'. *J. Glaciol.,* **3**, 765–772.
Whittow, J. B., 1966: 'The land forms of the Central Ruwenzori, East Africa'. *Geogr. J.,* **132**, 32–42.
Whittow, J. B., Osmaston, H. A., 1966: 'The deglacierization of East African mountains'. Paper to the Institute of British Geographers, Geomorphological Research Group, Occasional Paper No. 3, pp. 10–12.
*Whittow, J. B., Shepherd, A., 1959: 'The Speke glacier, Ruwenzori'. *Uganda Journal,* **23**, 153–161.
*Whittow, J. B., Shepherd, A., Goldthorpe, J. E., Temple, P. H., 1963: 'Observations on the glaciers of the Ruwenzori'. *J. Glaciol.,* **4**, 581–616.
Wichmann, H., 1889: 'Afrika'. *Petermann's Mitteilungen,* **35**, 182–184.
Wichmann, H., 1912: 'Afrika'. *Petermann's Mitteilungen,* **58**, 156–157.
Williams, M. A. J., Street, F. A., Dakin, F. M., 1978: 'Fossil periglacial deposits in the Semien highlands, Ethiopia'. *Erdkunde,* **32**, 40–46.
*Williams, M. A. J., Adamson, D. A., Williams, F. M., Morton, W. H., Parry, D. E., 1980: 'Jebel Marra volcano: a link between the Nile Valley, the Sahara and Central Africa'. pp. 305–337 in: Williams, M. A. J., Faure, H., eds., *The Sahara and the Nile*, Balkema, Rotterdam, 607 pp.
Winiger, M., 1981: 'Zur thermisch-hygrischen Gliederung des Mount Kenya'. *Erdkunde,* **35**, 248–263.
Winiger, M., 1979: 'Bodentemperaturen und Niederschlag als Indikatoren einer klimatisch-ökologischen Gliederung tropischer Gebirgsräume. Methodische Aspekte und Anwendbarkeit diskutiert am Beispiel des Mt. Kenya (Ostafrika)'. *Geomethodica,* **4**, 121–150.
Wollaston, A. F. R., 1908: *From Ruwenzori to the Congo*. Dutton and Co., New York, 308 pp.
World Meteorological Organization-ICSU, 1975: 'The physical basis of climate and climate modelling'. *GARP Publication Series*, No. 16, 265 pp.
World Meteorological Organization – United Nations Development Programme, 1974: *Hydrometeorological survey of the catchments of lakes Victoria, Kyoga and Albert*. 4 vols. RAF 66–025, Technical Report 1, Geneva, 925, 77, 576, 355 pp.
Wyss-Dunant, E., 1935: 'De la glaciation comparée des trois massifs de l'Est africain'. *Verhandlungen der Schweizerischen Naturforschenden Gesellschaft*, 116th session, Aarau, pp. 205–206.
Wyss-Dunant, E., 1938: *Mes ascensions en Afrique*. Payot, Paris, 255 pp.
Zeuner, F. E., 1949: 'Frost soils on Mount Kenya and the relation of frost soils to aeolian deposits'. *J. Soil Science,* **1**, 20–30.
Zienert, A., 1968: 'Gleiche Würm-Rückzugsstadien in den Gebirgen Mitteleuropas und Ostafrikas?'. *Eiszeitalter und Gegenwart,* **19**, 85–92.

AUTHOR INDEX

SUBJECT INDEX